Robust Sigma Delta Converters

ANALOG CIRCUITS AND SIGNAL PROCESSING

Consulting Editor: **Mohammed Ismail.** ***Ohio State University***

For other titles published in this series, go to
www.springer.com/series/7381

Robert H.M. van Veldhoven •
Arthur H.M. van Roermund

Robust Sigma Delta Converters

And Their Application in Low-Power Highly-Digitized Flexible Receivers

Springer

Dr. ir. Robert H.M. van Veldhoven
NXP Semiconductor, Central R&D
Mixed Signal Circuits & Systems
High Tech Campus, Building 32, room 3.24
5656 AE Eindhoven
The Netherlands
Robert.van.Veldhoven@nxp.com

Prof. dr. ir. Arthur H.M. van Roermund
Electrical Engineering
Mixed-signal Microelectronics Group
Technical University Eindhoven
Den Dolech 2
5612 AZ Eindhoven
The Netherlands
A.H.M.v.Roermund@tue.nl

Series Editors:
Mohammed Ismail
205 Dreese Laboratory
Department of Electrical Engineering
The Ohio State University
2015 Neil Avenue
Columbus, OH 43210
USA

Mohamad Sawan
Electrical Engineering Department
École Polytechnique de Montréal
Montréal, QC
Canada

ISBN 978-94-007-0643-9
e-ISBN 978-94-007-0644-6
DOI 10.1007/978-94-007-0644-6
Springer Dordrecht Heidelberg London New York

Library of Congress Control Number: 2011920957

Cover design: eStudio Calamar S.L.

Printed on acid-free paper

Springer is part of Springer Science+Business Media (www.springer.com)

Preface

The foundation of this book is based on my very exciting time at Philips Research Laboratories and later at NXP Research. After 14 years of research on Sigma Delta Converters and other IP blocks, I have learned that IC design and in particular the design of Sigma Delta Converters is not only a very interesting and multi-dimensional research topic, but is also a religion interpreted differently by each and every individual. In the book, I have tried to include my work, view and experience in the field of Sigma Delta Converters to give the reader a head-start in the design of this intriguing type of analog to digital converter.

The book rests on the experience of designing, and coaching others to design tens of different Sigma Delta Converters for numerous applications, like instrumentation, hearing aids, mobile phones, battery management, car radio, etc. In this book the application area of Sigma Delta Converters is limited to cellular and connectivity terminals, to limit the scope.

I partitioned the book systematically, by looking at what exactly determines the quality of a system. The found so-called quality indicators (Algorithmic accuracy, Robustness, Emission, Flexibility and Efficiency) are used as a framework throughout the book. The book shows all different aspects in the design of Sigma Delta Converters: system level specification and design, IP architecture, circuit implementation and layout are all subjects of this book. Also the verification of theory and silicon implementations by measurements is included. The book assumes some background on receiver architectures, Sigma Delta Converter theory, and IC design.

I very much hope you enjoy reading the book, and while reading, please do not forget: Analog design is beautiful, digital design is just a time discrete and quantized portion of it.

Valkenswaard
January 2011

Robert H.M. van Veldhoven

Contents

1 Introduction 1
1.1 Advanced, Multi-standard Cellular and Connectivity Terminals for the Mass Market 2
1.1.1 Complexity: Mobile Phone Trends, Its Impact on the Transceiver and the Quest for Integration 3
1.1.2 Transistor Scaling: VLSI and Moore 7
1.1.3 Smarter Circuits: $\Sigma\Delta$ Modulators for Mobile Applications . 9
1.2 Book Aims 10
1.3 Book Scope 11
1.4 Outline 11

2 System Quality Indicators 13
2.1 The System Function and Its In- and Outputs 14
2.2 System Quality 15
2.2.1 Accuracy 16
2.2.2 Robustness to Secondary Inputs 16
2.2.3 Flexibility 16
2.2.4 Efficiency 17
2.2.5 Emission of Secondary Outputs 17
2.3 The Digital Revolution 18
2.3.1 The Analog-Digital Interface 19
2.3.2 Digital Systems and the Quality Indicators 20
2.4 Conclusions 23

3 Integrated Receiver Architectures for Cellular and Connectivity 27
3.1 Wireless Receiver Architectures for Digital Communication 27
3.2 Receiver Architecture and the Quality Indicators 30
3.3 Conclusions 31

4 Specifications for A/D Converters in Cellular and Connectivity Receivers **33**
4.1 IF Choice 34
4.1.1 Image Rejection 34
4.1.2 Zero IF Architecture 35
4.1.3 Near Zero and High IF Architecture 37
4.1.4 IF Assessment 38
4.1.5 DC Offset and 1/f Noise 38
4.1.6 RF Front-End and ADC 1/f-Thermal Noise Corner Frequency 41
4.2 Top-End of the ADC DR 45
4.2.1 Signal Levels, Selectivity, and Maximum ADC Input Signal 46
4.2.2 Crest Factor 48
4.3 Receiver Gain 49
4.3.1 Narrow vs. Broad Band AGC 50
4.4 Bottom-End of the ADC DR 50
4.4.1 Receiver SNR Requirement 50
4.4.2 Receiver Noise Figure and ADC Noise Floor 51
4.5 DR of the ADC 53
4.5.1 DR of a Quadrature ADC 53
4.6 RF Front-End and ADC Linearity Requirements 54
4.6.1 Second and Third Order Harmonic Distortion 54
4.6.2 Second and Third Order Intermodulation and IP2 and IP3 . 55
4.6.3 Third Order Cross-Modulation 58
4.6.4 Distortion in a Quadrature ADC 59
4.7 Example Receiver Partitioning: Receiver for a GSM Mobile Phone 61
4.7.1 IF Choice and Image Rejection 62
4.7.2 Top-End of the ADC Dynamic Range 63
4.7.3 Receiver Sensitivity Requirement and the Bottom-End of the ADC Dynamic Range 65
4.7.4 Receiver Linearity Requirement and ADC Linearity 66
4.8 ADC Requirements, the System Quality Indicators and $\Sigma\Delta$ Modulators as the ADC Architecture 67
4.9 Conclusions 70

5 $\Sigma\Delta$ Modulator Algorithmic Accuracy **71**
5.1 $\Sigma\Delta$ Modulators with 1-bit Quantizer and 1-bit DAC 73
5.2 $\Sigma\Delta$ Modulators with b-bit Quantizer and b-bit DAC 78
5.3 $\Sigma\Delta$ Modulators with 1.5-bit Quantizer and DAC 79
5.4 $\Sigma\Delta$ Modulators with Multiple Quantizers and 1-bit DAC 80

5.5 ΣΔ Modulators with Additive Error-Feedback Loops 82
5.6 Cascaded ΣΔ Modulators 87
5.7 Conclusions 88

6 ΣΔ Modulator Robustness 91
6.1 Portable, Technology Robust Analog IP and Time-to-Market . . . 92
6.1.1 Technology Scaling and Its Impact on Analog Design Parameters 93
6.1.2 A Design Methodology to Increase the Portability of Analog IP 94
6.2 Continuous Time vs. Discrete Time Loop Filter 97
6.3 Feed-Forward vs. Feedback Loop Filter 99
6.4 Gain Accuracy 101
6.4.1 ΣΔ Modulator with 1-bit Quantizer and 1-bit DAC 101
6.4.2 ΣΔ Modulator with b-bit Quantizer and b-bit DAC 101
6.4.3 ΣΔ Modulator with Multiple Quantizers and 1-bit DAC . . 101
6.4.4 ΣΔ Modulator with Additive Error Feedback Loops 102
6.4.5 Cascaded ΣΔ Modulators 104
6.5 Circuit Noise of the Modulator's Input Stage and DAC 105
6.5.1 RC Integrator Input Stage and SI Feedback DAC 105
6.5.2 RC Integrator Input Stage and SR Feedback DAC 106
6.5.3 RC Integrator Input Stage and SC Feedback DAC 106
6.5.4 Impact of Supply Voltage on the Circuit Noise Requirements 107
6.6 Non-linearity 108
6.6.1 Non-linearity in the Input Stage 108
6.6.2 Non-linearity in the Quantizer Decision Levels 111
6.6.3 Inter-Symbol-Interference in the Feedback DAC 112
6.6.4 Non-linearity in the Output Levels of the Feedback DAC . . 113
6.7 Aliasing in ΣΔ Modulators 116
6.7.1 Aliasing in the Quantizer 116
6.7.2 ΣΔ Modulator with an SI Feedback DAC 117
6.7.3 ΣΔ Modulator with an SR Feedback DAC 120
6.7.4 ΣΔ Modulator with an SC Feedback DAC 122
6.8 Excess Loop Delay 127
6.8.1 Excess Time Delay Compensation 128
6.8.2 Excess Phase Compensation 129
6.8.3 DAC Feedback Pulse Shape and Delay 131
6.9 Clock Jitter in CT ΣΔ Modulators 132
6.9.1 The TAJE Model 133

6.9.2 The TPJE Model: Sine Wave Induced Jitter 139
6.9.3 The TPJE Model: Substitution of White Noise Jitter in the Sine Wave Induced Jitter Model 151
6.9.4 The TPJE Model: SI Versus SC Feedback DAC 161
6.9.5 The TPJE Model: An Application Driven Choice Between SI Versus SC Feedback DAC 162
6.10 Conclusions 169

7 ΣΔ Modulator Flexibility 175
7.1 Receiver Dictated Flexibility Requirements 175
7.2 ΣΔ Modulator Clock Flexibility 177
7.2.1 Receiver Architecture with LO-Dependent ADC Clock . . . 178
7.2.2 Receiver Architecture with a Flexible and Independent Clock for the ADC 179
7.2.3 Receiver Architecture with Fixed, Independent ADC Clock 180
7.2.4 Choice of Clock Strategy 182
7.3 Input Stage and DAC Flexibility 183
7.4 Loop-Filter Flexibility 183
7.5 Quantizer Flexibility 185
7.6 Conclusions 186

8 ΣΔ Modulator Efficiency 189
8.1 Power Efficiency FOM: FOM_{DR} 191
8.1.1 Benchmarking with FOM_{DR} 193
8.2 Power Efficiency FOM: $\text{FOM}_{eq,th}$ 194
8.2.1 Benchmarking with $\text{FOM}_{eq,th}$ 197
8.3 Distortion FOM: FOM_{HD3D} 200
8.3.1 Benchmarking with FOM_{HD3D} 202
8.4 Area FOM: FOM_{area} 205
8.4.1 Benchmarking with FOM_{area} 209
8.5 Conclusions 211

9 ΣΔ Modulator Implementations and the Quality Indicators 213
9.1 Digitization at System/Application Level: ΣΔ Modulators for Highly Digitized Receivers 214
9.1.1 A 1.5-bit ΣΔ Modulator for UMTS 215
9.1.2 A Triple-Mode ΣΔ Modulator for GSM-EDGE, CDMA2000 and UMTS 222
9.1.3 An Extremely Scalable ΣΔ Modulator for Cellular and Wireless Applications 232

9.1.4 Multi-mode Modulator Clock Strategy 238
9.2 Digitization at Analog IP Architecture Level: A Hybrid, Inverter-Based ΣΔ Modulator 240
9.2.1 ΣΔ Modulator Architecture 241
9.2.2 Circuit Design 242
9.2.3 Experimental Results 245
9.2.4 Conclusions 248
9.3 Digitization at Circuit and Layout Level: Technology Portable ΣΔ Modulators 249
9.3.1 ΣΔ Modulator Architecure 250
9.3.2 Circuit Design and Layout 251
9.3.3 Conclusions 256
9.4 Implementations Judged on the FOMs and Quality Indicators . . . 257
9.5 Conclusions 260

10 Conclusions **263**

A Harmonic and Intermodulation Distortion in an I&Q System **265**
A.1 Double Sided Spectrum of Second and Third Order Distortion of a Complex Signal 265
A.2 Double Sided Spectrum of Second and Third Order Distortion in a Complex System 266

B Distortion of a Differential Input Transistor Pair Biased in Weak Inversion **269**

C Fourier Series Expansion and Return-to-Zero **271**

D Clock Jitter in an I&Q System According to the TPJE Clock Jitter Model **273**

References **275**

Index **289**

List of Abbreviations

AA	anti-alias
AAD	anti-alias distance
A/D	analog-to-digital
ADC	analog-to-digital converter
AGC	automatic gain control
AWG	arbitrary waveform generator
BER	bit error rate
BPSK	binary phase shift keying
BT	bluetooth
CDMA	code division multiple access
CM	cross-modulation
CMOS	complementary metal oxide semiconductor
CNR	carrier-to-noise ratio
CT	continuous-time
CW	carrier wave
D/A	digital-to-analog
DAC	digital-to-analog converter
DC	direct current
DEM	dynamic element matching
DPSK	differential phase shift keying
DR	dynamic range
DSP	digital signal processor
DT	discrete-time
DUT	device under test
DVB	digital video broadcasting
DVB-H	digital video broadcasting hand-held
DVB-T	digital video broadcasting terrestrial
DWA	data weighted averaging
EDGE	enhanced data rates for GSM evolution

EMC	electro-magnetic compatibility
ENOB	effective-number-of-bits
FM	frequency modulation
FOM	figure-of-merit
GMSK	Gaussian minimum shift keying
GPRS	general packet radio service
GPS	global positioning system
GSM	global system for mobile communication
IC	integrated circuit
IEEE802.X	institute of electrical and electronics engineers local area network standards
IF	intermediate frequency
IL	implementation loss
IM	intermodulation
IIP	input intercept point
IP	intercept point *or* intellectual property
I&Q	in phase and quadrature phase
IR	image rejection
IRR	image rejection ratio
ISI	inter-symbol interference
ISSCC	International Solid-State Circuits Conference
JSSC	Journal of Solid-State Circuits
LCD	liquid crystal display
LNA	low noise amplifier
LO	local oscillator
LSB	least significant bit
LTE	long term evolution
MIMO	multi-input multi-output
MP3	MPEG-1 audio layer 3
MPEG	moving pictures experts group
MSB	most significant bit
NF	noise figure
NRTZ	non-RTZ
NTF	noise transfer function
NZIF	near-zero intermediate frequency
OIP	output intercept point
OSR	over-sampling ratio
OTA	operational transconductance amplifier
PA	power amplifier
PC	personal computer

PCB	printed circuit board
p-cell	parameterized cell
PDA	personal digital assistant
PLL	phase locked loop
PMU	power management unit
POD	performance on demand
PSK	phase shift keying
QAM	quadrature amplitude modulation
QPSK	quadrature phase shift keying
RF	radio frequency
RMS	root mean square
RTZ	return-to-zero
Rx	receiver
S2P	single-ended to parallel
SC	switched capacitor
SD	sigma-delta ($\Sigma\Delta$)
SI	switched current
SDR	signal-to-distortion ratio
SNR	signal-to-noise ratio
SNDR	signal-to-noise-and-distortion ratio
SoC	system-on-chip
SQNR	signal-to-quantization-noise ratio
SR	switched resistor
STF	signal transfer function
TAJE	time-to-amplitude-jitter-error
TD-SCDMA	time division synchronous code division multiple access
THD	total harmonic distortion
TPJE	time-to-phase-jitter-error
TV	television
Tx	transmitter
UMTS	universal mobile telecommunications system
USB	universal serial bus
VGA	variable gain amplifier
VHDL	VHSIC hardware description language
VHSIC	very high speed integrated circuit
VLSI	very-large-scale integration
WIBRO	wireless broadband
Wi-Fi	wireless fidelity
WIMAX	worldwide interoperability for microwave access

WLAN	wireless local area network
ZIF	zero intermediate frequency

Terminology

Adaptability
The ability of a system to change or be changed to fit a changing outside world. To be able to respond the system needs inputs which are a measure of the changes in outside world parameters.

Co-existence
The ability of two or more systems to operate at required performance being active at the same time.

Co-habitation
The ability of two or more systems to operate at required performance being in the same package or volume.

Durability
The property of a system being intensively used without degradation of the system quality.

Efficiency
The ratio between system performances and used resources. Efficiency should be as high as possible.

Flexibility
A combination of re-configurability, scalability, and adaptability.

Portability
The ease with which the system function can be transformed from one form to another. For instance a change of material or technology.

Re-configurability
The quality of the system to change from one system function into another system

function, by changing the order, or position of the different sub-systems of the main system.

Reliability
Reliability is the probability that a system will operate at its required performance with changing outside world influences over time.

Reproducibility
The quality of being reproducible. Reproducibility is a measure how sensitive the system function is to the imperfections and variations of the production process.

Robustness
The property of strong constitution to outside influences e.g. to temperature, humidity, radiation, force, interference, imperfections and variations of the production process.

Re-usability
The quality of a system to be re-used in the same or in a different system.

Scalability
The ability to scale or trade the system parameters to meet the requirements of the current system function application, by re-programming the system's parameters.

Simplicity
The quality of using minimum resources to achieve the maximum system functionality and performance.

Testability
The ease with which the system performance can be verified after manufacturing.

Variability
A collection of phenomena characterized by uncontrolled parameter variation between individual unit components. This collection is populated with a large number effects ranging from offset mechanisms to reliability aspects. Variability effects can be subdivided along three main axes: Time independent versus time variant effects. Global variations versus local variations. Deterministic versus stochastic (statistical) effects.

List of Symbols

A_{V_T}	threshold mismatch process parameter	V·m
A_x	area of block x	m^2
A_x	amplitude of signal x	- or V
AAD	anti-alias distance	-
b_n	modulator resonator coefficient n	-
B	bandwidth	Hz
B_{eff}	effective bandwidth	Hz
BER	bit error rate	%
CNR	carrier to noise ratio	Hz
D	duty cycle	-
DR	dynamic range	-
E_b	energy per bit	J
$ENOB$	effective-number-of-bits	bits
FOM_{DR}	conventional ADC power efficiency FOM	J/conversion
$FOM_{eq,th}$	power efficiency FOM based on ADC supply load and noise impedance	-
FOM_{HD3D}	third order distortion FOM	-
FOM_{area}	area FOM	W/m^2
f_{in}	input signal frequency	Hz
f_c	1/f-thermal noise corner frequency	Hz
f_s	sample frequency	Hz
G	gain	-
G_{Vl}	loaded voltage gain	-
F_x	gain of a filter of order x	-
gm	transconductance of a transistor	A/V
Gm	transconductance of a differential pair	A/V
$HD2$	second order harmonic distortion	-
$HD3$	third order harmonic distortion	-
$HD2D$	second order harmonic distortion distance	-

$HD3D$	third order harmonic distortion distance	-
I_D	drain current of a MOS transistor	A
i_n	normalized modulator integrator coefficient	rad
$IIP2$	second order input intercept point	-
$IIP3$	third order input intercept point	-
IL	implementation loss	-
$IM2$	second order intermodulation	-
$IM3$	third order intermodulation	-
$IM2D$	second order intermodulation distance	-
$IM3D$	third order intermodulation distance	-
$IR_{cochint}$	co-channel interference ratio	dB
IRR	image rejection ratio	-
j_n	normalized modulator feed-forward coefficient	-
k	Boltzmann's constant, $1.38 \cdot 10^{-23}$	J/K
k_n	normalized modulator resonator coefficient n	-
L	channel length of a MOS transistor	m
L	loop filter order	-
l_n	maximum signal swing of integrator n	-
m_n	normalized modulator stability coefficient	-
b	number of bits	bits
N	number of levels	levels
N_x	integrated noise power of x	V^2
NF	noise figure	-
N_0	single-sided noise spectral density	W/Hz
$NRTZ$	non-return-to-zero	-
OIP	output intercept point	-
OSR	over-sampling ratio	-
P	power	W
p_x	sub-block x area percentage out of the total area of an IP block	-
Q	charge	C
R	resistance	Ω
$R_{\frac{1}{f},th}$	1/f-thermal integrated noise power ratio	-
R_b	transmission bit-rate	bits/s
$R_{eq,th}$	equivalent ADC noise impedance	Ω
R_{load}	equivalent ADC supply load impedance	Ω
$R_{n,RF,ADC}$	RF front-end – ADC integrated noise contribution ratio	-
$R_{x,y}$	ratio between x and y	-

RTZ	return-to-zero	-
s_T	(effective) oxide scaling factor over succeeding technology generations	-
S_x	spectral noise power density of x	V^2/Hz
$S_{T,i}$	expected technology scaling factor of IP block i	-
SDR	signal-to-distortion ratio	-
SNR	signal-to-noise ratio	-
$SNDR$	signal-to-noise-and-distortion ratio	-
$SQNR$	signal-to-quantization-noise ratio	-
$SJNR$	signal-to-jitter-noise ratio	-
T	temperature	K
T_p	DAC output pulse width	s
T_s	sampling period	s
t_{ox}	(effective) oxide thickness	m
V_{GS}	gate-source voltage of a MOS transistor	V
V_{GT}	overdrive voltage of a MOS transistor	V
$V_{in,rms}$	rms value of the input voltage	V_{rms}
$V_{n,rms}$	rms value of the noise voltage	V_{rms}
V_{pp}	peak-to-peak voltage	V_{pp}
V_T	MOS transistor threshold voltage	V
$v_{x,in,max,rms}$	maximum input voltage of block x (x is ADC, RF etc.)	V_{rms}
W	channel width of a MOS transistor	m
$\beta_\square$	current factor	A/V^2
Δ_i	input signal to jitter tone ratio	-
Δ_s	clock carrier to jitter tone ratio	-
π	pi, 3.141593	-
σ_x	standard deviation on variable x	same as variable x
τ	time constant	s

Nomenclature

- A variable in dBx is referenced to x. For example:

dBV	unit in dB's with respect to 1 V
dBμV	unit in dB's with respect to 1 μV
dBm	unit in dB's with respect to 1 mW dissipated in a pre-defined reference resistor R

- To indicate that a variable is defined in decibels, the superscript dB is added to the variable's symbol. For example:

V	V is defined in [V]
V^{dBV}	V is defined in [dBV]
$V^{dB\mu V}$	V is defined in [dBμV]
P	P is defined in [W]
P^{dBm}	P is defined in [dBm]

- In Chap. 6 the TPJE jitter model is introduced. To separate time jittered signals from signals without time jitter, the superscript $\sim$ is introduced. A few examples:

t_x	time instant x of a clock without jitter
$t_x^{\sim}$	time instant x of a clock with jitter
T_s	sample period of the clock of frequency f_s
$T_s^{\sim}$	sample period of the jittered clock of frequency $f_s^{\sim}$

- Dynamic range (DR), Signal-to-Noise-ratio (SNR), Signal-to-Quantization-Noise-ratio (SQNR), Signal-to-Jitter-Noise-ratio (SJNR) are always voltage division when not specified in dB's.

- If a variable is dimensionless this is indicated by [-].

Chapter 1

Introduction

The introduction of IC technology has led to a revolution in the integration of electronic systems. Nowadays millions of transistors can be put in a very small volume, together forming complex functions. This has opened up the road to many new products, like today's personal computers, digital TV, and many battery powered products, like PDAs, advanced wrist watches, MP3 players and the mobile phone. In this book the mobile phone will be further investigated. More specifically, this book will zoom-in on the ADC in the receive path of a mobile phone.

In this chapter the work presented in this book will be motivated by making an inventory of important trends in the mobile phone industry, from the application and the technology side. These trends will be related to the implications these trends put on the ADC used in the transceiver of a mobile phone.

First, the market trends for cellular and connectivity terminals will be identified. These trends will be translated into system quality indicators, which will be used for the system specification and qualification. From the implementation side, Moore's law will be used to explore the technology trends of mainstream CMOS technologies used for the integration of the transceiver, and the implications these have on the implementation of circuits on silicon. Furthermore, Shannon's theorem will show that time resolution is in favor to amplitude resolution, which pleads for the use of a $\Sigma\Delta$ ADC architecture for the transceiver's ADC, the $\Sigma\Delta$ converter.

This chapter will end with the presentation of the book aims and scope and the outline of the book.

R.H.M. van Veldhoven, A.H.M. van Roermund, *Robust Sigma Delta Converters*,
Analog Circuits and Signal Processing, DOI 10.1007/978-94-007-0644-6_1,

1.1 Advanced, Multi-standard Cellular and Connectivity Terminals for the Mass Market

One of the markets boosted by IC technology is the mobile phone market. Due to the ability to integrate electronic systems in a single IC, the mobile phone has developed from a large device with only a phone call function, to a slim and multi-feature device. This has led to the mass market introduction of the mobile phone in the 1990s, and at the end of 2009 the number of mobile phone subscriptions has exceeded 4 billion. With an earth population of about 7 billion, that means at least half a subscription per person, an indication of the size of the market. And development has not stopped yet. The mobile phone has developed itself to more than a phone. Multiple features are added to the phone making it more compelling than the phone offered by the competitor. Larger touchscreen displays, FM radio, GPS navigation, electronic compass, cameras, mp3 player, voice recorder and games are being added to the phone without increasing and often even decreasing the phones size. Co-existing services like making a GSM phone call while looking up details on the internet or browsing your PC looking for documents should work seamless on these advances devices, and even the possibility of watching live video streams should be available. The increase of phone complexity which

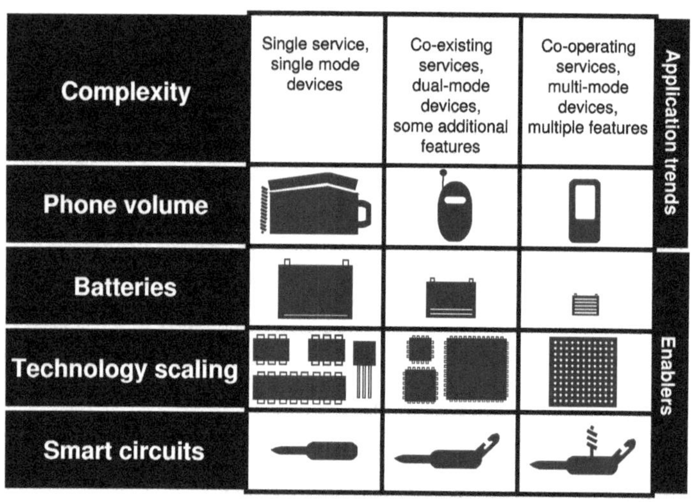

Figure 1.1: *Convergence of multiple, single application phones into single flexible, multi-mode phones*

at the same time has to fit in a smaller volume requires miniaturization of the technologies the phones are built with. Amongst other technologies, this means smaller batteries, more integrated functionality in less silicon area, and smarter systems and circuits (Fig. 1.1). The impact of IC developers on battery and IC technology miniaturization is very limited. However, IC developers can have a huge impact by coming up with smarter system and circuit solutions exploiting the advantages of transistor technology miniaturization. Only this way the increased functionality can fit the limited volume and power budget available. In the next few sections, it will be shown how the increase in phone complexity impacts the transceiver and the A/D converter in the receiver, and how transistor technology can be exploited by taking a smart system and circuit choice for the A/D converter architecture.

1.1.1 Complexity: Mobile Phone Trends, Its Impact on the Transceiver and the Quest for Integration

A typical block diagram of a mobile phone is shown in Fig. 1.2. The core of the phone is a DSP, which is surrounded by interface circuitry. The DSP is connected to the cellular network with a transceiver, to a user through the audio codec, keyboard and displays, to a PC with a USB, WLAN, and/or Bluetooth transceiver, to memory for data storage and to a battery being the energy source. Additional features included are a digital camera and an FM radio. The thick line outlines

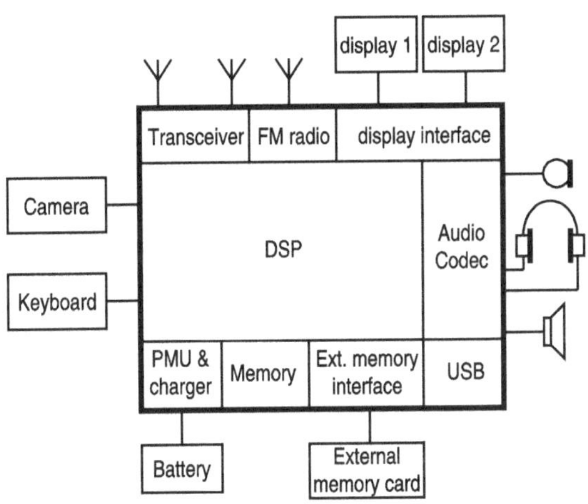

Figure 1.2: *Block diagram of a mobile phone with an extremely high degree of IC integration*

a possible IC boundary. In this example every functional block which can be integrated in a standard IC technology, is integrated on the same chip. If done so, the integrated system complexity will be huge, and possible co-existence issues between the systems integrated on the same chip have to be identified before the actual IC design starts. In most phones available on the market at this moment, functional blocks are separated in multiple ICs, to reduce the integration complexity, which is contradictory to a form factor decrease and a functionality increase. This asks for convergence of stand-alone, single application ICs into scalable, programmable, re-useable, and platform based ICs, to be able to create sophisticated devices in the limited volume available. This convergence is the driving force to integrate the mobile phone's transceiver as much as possible, using as little external components as possible. At the same time the radio design complexity is increased by four radio technologies arising, which are multi-standard (multi-mode) radios, Multi-Input Multi-Output (MIMO) radios, software defined radios and cognitive radios. These radio technologies ask for clever system and circuit solutions, which makes the transceiver one of the most challenging parts to integrate on an IC.

A multi-standard radio has the ability to be used anywhere around the globe and to connect to any cellular and connectivity communication network, which requires a very flexible transceiver. Figure 1.3 gives an overview of the most popular globally used standards, with their channel bandwidths. The figure shows that the channel bandwidth can vary between 200 kHz and 28 MHz which will put requirements on the flexibility of the receiver chain. Furthermore, the input signal dynamics and frequency content at the antenna will be different in each standard, increasing the flexibility requirements on the receiver further. The trend to more flexible receivers is confirmed by modern standards like WIMAX and LTE which already expect flexibility of the receiver, as the channel bandwidth is adaptable to the service to be delivered. MIMO radios have multiple radios integrated on a single IC to add diversity, increase sensitivity, or to be able to connect to multiple (different) communication channels at the same time. A typical use case could be one receive path for the connection to the GSM cellular network, one for the Bluetooth connection to a wireless headset, one as an FM radio, and one to search for documents on your personal computer via a WLAN connection. Next to that the receiver paths have to be reusable for different cellular (GSM, CDMA, UMTS, etc.), connectivity (Bluetooth, WLAN, WIMAX) or radio (FM) standards.

Software defined radios add performance parameter programmability to the receiver. The performance parameters are programmed according to the system's requirements to cover different standards. To increase the power efficiency of the radios, performance-on-demand (POD) is added to the receiver. The radio monitors the signal dynamics at the antenna and adapts the performance delivered and the proportional power consumed by each block accordingly.

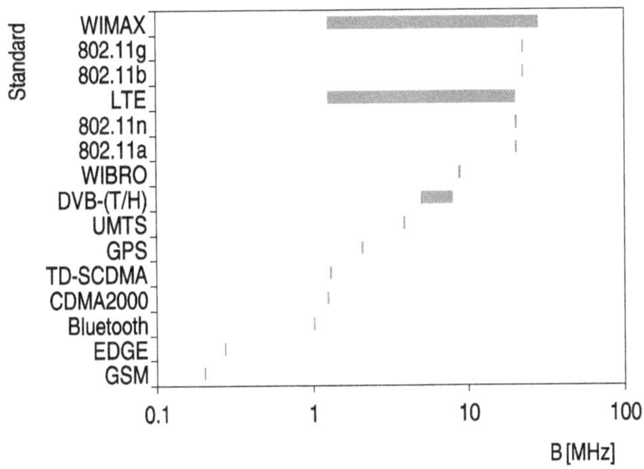

Figure 1.3: *Channel bandwidths of the different communication and connectivity standards*

Cognitive radios further improve the efficiency of the transceiver by adding spectrum sensing and flexible spectrum allocation to the radio. This will not only ask for flexibility in the transceiver, but also for flexibility in the services the radio networks provide.

A possible block diagram of such a multi-standard, MIMO, software defined and cognitive radio is presented in Fig. 1.4. The radio has multiple receive and transmit pipes, has reconfigurable blocks to implement POD, and has a radio resource manager, which programs the performance of the different blocks as required, and includes the spectrum sensing and allocation algorithm, to adapt to the transceivers environment. These new radio technologies ask for more adaptability and flexibility at every abstraction level of the transceiver, which does not come for free and will increase the integration complexity and design time of radios on a single IC. But it is not only these radio technologies which ask for more adaptability and flexibility of the transceiver. It is also the increased competition in the mobile market which forces phone manufactures to come up with clever, reusable system blocks, to reduce system design time, and to be able to set products on the market more quickly. The decreasing time-to-market in combination with the increasing complexity requires consolidation of the radio IC manufacturing industry, to reduce the development costs and time of these advanced radios.

In this book the focus will be on the receive path of the mobile phone's transceiver, and more specifically on the ADC in the receiver chain. The question arises which

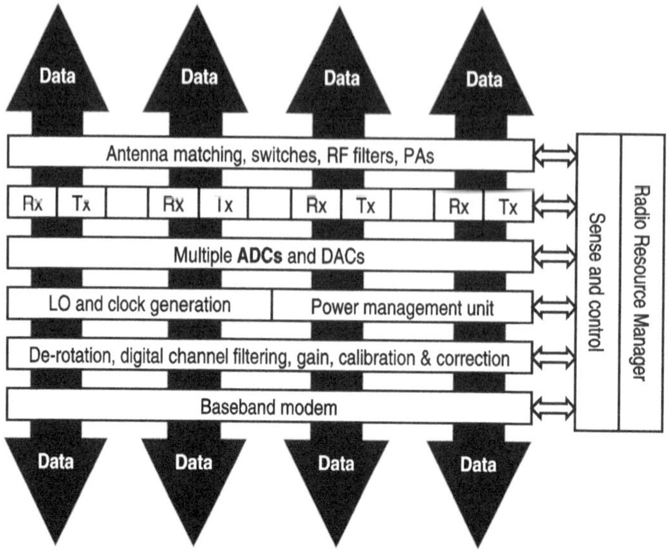

Figure 1.4: *Block diagram of a multi-standard/MIMO/software defined/cognitive radio*

receiver architecture is the best fit on the requirements of these modern radios, with the boundary of a reasonable ADC power consumption. The adaptability and flexibility of the receive paths ask for a digitized receiver architecture in which most of the adaptability and flexibility can be done in the digital domain. This moves the A/D converter closer to the antenna which will have a major impact on the required ADC accuracy and bandwidth, the ADC being the main subject of this book.

1.1.1.1 Implications of Trends on the ADC Specification Generalized in Quality Indicators

In the previous section, it has become clear that the A/D converter needs to have a high accuracy being close to the antenna, needs to be small as it houses in a mobile phone and needs to be power efficient as the phone is battery powered. As the phone life-time decreases, time-to-market becomes more important. The A/D converter needs to be flexible as it is used in a reconfigurable receiver which is used for different communication systems. Finally, the A/D converter should be robust to interference and of course should not generate interference, as the A/D converter has to coexist in a complex environment with other electronic systems in the same housing. The requirements mentioned above are captured in five quality indicators:

1. Accuracy
2. Robustness to secondary inputs
3. Flexibility
4. Efficiency
5. Emission of secondary outputs

In Chap. 2 these quality indicators will be founded.

In general, these quality indicators can be used to specify and qualify analog IP. In this book these quality indicators will be applied to $\Sigma\Delta$ modulators, except for emission which is out of the scope of this book.

1.1.2 Transistor Scaling: VLSI and Moore

In 1965 Moore predicted that the number of transistors the industry would be able to place on a chip would double every year [1]. In 1975, he updated his prediction to once every two years [2]. It has become the guiding principle for the semiconductor industry to deliver ever-more-powerful chips while decreasing the cost of electronic systems.

The development of modern CMOS technologies is mainly driven by the digital processor industry. The more transistors that can be put in the same area, the more powerful the digital processing per area will be, which requires technology scaling, and has led to a digital circuit specific technology optimization.

In ICs which interface with the analog world surrounding us, like the cellular and connectivity transceivers of Sect. 1.1, analog-digital and digital-analog interfaces are required, which have to be designed in the same digitally optimized process. This requires a strategy on how to exploit the advantages of the digital technology for analog circuit design, while dealing with the technology's disadvantages.

To make a quantitative inventory of pro's and con's, a transistor feature size scaling factor s_T is introduced. The transistor feature size scaling factor is defined by

$$s_T = \frac{L_{min,new}}{L_{min,old}} \quad [\text{-}] \tag{1.1}$$

and represents the scaling of the minimum L of transistors available within a new technology compared to the minimum L of transistors in the current technology node. For two succeeding CMOS technologies $s_T \approx 0.7$.

When constant field technology scaling is assumed (which was valid from the year 1990 to 2000), the advantages of technology scaling are [3, 4]:

- Speed increases with $1/s_T$
- Area decreases with s_T^2
- Dynamic power consumption decreases with s_T^2
- gm/I increases slightly (more gm for the same transistor bias current)

The disadvantages of technology scaling are:

- Power supply voltage decreases with s_T
- Noise margin of digital circuits decreases s_T
- Cross-talk increases with $1/s_T$
- Static power dissipation increases
- Transistor output impedance decreases

The scaling pro/con comparison shows that for both analog and digital functions the opportunities lie in the increasing speed of new technologies. The combined area and speed scaling of digital circuits make digital circuits $(1/s_T^3)$ times more powerful in the same area, when fabricated in a next generation technology for constant field scaling. Furthermore, the power-delay product decreases, which makes the digital circuits more efficient in the next technology node (in the constant field technology scaling period, power efficiency increased with $1/s_T^3$). This makes it attractive to shift analog functions into the digital domain where possible[1].

This means that the ADC and DAC converters at system level are shifted closer to the out-side-world, and analog signal conditioning is eliminated as much as possible. The scaling disadvantages for the remaining analog functions should be solved by choosing smart analog function architectures, or even better, by assistance of digital circuits.

[1]It has to be noted here that for constant voltage scaling in the period 2001 to date, the efficiency of digital circuits still increases, but at a reduced pace of $1/s_T$ [4].

1.1.2.1 Transistor Scaling in the Context of Shannon's Channel-Capacity Theorem

Shannon's channel-capacity theorem relates the systems' bandwidth and signal to noise ratio into the system's channel-capacity. The higher the available bandwidth and SNR, the more information bits/s can be put through the system reliably, without information loss. When both the signal source and the channel noise sources have a Gaussian distribution, the capacity of a channel can be calculated by:

$$\text{Channel capacity} = B \cdot \log_2(1 + \text{SNR}) \;\; \text{[bits/s]} \tag{1.2}$$

The information sent over the channel can be put in the amplitude/resolution (SNR) or in the time (bandwidth) domain. As can be seen from Eq. 1.2, the cost of increasing the channel capacity is easier in the time domain[2]. This maps onto the scaling advantages of deep sub-micron CMOS technologies, as the speed of a new technology generation increases, making it future proof. In the amplitude domain, technology scaling trends predict that the ratio $V_{supply}(s_T)/\sigma_{V_T}(s_T)$ is constant, which means that the performance at best remains the same.

1.1.3 Smarter Circuits: $\Sigma\Delta$ Modulators for Mobile Applications

The choice of the architecture of the A/D converter of Sect. 1.1 should be driven by the receiver application of the A/D converter, and by the speed advantages of (future) deep submicron technologies, while being robust to the disadvantages.

Sigma Delta modulators trade amplitude resolution for time resolution, by using over-sampling in combination with noise shaping. In particular 1-bit $\Sigma\Delta$ modulators only use a 1-bit quantizer and DAC, making the modulator insensitive to transistor mismatch, and inherently linear.

Traditionally, a continuous-time (CT) $\Sigma\Delta$ modulator has a better power efficiency compared to switched capacitor (SC) modulators, as a SC $\Sigma\Delta$ modulator needs high bandwidth filter circuits. Furthermore, CT $\Sigma\Delta$ modulators take advantage of the CT nature of the loop filter as it provides anti-alias filtering, which is of great merit when the $\Sigma\Delta$ modulator is used in a receiver architecture, because the $\Sigma\Delta$ modulator is more robust to interference at the input of the ADC. A SC implementation of the filter loses the advantage of a built-in anti-alias filter, as it has a sampler at the input. Furthermore, SC $\Sigma\Delta$ modulators are more prone to emit and receive interference, as they use switching everywhere in the filter.

---

[2]Note that the SNR is related to bandwidth. Therefore, this remark is not so straightforward.

A disadvantage of a CT ΣΔ modulator is its sensitivity to time jitter on the clock. This effect can be reduced by using a SC instead of a switched current (SI) feedback DAC.

ΣΔ modulators exploit Shannons' bandwidth parameter to increase channel capacity, and use the speed advantages of future deep submicron technologies. Furthermore, ΣΔ modulators using a CT loop filter and a SC feedback DAC, combine the advantages of CT and SC ΣΔ modulators. For these reasons ΣΔ modulators with a CT loop filter in combination with a 1-bit SC feedback DAC are chosen as the basis of most of the presented ΣΔ modulators in this book.

1.2 Book Aims

As seen in the introduction the trends in an application can have a major impact on the requirements of the IP to implement the application. This book studies how to deal with the ever increasing requirements on such IP, and how the technology advances of the technology the IP is manufactured in can be exploited choosing the right design methodology.

It is the objective of this book to explore possibilities to implement high quality ΣΔ modulators. Key steps in this process are:

1. Find quality indicators which can be used to qualify a signal processing system and the analog IP blocks with which it is built (Chaps. 1 and 2)

2. Define a general design methodology for high quality analog IP blocks (Chap. 2)

3. Derive requirements for a ΣΔ modulator used in a low-power, multi-standard, highly digitized wireless receiver (Chap. 3 through Chap. 4)

4. Contribute to ΣΔ modulator theory and categorize this theory along the presented quality indicators (Chap. 5 through Chap. 8)

5. Apply the presented design strategy and theory to ΣΔ modulators and implement them on silicon (Chap. 9)

6. Judge the implemented ΣΔ modulators on the quality indicators to see whether the chosen design strategy was successful (Chaps. 9 and 10)

1.3 Book Scope

The design methodology for high quality analog IP blocks will be applied to ADCs in (N)ZIF receiver architectures only. Chosen ADC architecture is the $\Sigma\Delta$ modulator. The $\Sigma\Delta$ modulator presented in this book will either have a CT or a partly CT—partly digital loop filter; no SC loop filters will be used. The number of feedback levels used in the $\Sigma\Delta$ modulator feedback path will be limited to 2 or 3. These choices will be reasoned during the book.

As the modulators presented were to be part of a receiver SoC with a large amount of digital processing on board, the design technologies used to implement the modulators presented in this book are all standard digital CMOS technologies. No additionally available non-standard process options were used in these technologies. Native technology supply voltage was used in all cases.

1.4 Outline

The outline of the book is summarized in Fig. 1.5. Chapter 1 summarizes the

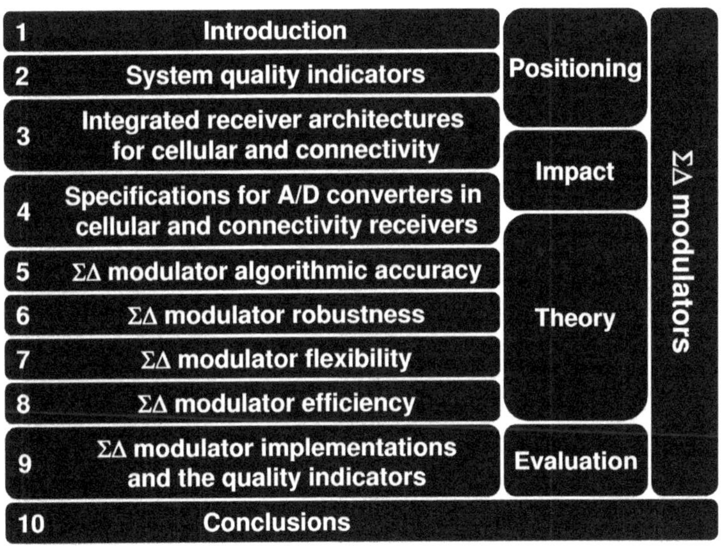

Figure 1.5: *Outline of the book*

trends in transceivers for cellular and connectivity and the impact these trends can have on the analog IP blocks these transceivers are built with. Chapter 2 will describe how we can categorize this impact in quality indicators, and will explain why it can be advantageous to increase the level of digitization in your system. Chapter 3 will show an example of such a digitization process at system level. A receiver architecture will be shown for different levels of digitization, by shifting the ADC closer and closer to the antenna, and will briefly elaborate on the impact on the ADC. The specification of the ADC in such receivers and the interchange between RF front-end and ADC performance are derived in Chap. 4. Based on the specification outcome and using the quality indicators, a choice for the ADC architecture will made in this chapter, which is the $\Sigma\Delta$ modulator architecture. In Chaps. 5 to 8 $\Sigma\Delta$ modulator theory is derived and extended and is categorized along the quality indicators presented in Chap. 2. In each of these four chapters, the properties of $\Sigma\Delta$ modulators will be tested on the quality indicator presented in that chapter, and will show how the score on this quality indicator can be improved. Chapter 9 presents all implemented modulators. In this chapter $\Sigma\Delta$ modulators will be shown, which are subject of digitization at different abstraction levels. Furthermore, the implemented modulators will be benchmarked with state-of-art $\Sigma\Delta$ modulators published in literature and will be tested on the quality indicators (except for emission Sect. 1.1.1.1). At the end, the conclusions will be presented in Chap. 10.

Chapter 2

System Quality Indicators

The integration of systems on a chip, has led to a revolution in the electronic industry. Large, complex system functions can be integrated in a single IC, paving the road to many battery powered portable applications like the cellular phone, wireless products, MP3 players and so on. The constant drive to improve these applications and to include extra features has enormously increased the pace with which new generation portable products are introduced on the market. Keeping its main function, extra demands are put on the system realizing this function. Smarter integrated system solutions, which are cheaper, smaller, more power efficient, robust to interference, more flexible, etc. are required. In this chapter these additional system requirements are captured in five quality indicators which indicate the quality of the integrated system, and which help to structure the analysis complex systems. The five quality indicators used are: accuracy, robustness, efficiency, flexibility, and emission. The system and its quality indicators are presented in Sects. 2.1 and 2.2 respectively.

In Sect. 2.3 the quality indicators are used to motivate why it can be advantageous to shift analog functionality into the digital domain which implicates the need for high dynamic range and high bandwidth analog-digital interfaces. In Chap. 3, the quality indicators are used to find a power efficient receiver architecture for use in a mobile phone. The influence of system partitioning on the quality indicator requirements of the analog-digital interface used in such receiver is postponed to Chap. 4. The quality indicators are used to determine the quality of the analog-to-digital interface in Chaps. 5 to 8. In a later stage in this book (Chaps. 8 and 9), the quality indicators are used to compare the analog-to-digital interfaces presented in this book to the quality of analog-digital interfaces presented in literature with the help of a benchmark. In this benchmark, the same or similar analog-digital

R.H.M. van Veldhoven, A.H.M. van Roermund, *Robust Sigma Delta Converters*,
Analog Circuits and Signal Processing, DOI 10.1007/978-94-007-0644-6_2,

interfaces are compared, on their quality indicators as these indicators can be a key differentiator to a customer.

2.1 The System Function and Its In- and Outputs

A system could be defined as a group of interacting, interrelated, and interdependent elements executing a function. A system function has one or more input(s) X, which are processed in some way by the system function F, yielding one or more output(s) Y. This is schematically shown in Fig. 2.1. The system inputs

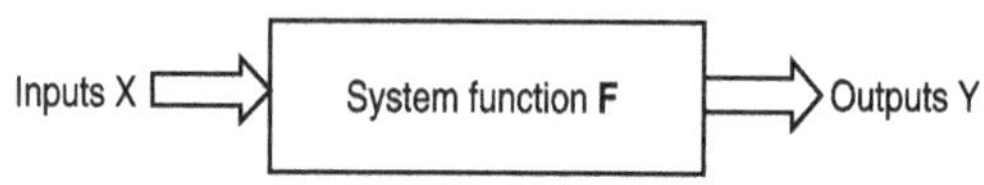

Figure 2.1: *System function with its inputs and outputs*

can be sub-divided in 2 categories, namely the primary inputs and the secondary inputs. The primary inputs are the wanted inputs, which have to be transferred by the system to the wanted outputs, a process which is called the primary process. The secondary inputs are inputs, which are unavoidable in some way, when implementing the system.

The secondary inputs are split up in 3 categories:

- Resources
- Outside world influences
- System interface

The first category describes the resources that are required for the systems' primary process (e.g. power source, material, design effort). The second category comprises the outside world influences, which describe inputs imposed by the outside world onto the system and can degrade the quality of the primary process (e.g. temperature, interference, manufacturing imperfections, noise). The last category represents inputs, which are required by the user or the system itself, to adapt and change the properties of the primary process to the current system application (e.g. volume control, or tuning function). The secondary inputs are shown in Fig. 2.2. The outputs of the system can also be sub-divided in the primary and secondary categories. The primary output is the output the system was designed for, the wanted output. A secondary output is an output, which was not

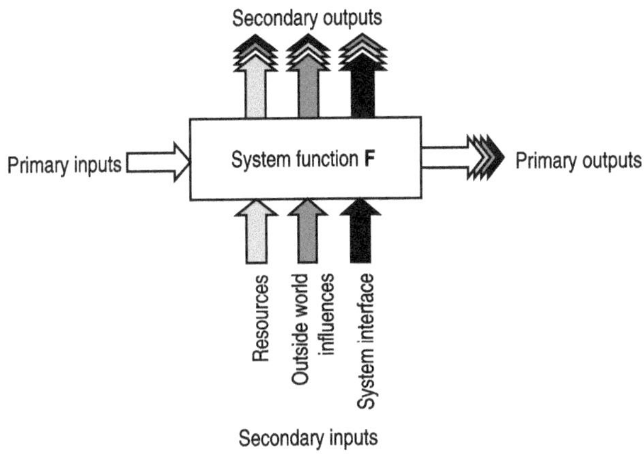

Figure 2.2: *System with its primary and secondary inputs and outputs*

intended to be an output of the system function, like the heat or interference generated by the system. The primary output might be a function of the secondary inputs. Next to that, the combination of primary and secondary inputs might cause cross-correlated secondary outputs. The different in- and outputs are shown in Fig. 2.2. It is very likely that some of the cross-correlation factors of F are zero. Of course there is also wanted correlation between inputs and outputs, examples are: primary input to primary output, secondary system interface input to primary output.

2.2 System Quality

An ideal system has infinite accuracy, uses its resources 100% efficiently, is unaware of influences from the outside world and is re-usable for different applications. However, during the system implementation phase, it will show, that there are limits to the accuracy and efficiency the system can achieve, including flexibility turns out to have its cost, the system will be susceptible to the outside world, and the system will generate secondary outputs which might interfere with the system itself or neighboring systems.

To determine the quality of a system it is judged on several quality indicators, which are divided into five groups:

1. Accuracy
2. Robustness to secondary inputs
3. Flexibility
4. Efficiency
5. Emission of secondary outputs

The quality indicators will be explained by the following sections.

2.2.1 Accuracy

The accuracy is the precision with which the primary system function can be fulfilled. The accuracy or performance of the system is measured on the quality of its primary outputs, compared to the quality of the primary inputs, and is determined by system choices.

2.2.2 Robustness to Secondary Inputs

Another measure to judge the quality of a system is the systems' robustness. The outside world can distort the primary function of the system in some way due to implementation aspects. The more insensitive the system is for influences from the outside, the more robust the system is. Examples of outside world influences are temperature, humidity, interference, noise, force, process spread and material imperfections. A few examples of different measures to quantify the systems' insensitivity to the outside world are durability, reliability, reproducibility and portability (technology independence).

2.2.3 Flexibility

The flexibility of a system indicates the re-configurability, adaptability and scalability a system, to meet changing requirements, or circumstances. It measures the extent in which (parts of) the system function can be changed into different system functions, for instance with a different accuracy. An adaptable system has the ability to respond to a changing outside world. To be able to respond, the system needs inputs measuring the changes in the outside world. Scalability describes the ability to scale or trade system parameters to meet the requirements of the current system function application. A re-configurable system is able to change from one system function into another system function, by changing the order or position of the different sub-systems of the main system.

The requirements on the flexibility of a system are often identified by use-case studies. It makes an inventory of expected human behavior and the way a system is expected to be used. To make an inventory of use-cases, marketing research has to be done.

2.2.4 Efficiency

Efficiency indicates how economical a resource is spent. Important efficiencies are power and area efficiency, as nowadays feature rich, battery powered and portable applications require low power consumption and small form factor. Other relevant efficiencies are testability, re-useability and design effort.

Testability describes the ease with which the required system accuracy of a system can be verified after manufacturing. Re-useability describes the extent in which parts of the system can be re-used for other systems. Sub-system functions can be categorized in libraries with clearly defined input and output conditions. In this way new system functions can be created with of the shelve parts coming out of the library, decreasing time to market, and reducing maintenance of different products as they share parts from the same library. Design effort describes the effort to build the system and is a resource which should be spent with great care, as it is costly and scarce.

Benchmarking is used to quantify the efficiency of a certain system. In a benchmark different system implementations, which have the same or similar system functionality are compared on their efficiencies. The efficiencies are bounded by fundamental limits (like thermal noise, maximum technology speed and availability of man power), but as the implementation of a system has additional cost, the maximum system efficiencies achievable are determined by the current state-of-art.

2.2.5 Emission of Secondary Outputs

Another system quality indicator is the amount in which the system generates secondary outputs. It is important to make an inventory of the secondary outputs the system emits as these outputs can distort the primary process of the system itself, or the primary process of other systems. Examples are heat, and electrical and magnetic interference.

2.3 The Digital Revolution

The quality indicators presented in the previous section, make the introduction of digital circuitry in nowadays integrated system functions unstoppable. A digitally implemented system is greatly in line with the quality indicators as will be shown in the next section, something that is not so obvious for the same system function implemented with analog circuits.

The application of digital enhancements to system functions is numerous. Below several examples are given of systems, which use digital functionality to implement tasks, which are very difficult to implement with analog circuits, if possible at all.

- The reliability of wireless transmission of speech and video streams is greatly improved by the introduction of digital data transmission. The digital modulation techniques used in these wireless links are much more robust to interference than completely analog modulation schemes. Digital error correction algorithms further improve the reliability of the wireless link.

- In medical imaging applications an A/D converter converts the sensor outputs of medical imaging equipment into the digital domain. The digital signal processing following the A/D converter for example allows for the construction of 3D images of the human body, leading to a diagnosis of better quality and potentially a longer life.

- A digital photo camera turns something visible into a digital representation using a photo sensor and A/D converter. After transferring the data to a PC, further image processing and retouching (like red-eye reduction) can easily be done in software.

The digital world is penetrating daily life everywhere. But it is not only the digital processing, which facilitates this increased quality of life and number of features; an interface is required between the analog and digital world. Although the outside world is analog, it is seems much easier to do advanced signal processing in digital hardware or software. It must be noted though that this shift from analog to digital signal processing, does not come for free and is bounded by the performance-cost ratio of A/D and D/A functions and by the speed limits of the technology the digital processing is made in.

2.3.1 The Analog-Digital Interface

Because the outside world is still analog and the processing is preferably done digitally, the introduction of analog-to-digital and digital-to-analog converters has been inevitable. Figure 2.3 shows a generalized implementation of a system function which has been (partly) digitized. From the figure it is evident that the quality of the A/D and D/A converters used in the signal path can be a quality determining factor in the overall system function. This opposes challenges on the design of the A/D and D/A converters. The more our world is captured digitally, the more we must convert from analog to digital and reconstruct from digital to analog, which implicates a trade-off between the amounts of analog and digital functionality and their implementation cost. The system of Fig. 2.3 is split into a receiver, a process-

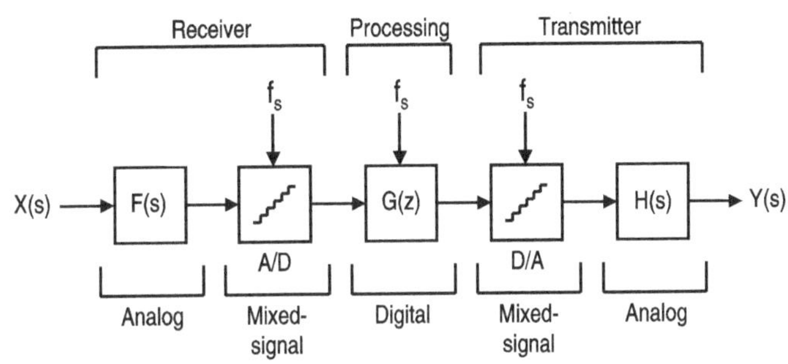

Figure 2.3: *Partially digitized system*

ing unit and a transmitter. The receiver receives an analog input X from a sensor, e.g. a microphone, a temperature sensor or an antenna. X is conditioned by F, which can include both gain and filtering, such that it most efficiently fits the input DR of the ADC. The ADC converts the analog input signal into its digital representation at a clock rate f_s. In the digital domain G represents the required digital signal processing which implements the task which is more efficient or powerful in digital hardware, or maybe even software. The output of the processing unit is connected to a D/A converter, which outputs the analog signal, which again is conditioned to the right amplitude and frequency content, yielding the desired output Y.

The more of the analog functionality represented by F and H is shifted into G, the more demands will be put on the A/D and D/A converter. This requires a system optimization which leads to realistic A/D and D/A converter requirements, which are in line with what is dictated by a benchmark of converters with state-of-the-art performance. This process will be described in Chap. 4.

Before going into the converter function, first it will be motivated why it is advantageous to replace as much as analog functionality by digital functionality. This is done in the next section, where digital functionality will be tested for its compliance to the quality indicators.

2.3.2 Digital Systems and the Quality Indicators

The advantages of digital signal processing compared to analog signal processing are clear. Once in the digital domain, the signal processing is much more powerful, and advanced features can be added in the signal processing path much easier. In this section the match between digital circuits and the quality indicators will be explored.

2.3.2.1 Accuracy

One of the primary advantages of digital circuits is the accuracy of digital circuits is 100% when operating well within the noise margin [5] and below the maximum speed of the technology. The maximum switching frequency of the technology chosen sets an upper bound for the sample frequency that can be used for the digital processing unit. If digital circuits are designed on the edge of the speed boundary of the technology and are processed in a slow technology corner, timing errors might occur leading to faulty outputs.

For analog circuits the accuracy analysis is much more difficult. The accuracy of the analog circuits is much more dependent on bias conditions and transistor parameters. Furthermore, once introduced, the offset, noise and distortion introduced by the analog circuits accumulates along the signal path, whereas in digital circuitry the accuracy is independent of transistor offset, distortion, circuit noise and interference, when operating well within the noise margin and below the maximum technology speed.

Because digital circuits are 100% accurate within the noise margin, they can be captured in a high level descriptive language. The mapping of the VHDL code functionality on the functionality extracted from the layout of the digital system normally is 100% when the digital circuits operate well within the maximum technology speed and noise margin. The maximum achievable accuracy is set by the sample frequency of the digital system, and the number of bits used for the calculations. If the required accuracy is proven by simulation, the hardware implementation of it will show exactly the same performance, under ideal outside world circumstances.

2.3.2.2 Robustness

The noise margin and maximum technology switching speed of digital circuits are subject to outside world influences, like process spread, process corners (slow, typical and fast processing), power supply variations (typical +/−10% of the nominal technology supply voltage), temperature (typical −40 and 125°C). To characterize the influences of these conditions on the noise margin and speed of the technology, several standard digital cells are exposed to these conditions. The outcome of this characterization can than be generalized to define the performance of the technology. At the end of the design trajectory of a digital system, timing verification is done to verify if the accuracy is guaranteed by the system when exposed to these conditions. The extraction of the noise margin from the characterization of different digital cells, will lead to a general substrate, power supply and decoupling strategy.

For analog circuits a generalization of the design strategy is much more difficult. As the errors introduced by the outside world influences mentioned above accumulate along the signal path.

Due to the robustness of digital systems, they are almost push-button portable to newer technologies, which adds more flexibility to the system. Once available in VHDL code, the layout of a digital system can be ported from one technology to another in only limited amount of time, with a high degree of automation.

Although in the discussion above digital circuits seem very robust, the technology scaling of digital circuits predicts that interference within the digital system is an increasing threat. As the accuracy in lithography scales with s_T, wires are closer to each other, increasing mutual crosstalk. Furthermore the impedance of supply lines is increasing, which together with an increase of the current density per area increases the supply bounce. With the increasing number of switching transistors per area the dI/dt increases per area which causes the ground bounce to increase. Next to that the noise margin will become smaller as supply voltage and V_T are decreasing. This means that shifting analog functionality into the digital domain does not come for free, and noise margin, supply and substrate bounce, and decoupling strategy will become more and more important. As in this book the digital circuits which are used to replace the analog functions are comparably small in area, the (influence on the total digital) interference problem is only small.

2.3.2.3 Flexibility

As the performance overhead in the noise margin of digital circuits allows for a high abstraction level description (like VHDL) of digital systems, the flexibility potential of digital circuits is enormous. As analog design is mostly custom design it is much more difficult to make flexible. Moreover, adding flexibility to analog circuits introduces parasitic behavior which can even limit the maximum achievable accuracy of the analog circuit.

The VHDL code describing a digital system can be set up in a scalable way by using parameterization, to be able to program the systems' performance in line with the current application requirements. If the VHDL code describing a digital sub-system is set-up in a scalable way, with clearly defined input and output conditions, the main system function can easily be re-configured to a different system function re-using sub-system functions in a different way or order. A digital system function can be made adaptable to changing outside world circumstances, by reprogramming of the coefficients of the input-output matrix defining the system. To be able to respond to changes in the outside world, the digital system should be supplied with inputs which represent the changes in the outside world.

2.3.2.4 Efficiency

The power consumption of digital circuitry is related to $P = C \cdot V_{supply}^2 \cdot f_s$. As V_{supply} scales with s_T (for constant field scaling [4]), and C also scales with s_T, the consumed power of a digital circuit switching at a constant f_s scales with s_T^3 (for constant voltage technology scaling, consumed power scales with s_T). This makes it attractive to shift analog functionality in the digital domain, because power consumption of analog circuits[1] at best remains constant when scaled into to deep submicron technologies.

The area of digital circuitry scales with s_T^2 as the minimum gate length of the smallest transistor that can be used in logic cells, scales with s_T. As with power, the area of analog functions at best remains the same when scaling an analog function into deep submicron technologies. Looking into the future, the scaling of digital systems in deep submicron technologies shows promising area and power advantages compared to the scaling of analog systems.

[1]Note that the focus of this book is on A/D converters. For other analog circuits like for example oscillators, technology scaling also provides some advantages. The exact analysis of these advantages however is without the scope of this book.

Although difficult to measure, the effort to design a certain function (e.g. a channel filter) with analog circuits is more time consuming compared to the design of the same functionality with digital circuits. Moreover, for digital circuitry the generation of layout is automated to a great extent. Analog layout often still is handcraft, for sure for high-end analog functions. For analog functionality some design and layout automation methods have been published ([6, 7] and many more), but are often limited to a specific analog function.

To test high performance analog functionality, expensive equipment is required to be able to generate and qualify the analog signals going in or coming out of the analog block respectively. Complicated and difficult to generalize tests with high quality input signals have to be carried out, to be able to completely check if the analog system achieves the required performance under all conditions. The qualification of the system accuracy is difficult because it is degraded by the noise, distortion and interference introduced along the analog signal path.

In digital circuits test chains are introduced to verify the systems' performance. A pattern generator generates input vectors which sufficiently cover the system functionality. The output vectors of the system are either wrong or right. In general the testing of digital systems is much easier as the behavior of digital circuits is much more predictable, and the results are easier to interpret.

2.3.2.5 Emission

A drawback of digital circuits is the fact that they are notorious for their emission of interference to supplies and substrate. This asks for a good supply, substrate, and decoupling connection strategy. This way the interference generated by the digital system can be kept under control, and is no threat to the surrounding systems on the same chip or in the same application. As nowadays deep submicron technologies have a deep N-well technology option, at least the substrate bounce of digital circuits can be better shielded from other systems on the same chip.

2.4 Conclusions

The design of a system is not only about system functionality but also about system quality. The wish to create more efficient and flexible systems, insensitive to outside world influences, comes from the drive to get products faster to the market, at a lower price, and including more features in a smaller volume, making products more differentiating. This asks for the introduction of quality indicators, with which the quality of a system can be judged. In this chapter five quality indicators have been presented, which are: accuracy, robustness to secondary inputs, flexibility, efficiency, and emission of secondary outputs. Throughout the book these are used to judge a system's quality. Quality indicator emission is outside the scope of this book. The quality indicators are shown in Fig. 2.4. In Chap. 1

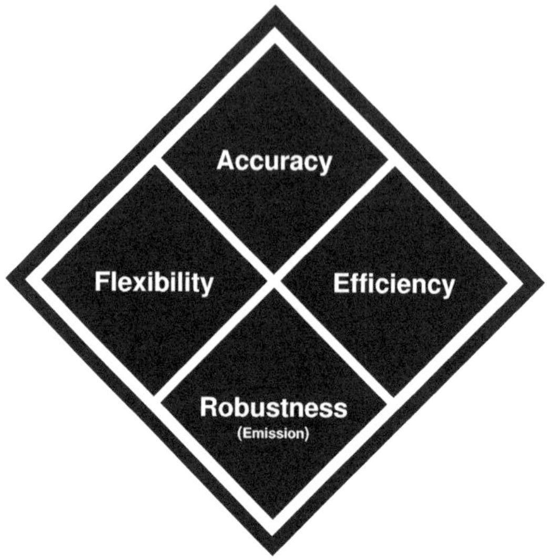

Figure 2.4: *The quality indicators introduced in this book*

it was shown that Moore's law predicts that if a digital function is ported to the next technology node, clear technology advantages like area scaling (s_T^2), increase of power efficiency ($1/s_T^2$) and speed increase ($1/s_T$) become available. Furthermore, as digital circuits have built-in performance overhead in their noise margin, a high degree of automation to do the port to the next technology node is possible. Next to that, digital circuits can be made re-configurable very easily as they are captured in a descriptive language like VHDL.

For a fixed analog function the area scaling in the next technology node is not that evident. The change of analog design parameters like power supply, V_T, etc., ask for a re-design of all the analog circuit blocks when going to the next technology node. This reduces the portability of these analog blocks and thus increases time-to-market. Next to that, analog circuits are much more difficult to make re-configurable.

Therefore, it is advantageous to increase the digitization of a system as digital circuits score high on the quality indicators. In this book the digitization process

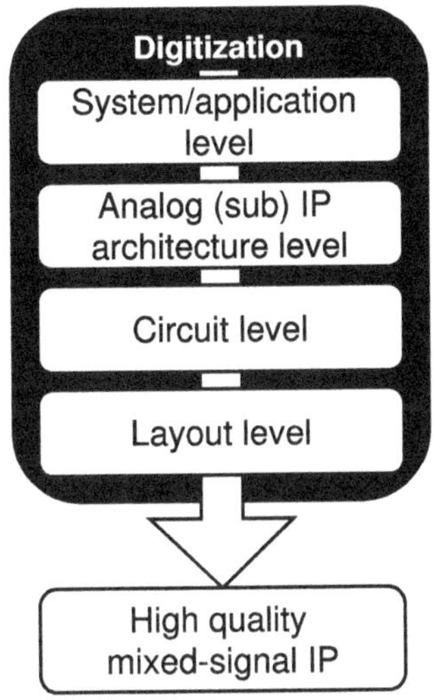

Figure 2.5: *Digitization of an analog system at different levels*

will be carried through four different abstraction levels, displayed in Fig. 2.5. At system level, this calls for an early introduction of the A/D and D/A conversion in the system pipe-line, which shifts the signal processing as much as possible into the digital domain. Once in the digital domain, the systems' accuracy is only determined by the accuracy described in the VHDL code when operating within the maximum achievable speed of the technology and within the noise margin. This makes the system robust to outside world influences. In the digital domain the signal processing is more powerful, can be setup in a flexible way more easily and shows increased power and area efficiencies in newer technologies, being future proof. However, shifting more of the signal processing in the digital domain, higher demands are put on the DR and bandwidth requirements of the ADC. It is the challenge to trade off analog and digital functionality with the ADC DR to come to a realistic but competitive system solution.

At analog IP architecture smart circuit choices should be made to reduce the amount of critical analog functions and replace or assist them with digital circuits as much as possible.

At circuit level, the circuits should be designed such that the analog blocks can be built up by a limited amount of unit cells, like in digital circuits. Due to the simplicity of the analog unit cells, the analog library can be ported to a next technology node very quickly, as its optimization process can be done by using simulation scripts for the analog simulator.

At layout level, each unit cell out of the analog unit cell library is turned into a parameterized layout (p-cell layout). Once these p-cell layouts are available, the routing tool normally used to layout digital circuits can be used, which reduces time-to-market tremendously.

This way digitally assisted systems and circuits are created which score high on the quality indicators.

Chapter 3

Integrated Receiver Architectures for Cellular and Connectivity

In wireless digital communication complex modulated signals containing user data are transmitted via the ether. A receiver is required to receive and detect the user data bits. Somewhere along the receive path, A/D converters are required to convert the analog signal into a digital signal. One of the trade-offs that has to be made during the receiver design is the degree of digitization. The quality indicators presented in Chaps. 1 and 2 predicted it could be advantageous to shift as much of the analog signal processing into the digital domain as possible, but this will come at the cost of increased requirements for the ADC.

In this chapter a receiver architecture will be searched for, which scores high on the quality indicators and combines a high degree of digitization with a reasonable ADC power consumption.

3.1 Wireless Receiver Architectures for Digital Communication

One of the earliest forms of a receiver architecture for digital transmission, is the superheterodyne architecture presented in Fig. 3.1. It uses two IF frequencies, and lots of analog signal conditioning blocks as well as expensive high Q (ceramic) filters to reduce the incoming antenna signal to our wanted channel. At the output of the filter in front of the A/D converters, our wanted signal is at a

R.H.M. van Veldhoven, A.H.M. van Roermund, *Robust Sigma Delta Converters*,
Analog Circuits and Signal Processing, DOI 10.1007/978-94-007-0644-6_3,

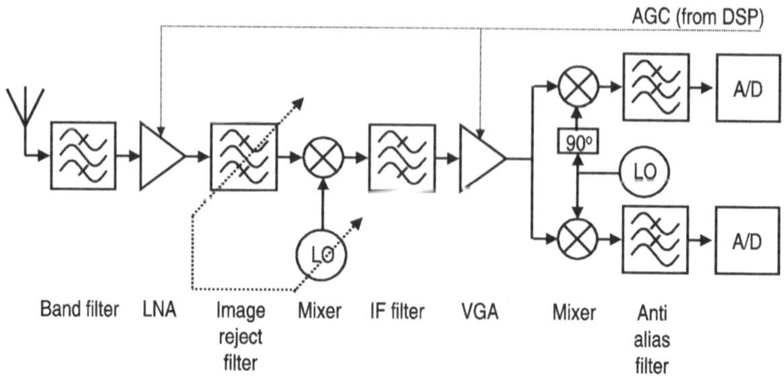

Figure 3.1: *Traditional receiver architecture*

low or zero-IF and is almost clear of interference, which means that the DR and linearity requirements of the A/D converter are very low, and the ADC power consumption will be smaller than 100 μW. Drawback of this architecture is that it requires a lot of (tuneable) analog filters, and although the A/D converter in this architecture is very simple, the analog design of this kind of receivers is very time consuming, hard to port from one IC technology to another and difficult to make re-configurable, as it needs multiple (tunable) analog filters, and mixers.

The continuous development of modern IC technologies and increasing availability of digital processing power has paved the road to more digitized receiver architectures. The first example is a digitized IF receiver architecture which simplifies the architecture above by removing the second mixer from the receiver. Only one (bandpass) ADC is required to convert the wanted channel into the digital domain. A drawback of this architecture is the tune-able image filter required, which is difficult to design. Because the signal at the input of the ADC is at a high IF, the clock frequency of the ADC should at least be twice the IF frequency to avoid aliasing. This means that the bandwidth requirements of the circuits used in the ADC will be high, which increases ADC power consumption. In the receiver architecture proposed in [8], the aliasing is used to the advantage of the receiver, in this case it replaces a mixer.

In a direct conversion receiver (Fig. 3.2) a complete band of channels is directly down converted to a zero IF (ZIF) or Near Zero IF (NZIF), to reduce the circuit bandwidth requirements for the ADC. The LO is tuned to the wanted signal frequency which is mixed down to the low IF frequency, at which it is converted into the digital domain. To reduce image interference, an I&Q mixer is used. After

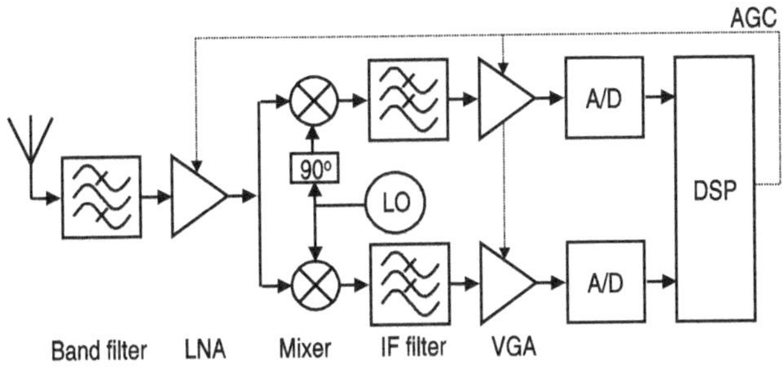

Figure 3.2: *Direct conversion receiver architecture*

down mixing, the analog filter reduces the input power to the ADC and removes frequency components, which can cause aliases due to the sampling operation of the ADC. The VGA conditions the signal such, that it uses the ADC dynamic range in the most efficient way. The signal is digitized by the quadrature A/D converter and finally the channel filtering and demodulation is done in the DSP. In modern receiver architectures for mobile phones the concept of direct conversion has become very popular. It only uses one IF frequency and does not need an expensive high Q ceramic IF filter compared to the traditional super heterodyne receiver. Instead it uses a quadrature mixer and two ADCs. This way, part of the channel filtering can be shifted into the digital domain. Implemented in the digital domain, the channel filtering can be made scalable more easily to satisfy the different filtering requirements depending on the communication standard the receive pipe is used for. The question is how much of the analog filtering if not all can be shifted in the digital domain. This imposes a trade-off between analog and digital filtering and ADC dynamic range. If only limited analog filtering is assumed in front of the ADC, it will be shown later that a state-of-art ADC power consumption is in the 100 μW–10 mW range.

The holy grail of receiver architectures is the digitized RF receiver architecture. It is shown in Fig. 3.3. It only has limited selectivity at RF to select the RF band

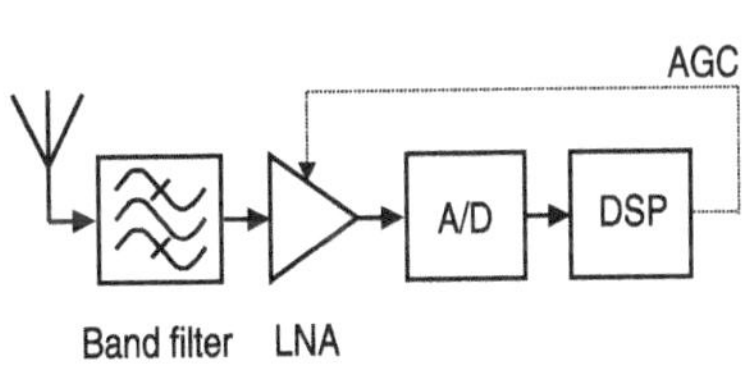

Figure 3.3: *RF A/D conversion receiver architecture*

of interest. The RF band is directly digitized by an A/D converter at RF and channel selection is completely done in the digital domain. The signal at the input of the ADC now contains all channels of e.g. the GSM system. Although this architecture enables new features such as multi channel reception, it will increase the dynamic range of the ADC dramatically and thus its power consumption. In a GSM receiver for example, the signal dynamics and the noise requirement leads to an SNR close to 100 dB (Sect. 4.7). Furthermore, the GSM channels at RF cover a bandwidth of 35 MHz. If this would be digitized by an ADC processed in current state-of-art technology, its estimated power would be larger than 1 W, which is way too much for portable applications like the mobile phone. Next to that as the RF channels are positioned in the GHz range and no mixer is used, a multiple GHz sample frequency will be required, which will lead to even higher power consumption for the ADC, let alone the power consumption of the digital channel filters following the ADC. It is believed to be a power inefficient system if possible at all in modern state-of-art mainstream IC technologies.

3.2 Receiver Architecture and the Quality Indicators

Figure 3.4 summarizes the receiver architectures discussed. The ADC in a digi-

Receiver architecture	ADC			
	Accuracy	**Robustness**	**Flexibility**	**Efficiency**
Superhetrodyne	Low	Wanted signal only	Too much analog blocks (almost no digitization)	Extensive signal conditioning (P_{ADC}~100μW)
Direct conversion	Moderate to high	Wanted plus neighboring channels	Limited number of analog blocks (highly digitized)	Limited signal conditioning (P_{ADC}~0.1-10mW)
Digitized RF	Extremely high	Complete band (all channels)	Completely digitized	No signal conditioning (P_{ADC}~1W)

Figure 3.4: *Summary of receiver architectures*

tized RF receiver requires 1 W of power, which is already higher than the compete receiver power budget, and therefore is no option. The superhytrodyne receiver is no option as the number of analog circuits is too high. If such a receiver would be made flexible, it would lead to a very bulky solution. The (N)ZIF direct conversion receiver has a limited number of analog blocks, and a high degree of digitization. Therefore, it benefits from the advantages transistor technology scaling brings (Sect. 1.1.2). Furthermore, part of the flexibility can be shifted into the digital domain, like the channel filters. The remaining analog filtering should reduce out-of-band interference such that the required ADC accuracy yields a realistic

ADC power consumption of 0.1–10 mW. The ADC converts the wanted signal into the digital domain, but because of the limited filtering in front of the ADC, neighboring channels will not be at a negligible level. This will put additional accuracy requirements (DR, distortion and aliasing) on the ADC. The exact analysis of the impact of the neighbors on the accuracy requirements of the ADC will be given in Sect. 4.7 for a direct conversion receiver for GSM.

3.3 Conclusions

In this chapter different receiver architectures have been shown with their pro's and con's. The degree of digitization of these architectures has been balanced with the ADC power consumption. The highly but not completely digitized (N)ZIF direct conversion receiver architecture with limited analog filtering seems to be the best candidate for integration in nowadays state-of-art IC technologies. It combines a high degree of receiver digitization with a reasonable ADC power consumption. Because of the limited amount of analog blocks it can be made flexible more easy, at the cost of a high ADC DR. Furthermore, the variety of communication standards will put requirements on ADC bandwidth and clock frequency programmability.

The ADCs presented in this book are designed for a direct conversion receiver which can be used for multi-standard, MIMO, software defined and cognitive radios. Before the design of the ADC is discussed, Chap. 4 will relate the RF front-end and ADC performance parameters, to come to realistic specification balance for the RF front-end and ADC.

Chapter 4

Specifications for A/D Converters in Cellular and Connectivity Receivers

The design of a receiver system asks for relations between system level choices, and their impact on the specifications of the different receiver building blocks, to come to a optimized system partitioning. This chapter will relate system level specifications to the RF front-end and ADC requirements. The implementation loss of the digital part following the ADC will be assumed to be zero.

The chapter starts with a discussion about the relation between the IF frequency choice and the receiver image rejection, DC offset, and 1/f noise requirements. Next, the ADC DR will be calculated by determining its top and bottom-end. The top-end of the ADC DR will be calculated as a function of antenna input signals, their crest factor, and the RF front-end selectivity. The gain of the receiver will be related to the maximum ADC input signal, and the maximum input swing possible within the ADC supply. The bottom-end of the ADC DR will be related to the modulation scheme, the wanted signal at sensitivity (the minimum signal the receiver should be able to receive), the receiver and RF front-end noise figure, and the receiver gain. Furthermore, various performance parameters of the RF front-end like 1/f-thermal noise corner, NF, and IP2/3 will be related to the ADC 1/f corner, DR and IM2/3 respectively. The last part of this chapter describes an example of the performance parameter trade-offs between RF front-end and ADC for a GSM receiver.

R.H.M. van Veldhoven, A.H.M. van Roermund, *Robust Sigma Delta Converters*,
Analog Circuits and Signal Processing, DOI 10.1007/978-94-007-0644-6_4,

4.1 IF Choice

The choice of the IF frequency in the direct receiver architecture has impact on the 1/f noise, offset and image rejection ratio (IRR) requirements of the receiver. This section will relate the RF front-end and ADC image rejection requirements. Furthermore, the influence of the choice of the IF frequency on the receiver 1/f noise requirements will be discussed.

The starting point is the RF band depicted in Fig. 4.1. A wanted channel and a neighboring channel are used to illustrate the impact of IF choice on image rejection, DC offset end 1/f noise requirements. To illustrate the impact, it suffices to consider the dark grey neighboring channel only. The wanted signal is down-

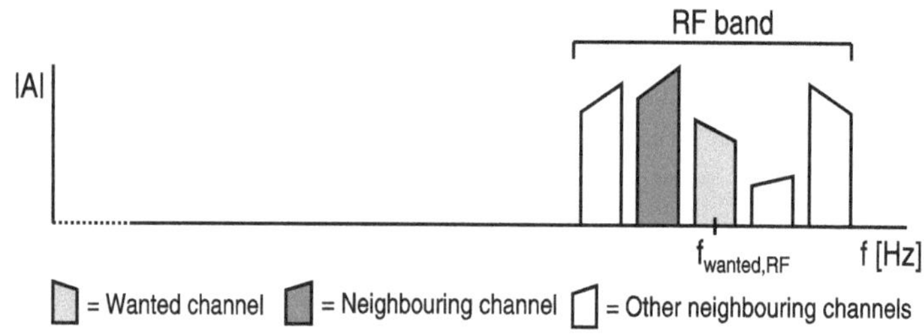

Figure 4.1: *Example RF band*

converted such that the IF frequency is in the middle of the signal bandwidth. Therefore, the wanted signal spans the bandwidth $f_{IF} \pm \frac{1}{2}B_{wanted}$.

4.1.1 Image Rejection

Figure 4.2 shows the I&Q mixer and ADCs. The mixer input signal X contains the wanted signal which is at a frequency $\omega_{wanted,RF}$, and the mixer frequency is

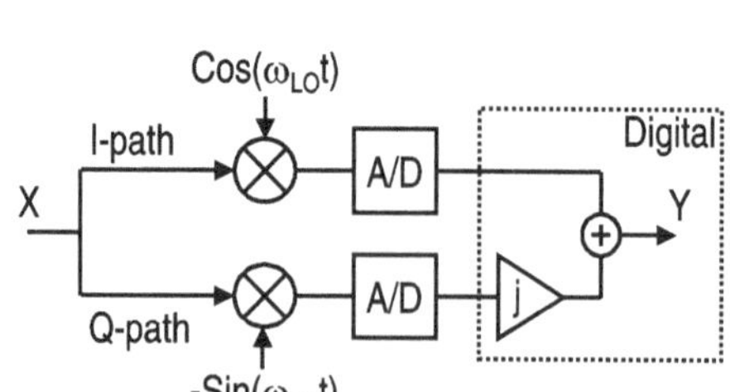

Figure 4.2: *Model to relate gain and phase matching to the IRR requirement*

at ω_{LO}. The LO frequency with which the wanted signal will be down-converted, is positioned in the middle of the wanted signal at RF plus the IF frequency, or $f_{LO} = f_{wanted,RF} - f_{IF}$.

If the mixers and ADCs have perfect amplitude and phase matching and the LO signals driving the mixer are perfectly in quadrature, the digital ADC re-combined output signal Y is:

$$\begin{aligned} Y &= e^{j\omega_{wanted,RF}t} \cdot (\cos(\omega_{LO}t) - j\sin(\omega_{LO}t)) \\ &= e^{j\omega_{wanted,IF}t} = e^{j(\omega_{IF}+\omega_{wanted})t} \end{aligned} \tag{4.1}$$

using equations:

$$\sin(\omega t) = \frac{e^{j\omega t} - e^{-j\omega t}}{2j} \quad \text{and} \quad \cos(\omega t) = \frac{e^{j\omega t} + e^{-j\omega t}}{2} \tag{4.2}$$

and in which f_{wanted} is defined from $-\frac{1}{2}B_{wanted}$ to $\frac{1}{2}B_{wanted}$ Hz.

If there is an amplitude mismatch Δ or a phase mismatch θ between the two mixers or ADCs, or if the LO signals are not perfectly in quadrature, an image will appear in the output signal Y at a frequency $-\omega_{wanted}$. The amplitude difference between the original signal and its image is called the image rejection ratio, and can be approximated by [9]:

$$\text{IRR}^{dB} \approx 10 \cdot \log_{10}\left(\frac{4}{\Delta^2 + \theta^2}\right) \ [\text{dB}] \tag{4.3}$$

The individual amplitude and phase mismatches in the mixer, LO signals, IF amplifiers and ADCs have to be added together to determine the receiver image rejection. The way the amplitude or phase errors should be added is determined by the nature of the errors. The error can be deterministic (e.g. unbalanced parasitics in layout) or stochastic (component mismatch) which require linear or rms addition of the errors respectively.

The mechanism which determines the IRR specification of the receiver will be subject to the choice of the IF frequency, which can be split up in three variants, namely: Zero IF (ZIF), Near Zero IF (NZIF), high IF.

4.1.2 Zero IF Architecture

In a ZIF architecture the wanted channel is mixed down to DC, or 0 Hz IF and spans the frequency range of +/− half the signal bandwidth. Figure 4.3 shows the

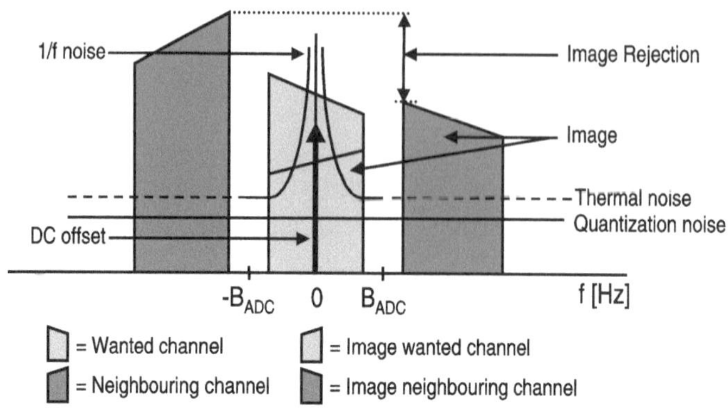

Figure 4.3: *ADC output signal in a ZIF configuration*

signal just after the I&Q ADCs. The figure shows the wanted and one adjacent channel. Furthermore, it shows the 1/f and thermal circuit noise and the ADC quantization noise. The ADC bandwidth is only plus and minus half the channel and low bandwidth ADC circuits can be used, which reduces power consumption. A disadvantage of zero-IF is that DC offset and 1/f noise are in the middle of the signal bandwidth, which distorts the signal. A DC cancellation loop can be introduced in the receiver path to reduce the DC offset (and part of the 1/f noise), but as these loops introduce a notch in the spectrum at DC, this might be unattractive to do in narrow band systems like GSM, as it creates a notch in the middle of the signal bandwidth. Lowering cut-off frequency to reduce the influence of the notch, will lead to an undesirable settling time. A solution to 1/f noise and DC offset in the receiver's circuitry is the introduction of chopping in for example the input stage of the ADC. Using chopping, the 1/noise and DC offset is modulated to an out-of-band frequency where it can be removed by filtering. A drawback of chopping is that the chopping is done on one of the most sensitive nodes in the ADC, and therefore very careful design is required to avoid non-linearities caused by the chopping switches and the switching process itself. When the DC offset is induced by second order distortion in the receive path, chopping does not help, and a shift from ZIF to NZIF is required.

Furthermore, the limited I&Q balance in the LO signals, mixers and ADCs will cause an image in the channel bandwidth. In a ZIF architecture the image is a mirrored copy of the wanted signal itself, and the co-channel interference ratio $\mathrm{IR}_{cochint}$ is determined by the receiver IRR only. The $\mathrm{IR}^{dB}_{cochint}$ requirement normally is only a few dBs more than the required minimum receiver SNR^{dB} for demodulation, which leads to an image rejection specification in the order of 20–30 dB for most digital modulation schemes.

4.1.3 Near Zero and High IF Architecture

In an NZIF direct conversion receiver, the wanted channel is mixed down close to DC. Advantages are that the DC offset ($f_{IF} > \frac{1}{2}B$) and (most of the) 1/f noise is out of the channel bandwidth, and can be removed in the digital domain. A disadvantage of NZIF is the higher bandwidth required for the circuits used in the ADC (real or complex) which makes this approach less power efficient compared to the zero-IF architecture. The signal just after the ADCs is shown in Fig. 4.4. Another

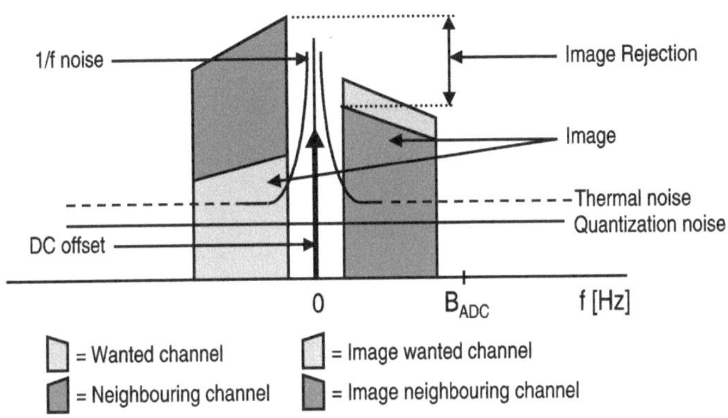

Figure 4.4: *ADC output signal in a NZIF configuration*

disadvantage of NZIF is that the co-channel image appearing in the wanted signal bandwidth is now coming from an adjacent channel. The magnitude of the image is not only dependent on I&Q balance, but also on the magnitude of the adjacent channels. In the NZIF architecture, the receiver IRR specification is defined by the combination of the adjacent channel powers, the exact IF frequency, and the co-channel interference ratio that can be tolerated according to the standard's specification. The image rejection is not only determined by the image appearing in the wanted signal bandwidth. The image powers appearing at the adjacent channel distances are limited by the power spectral mask, which defines the maximum power at each fixed distance to the wanted signal. The image interference should not exceed these powers, as these are used as a reference for the receiver design.

The higher the IF is chosen, the higher the IRR requirement will be, as the power of adjacent channels and blocker signals increases (Sect. 4.2.1) with the frequency offset to the wanted signal. This asks for a trade-off between 1/f noise and IF frequency, to come to an optimal IRR and 1/f noise specification. Next to that, increasing the IF frequency will increase the ADC design complexity.

The actual image requirement of the ADC will be subject to the IF filtering in front of the ADC, as the ADC's IRR is dependent on the power of the adjacent channels.

4.1.4 IF Assessment

In Table 4.1 the pro's and con's of the three IF architectures presented are displayed. ZIF seems to have the most advantages, but 1/f noise and DC offset are potential problems. ZIF also seems to be the most attractive architecture for a

ADC parameter	Intermediate frequency		
	ZIF	NZIF	High IF
Circuit bandwidth	+	−	−
Power consumption	++	+	−
Image rejection	++	−	−
DC offset	−	+	++
1/f noise	−	+	++

Table 4.1: *IF pro's and con's*

low power receiver, as it puts the lowest bandwidth requirements on the ADC. The 1/f noise and mismatch caused DC offset can be eliminated by chopping. If the DC offset is caused by second order distortion e.g. of an interferer an NZIF architecture is recommended.

4.1.5 DC Offset and 1/f Noise

The DC offset and 1/f noise introduced along the receiver chain can degrade the performance of a receiver. The DC offset is introduced by mismatch in the receiver circuits and self-mixing in the mixer. The 1/f noise is introduced by the switches of the mixers and the circuits of the IF stages and the ADCs.

The DC offset of the circuits can be reduced by increasing the area of the transistors to improve matching. Furthermore, a DC cancellation loop can be added to the receiver.

The contribution of the 1/f and thermal noise to the total noise is dependent on the IF frequency chosen. To demonstrate this effect a generic circuit block noise model shown in Fig. 4.5 is used. A relation between the 1/f and thermal noise contributions of a circuit block can be derived from Eq. 4.4, in which it is assumed

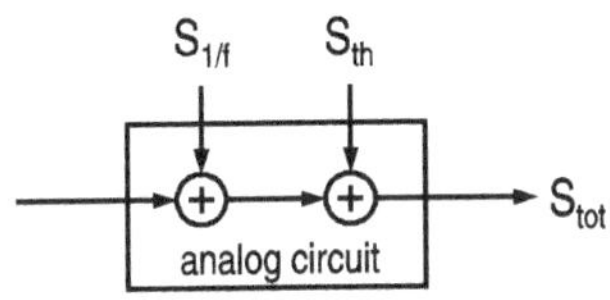

Figure 4.5: *Analog circuit noise model*

that 1/f noise has infinite power. In the equation, the 1/f noise level is related to the power density of the thermal noise $S_{th}(f)$.

$$\begin{aligned} N_{tot} = N_{th} + N_{\frac{1}{f}} &= \int_{f_1}^{f_2} \left(S_{th}(f) + S_{\frac{1}{f}}(f) \right) \cdot df \;\; [\mathrm{V}^2] \\ &= S_{th} \int_{f_1}^{f_2} df + S_{th} \int_{f_1}^{f_2} \frac{f_c}{f} \cdot df \;\; [\mathrm{V}^2] \\ &= S_{th} \cdot \left((f_2 - f_1) + f_c \cdot \ln \left(\frac{f_2}{f_1} \right) \right) \;\; [\mathrm{V}^2] \end{aligned} \tag{4.4}$$

The factor f_2/f_1 in Eq. 4.4 shows that the power of the 1/f noise is the same for each frequency octave or decade. Equation 4.4 also shows, that halving the corner frequency decreases the 1/f noise by 3 dB. It depends on the level of the thermal noise whether this reduction in f_c is beneficial.

In an NZIF receiver architecture the signal bandwidth and IF are often related by $f_{IF} = \frac{1}{2}B$. This means, the 1/f noise is integrated from 0 Hz, which would theoretically lead to an infinite 1/f noise contribution. The actual effective noise integration bandwidth is B_{eff} and is smaller than B. The exact B_{eff} is determined by system aspects like system frame rate, and DC offset calibration of which the exact explanation is out of the scope of this book. The influence of the effective integration bandwidth on the 1/f noise contribution is shown in Fig. 4.6 for $B_{eff} = 0.9B$ and $B_{eff} = 0.99B$, for $B = 1$ Hz. On the y-axis of Fig. 4.6, the total noise (normalized to the thermal noise density) is plotted. On the same y-axis the thermal noise limit is plotted. In all noise plots the noise is normalized to the thermal noise density S_{th}, as the absolute value of the noise is of no importance in these plots. The figure shows that the effective noise bandwidth has a great influence on the 1/f noise contribution when f_{IF} is 0.5 Hz, as the factor f_2/f_1 in Eq. 4.4 approximates infinity when B_{eff} is close to 1. If f_{IF} is increased above 0.5 Hz, f_2/f_1 reduces rapidly. For very large f_{IF} the 1/f noise is negligible, and the total noise approximates the thermal noise limit. For $B_{eff} = 0.9$ Hz and $B_{eff} = 0.99$ Hz the S_{th} normalized thermal noise limits, are $\sqrt{0.9}$ Hz and $\sqrt{0.99}$ Hz respectively.

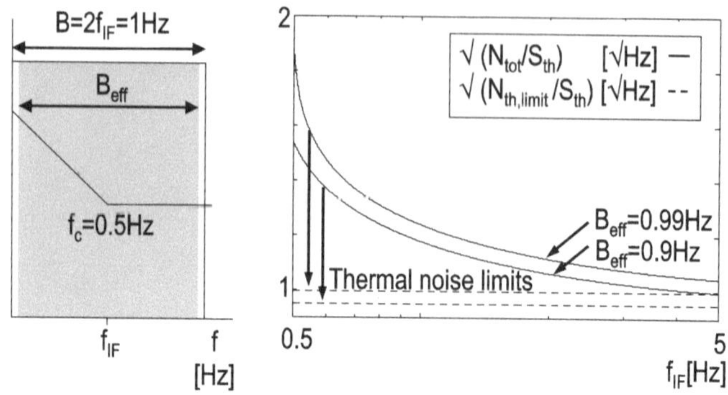

Figure 4.6: *Total integrated circuit noise (normalized to thermal noise density) as function of f_{IF} and B_{eff} ($B = 1$ Hz, $f_c = 0.5$ Hz) and corresponding thermal noise limits*

Another system partitioning parameter is the choice of the 1/f-thermal noise corner compared to the signal bandwidth. To show this effect of the IF frequency on the contributions of the 1/f and thermal noise to the total circuit noise, the ratio $R_{n,\frac{1}{f}/th}$ is introduced:

$$R_{n,\frac{1}{f}/th} = \sqrt{\frac{N_{1/f}}{N_{th}} = \frac{f_c \cdot \ln(\frac{f_2}{f_1})}{f_2 - f_1}} \quad [\text{-}] \tag{4.5}$$

In Fig. 4.7 the $R_{n,\frac{1}{f}/th}$ is plotted on the y-axis for corner frequencies 0.125, 0.5 and 2 Hz. For each corner frequency the IF is varied from 0.5–5 Hz. On the same y-axis the total noise N_{tot} normalized to S_{th} is displayed. In the example

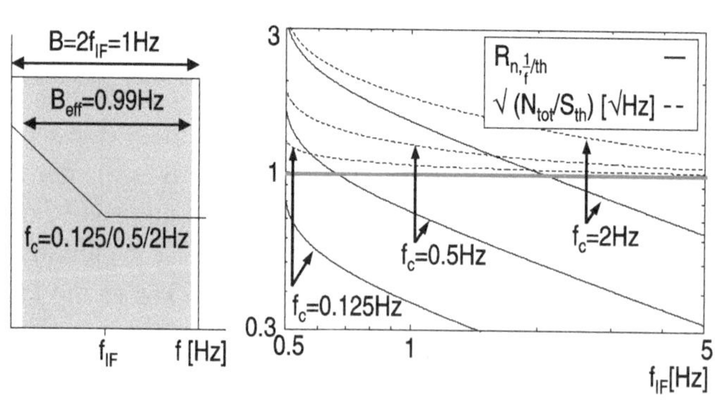

Figure 4.7: $R_{n,\frac{1}{f}/th}$ *and total noise (normalized to thermal noise density) for 3 different 1/f-thermal noise corner frequencies*

above B_{eff} is 99% of B ($B = 1$ Hz). In the corner frequency example of $f_c = 0.5$ Hz, it can be read from the plot that the 1/f-thermal noise contributions are equal at an IF of 0.65 Hz. Shifting the IF frequency from 0.5 Hz to 0.65 Hz reduces the total integrated noise in B_{eff} by 1.35 times. At IF frequencies much larger than 0.65 Hz, the total circuit noise will approximate the thermal noise limit of $\sqrt{0.99}$ Hz (grey line). The more f_c is lowered below the IF, the less it helps to lower the total integrated noise, and the more dominant the thermal noise becomes, which is a likely scenario for the more wide-band (MHz range) systems, like WLAN. In narrow bandwidth systems like GSM, 1/f noise will have a non-negligible contribution to the total noise, when the thermal-1/f noise corner frequency f_c of the IF receiver circuits is of the same magnitude as the IF and signal bandwidth. Although the 1/f noise can be solved by increasing the circuit area, this might lead to costly IC dimensions.

4.1.6 RF Front-End and ADC 1/f-Thermal Noise Corner Frequency

The contribution of the 1/f and thermal noise of the RF front-end and ADC to the total receiver noise is a trade-off between circuit area and power. The 1/f noise is inversely proportional to the circuit area $\sqrt{A_{circuit}}$, while the thermal noise is inversely proportional to power consumption of the circuit $\sqrt{P_{circuit}}$ as will be shown in Chap. 5. This will have its impact on RF front-end and ADC area (A_{RF} and A_{ADC} respectively) and power consumption (P_{RF} and P_{ADC} respectively).

In this section, the ratio $R_{n,RF/ADC}$ is introduced, which will relate the integrated noise contributions of the RF and ADC circuits, referred to the input of the ADC. Using this relation, the 1/f noise corner frequency of the receiver $f_{c,Rx}$ will be related to the individual corner frequencies of the RF front-end ($f_{c,RF}$) and ADC $f_{c,ADC}$. The section will prove that choosing a very high corner frequency for the RF front-end, might lead to an unrealistically low corner frequency of the ADC, and vice versa.

To derive these relations the receiver noise model of Fig. 4.8 is used. The model shows the signal in and outputs of the receiver and the input parameters $R_{n,RF/ADC}$, $G_{VI,RF}$ and $f_{c,Rx}$, of which the actual intermediate gain $G_{VI,RF}$ is determined in Sect. 4.3.

Equation 4.6 presents the ratio $R_{n,RF/ADC}$ when a bandwidth of (f_1, f_2) is assumed.

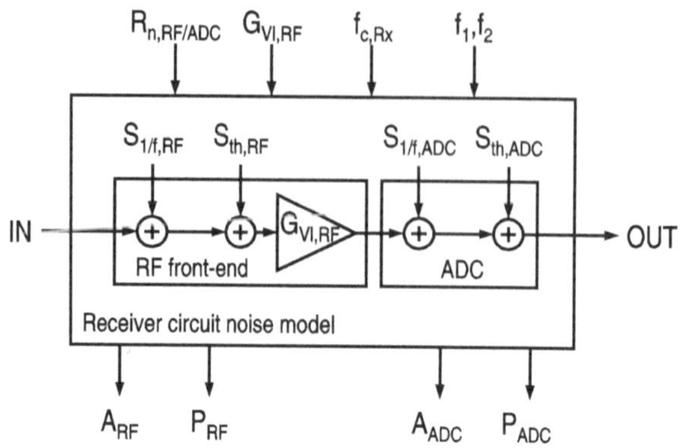

Figure 4.8: *Model for RF-ADC combined 1/f-thermal noise corner frequency calculation*

$$R_{n,RF/ADC} = \sqrt{\frac{G_{Vl,RF}^2 \cdot S_{th,RF} \cdot ((f_2 - f_1) + f_{c,RF} \cdot \ln(\frac{f_2}{f_1}))}{S_{th,ADC} \cdot ((f_2 - f_1) + f_{c,ADC} \cdot \ln(\frac{f_2}{f_1}))}} \quad [\text{-}] \tag{4.6}$$

The 1/f-thermal noise corner frequency of the RF front-end and ADC combination is:

$$f_{c,Rx} = \frac{S_{\frac{1}{f},RF}(f) \cdot G_{Vl,RF}^2 + S_{\frac{1}{f},ADC}(f)}{S_{th,RF} \cdot G_{Vl,RF}^2 + S_{th,ADC}} \quad [\text{Hz}] \tag{4.7}$$

Which can be rewritten to:

$$f_{c,Rx} = \frac{(f_2 - f_1) \cdot (f_{c,RF} + f_{c,ADC}) \ln(\frac{f_2}{f_1}) \cdot f_{c,RF} \cdot f_{c,ADC} \cdot (1 + R_{n,RF/ADC}^2)}{(f_2 - f_1) \cdot (1 + R_{n,RF/ADC}^2) + \ln(\frac{f_2}{f_1}) \cdot (f_{c,RF} + f_{c,ADC} \cdot R_{n,RF/ADC}^2)} \quad [\text{Hz}] \tag{4.8}$$

If the $f_{c,Rx}$, the bandwidth (f_1, f_2) and $R_{n,RF/ADC}$ is given, the ADC and RF front-end 1/f-thermal noise corner frequency relation can be calculated using Eq. 4.8. Note that $G_{Vl,RF}$ is eliminated from Eq. 4.8.

Figure 4.9 shows the RF corner frequency as a function of the ADC corner frequency for $R_{n,RF/ADC}$ equal to 1, and under the conditions $B = f_2 - f_1 = 1$ Hz and $f_{c,Rx} = 1$ Hz. The varying corner frequencies in the plot lead to different values for S_{th} and $S_{1/f}$ to keep the same total integrated noise. Although B is

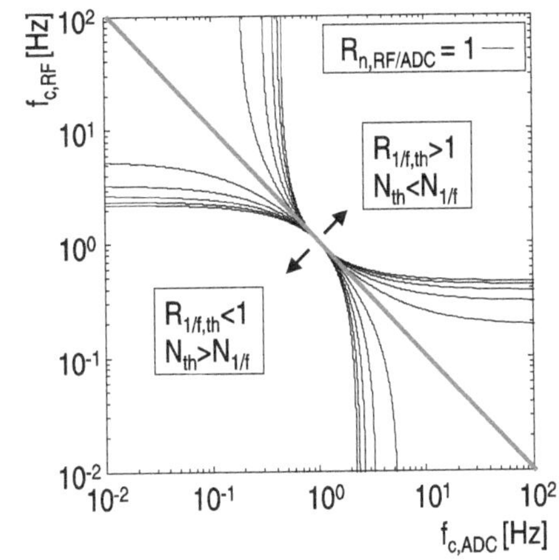

Figure 4.9: *RF front-end vs ADC 1/f-thermal corner frequency for $R_{n,RF/ADC}$ is 1*

constant for all curves, f_1 and f_2 are varied in the different plots (f_2/f_1 is not constant). Exactly on the grey line $R_{1/f,th}$ (Sect. 4.1.5) is 1, which means the thermal and 1/f noise contributions are equal. In all cases $R_{n,RF/ADC} = 1$. For the curves above the grey line, the receiver is 1/f noise dominated, below the grey line the receiver is thermal noise dominated. In Fig. 4.10 the noise contributions of RF and ADC differ by a factor of three ($R_{n,RF/ADC} = 1/3$ and $R_{n,RF/ADC} = 3$). In Fig. 4.10, four extreme situations for $R_{n,RF/ADC} = 3$ are indicated which are displayed in the frequency domain in Fig. 4.11. For these four different cases the dominant noise contributions will be described in more detail.

- 1. Thermal noise dominated design. RF dominates thermal noise, ADC dominates 1/f noise. Very low $f_{c,RF}$ required. High $f_{c,ADC}$ allowed.
- 2. 1/f noise dominated design. RF dominates thermal and 1/f noise. The $f_{c,RF}$ is close to $f_{c,Rx}$. High $f_{c,ADC}$ allowed.
- 3. 1/f noise dominated design. RF dominates 1/f noise, ADC dominates thermal noise. High $f_{c,RF}$ allowed. Low $f_{c,ADC}$ is required.
- 4. Thermal noise dominated design. RF dominates thermal and 1/f noise. The $f_{c,RF}$ is close to $f_{c,Rx}$. Low $f_{c,ADC}$ is required.

For 1/f noise in (N)ZIF receiver architectures it can be concluded from Sects. 4.1.5 and 4.1.6 that:

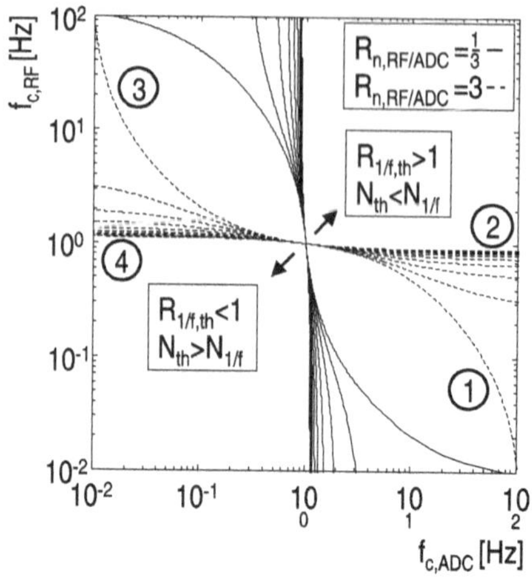

Figure 4.10: *RF front-end vs ADC 1/f-thermal corner frequency for $R_{n,RF/ADC}$ is 1/3 and 3*

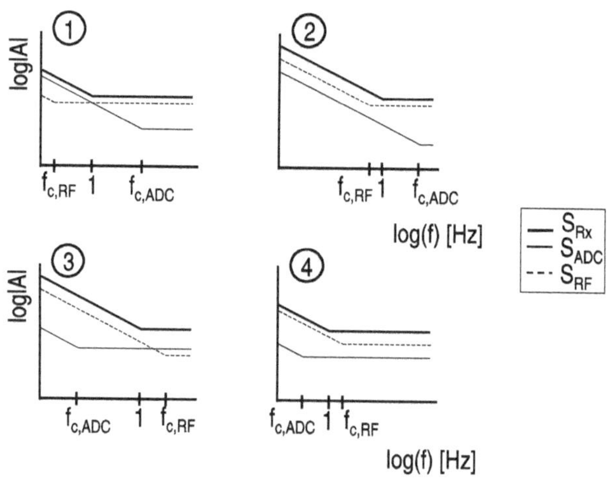

Figure 4.11: *Four extreme RF front-end and ADC corner frequency cases for $R_{n,RF/ADC} = 3$*

- It is important to know what the exact system bandwidth (B_{eff}) related to the IF frequency is, to be able to optimally design the corner frequency of circuits required in the RF front-end and ADC. An inaccurate specification of B_{eff}, might lead to over-specification of the analog blocks, which is bound to cost area and power.

- When the channel bandwidth B is close to $2f_{IF}$, the 1/f noise can have severe impact on the total noise contributions. A limited increase of the IF directly relaxes the 1/f noise requirement.

- A high corner frequency for the RF front-end might lead to an unrealistic low corner frequency for the ADC, and vice versa.

In Chap. 5 the relation between ADC 1/f-thermal noise corner frequency and ADC area and power consumption will be calculated. For the RF front-end the same should be done to come to an optimized system, but this is out of the scope of this book.

4.2 Top-End of the ADC DR

The top-end of the ADC DR is determined by the signals to be expected at the receiver antenna and the signal conditioning throughout the receiver. The trade-off between ADC DR and analog filtering (and thus the degree of digitization) of the receive path is shown by three examples. The input signal of the receive path consists of a weak wanted signal and a large unwanted signal. The wanted signal is in the grey area which indicates the signal bandwidth of interest. Figure 4.12 shows the most traditional implementation of the receive path. It uses high order

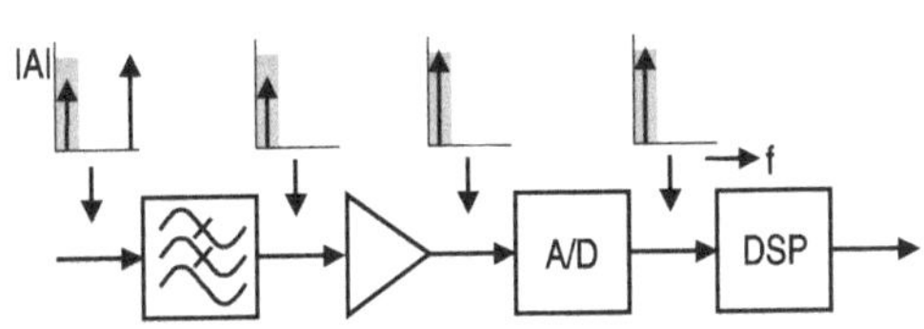

Figure 4.12: *The analog filter used for channel filtering resulting in a low DR requirement for the ADC*

analog filtering to remove the interfering signal and uses amplification to condition the wanted signal such that it exactly matches the maximum input signal of the ADC, relaxing its noise requirements. Because the interferer is attenuated to a negligible level, the required bandwidth is only dependent on the wanted signal, as alias products due to the sampling process of the ADC have not to be feared. Figure 4.13 shows a more advanced receive path. It reduces the analog filtering to a level where the interference is conditioned such that it is close or equal to the wanted signal level. Only low order analog filtering is required, and part of the filtering is shifted in the digital domain. The bandwidth requirement for the ADC has now increased to at least the unwanted signal frequency, to avoid aliases.

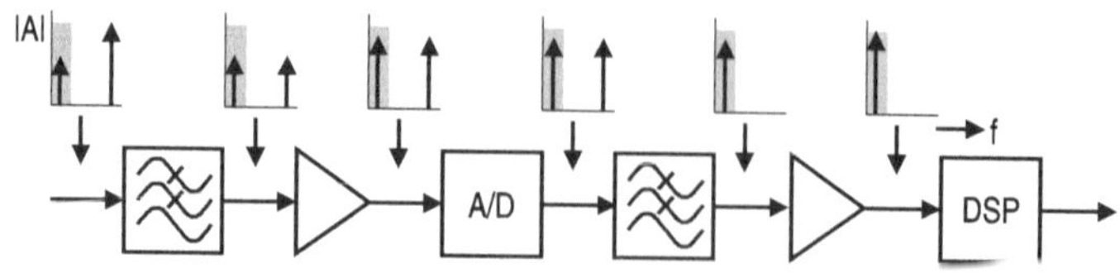

Figure 4.13: *The analog filter used for signal conditioning resulting in a medium DR requirement for the ADC*

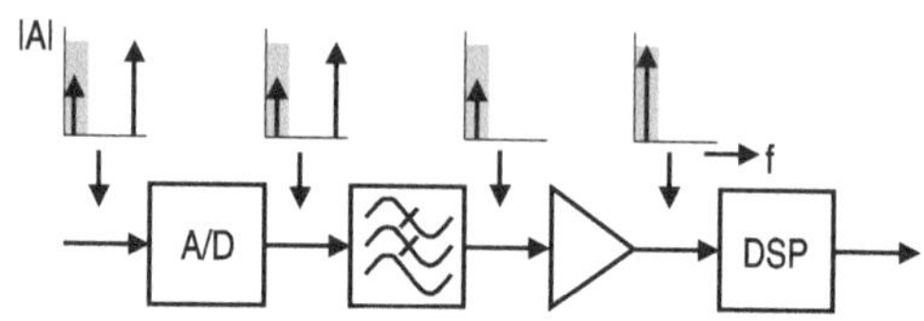

Figure 4.14: *No analog filtering resulting in a high DR requirement for the ADC*

Figure 4.14 shows the system with the highest degree of digitization. It shifts all the analog signal conditioning into the digital domain. Analog functionality is reduced to the ADC, at the cost of increased ADC DR and bandwidth requirements. It should be noted that the bandwidth at the input of the ADC is not limited at all, and the sample frequency should be chosen such that it is at least twice the highest unwanted signal frequency to be expected at the input of the ADC, to avoid in-band aliases. Removal of the unwanted signals can now completely be done in the digital domain. The highly digitized receiver architecture of Fig. 4.14 might open up the road to new applications such as multi-channel reception, as the wanted and unwanted signal might be two radio stations, which can be selected and filtered out in the digital domain.

In the more digitized receiver architectures of Figs. 4.13 and 4.14 there is only limited or no filtering in front of the ADC, which means it is very likely that the maximum input signal of the ADC is determined by out-of-band interfering signals. To come to a realistic ADC DR a trade-off between analog filtering and ADC DR has to be made. As discussed in Sect. 3.1, this trade-off is found in the direct conversion receiver.

4.2.1 Signal Levels, Selectivity, and Maximum ADC Input Signal

The in- and out-of-band signal levels are defined in the communication standard specification. Figure 4.15 shows an interferer/blocking/wanted signal spectral

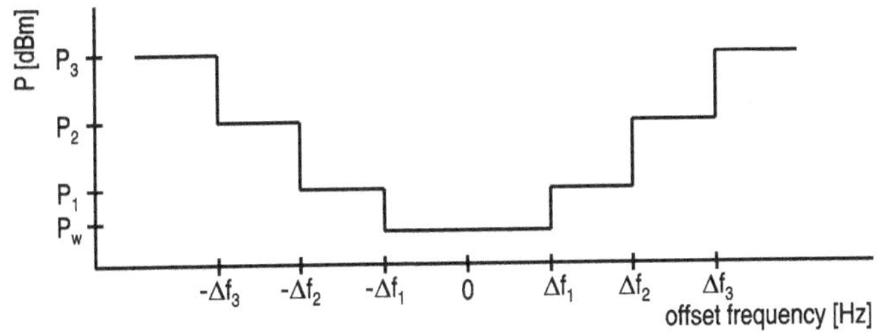

Figure 4.15: *Spectral mask model*

mask example. The x-axis displays the frequency offset to the wanted signal, the y-axis displays the signal power. The figure shows the signal levels that can be expected at the antenna of the receiver, of which the far-off signals are the strongest. Closer to the wanted signal the out-of-band signals gradually become smaller.

At the antenna, the interferers are normally specified in dBm. A power of 0 dBm, refers to 1 mW of power in a reference impedance R_{ref}, or:

$$P_{in}^{dBm} = 10\log_{10}\left(\frac{V_{in,rms}^2}{R_{ref}\cdot 1\ \text{mW}}\right) \quad [\text{dBm}] \tag{4.9}$$

The signal level can be calculated to dBμV$_{rms}$ by:

$$V_{in,rms}^{dBV} = P^{dBm} + 90 + 10\log_{10}(R_{ref}) = P_{50\ \Omega}^{dBm} + 107 \quad [\text{dBμV}] \tag{4.10}$$

if it is assumed that the RF front-end input impedance matches the antenna output impedance of $R_{ref} = 50\ \Omega$.

The maximum ADC input signal is determined by the signal levels at the antenna, together with the filtering in the RF front-end. In the direct conversion receiver the RF front-end filtering consists of a duplexer and the IF amplifier filter. In Fig. 4.16 the maximum ADC input signal is shown for a first (F_1) and a second (F_2) order IF amplifier filter. The filtering introduced by the duplexer, is the same in both cases. The maximum ADC input signal in case of F_1 and F_2 only differs about 6 dB, and is determined by an interferer at Δf_4 and Δf_2 respectively. A first order filter in the IF amplifier means 6 dB additional ADC top-end dynamic range compared to a second order filter. From the figure it becomes clear, that a higher order filter only has limited effect on the top-end, as the close by interferer (adjacent channel) is dominating the top-end of the ADC DR.

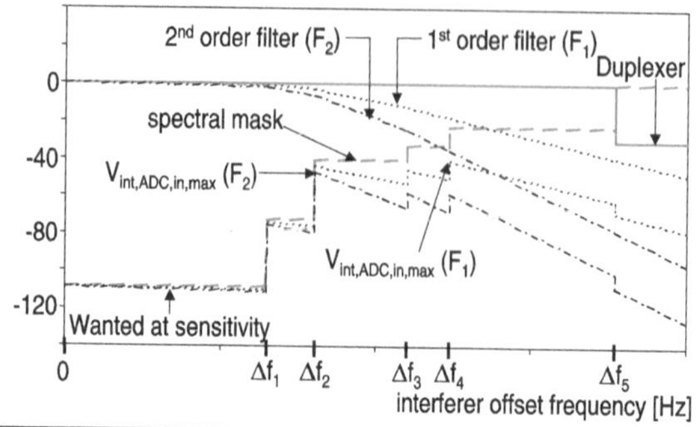

Spectral mask	dBm	F_1^{dB}	dB	$P_{int,ADC,in,max}^{dBm}(F_1)$	dBm
Duplexer gain	dB	F_2^{dB}	dB	$P_{int,ADC,in,max}^{dBm}(F_2)$	dBm

Figure 4.16: *Derivation of the ADC maximum input signal*

4.2.2 Crest Factor

The dynamic range of A/D converters is generally qualified with a sine wave input signal. The peak SNR of the ADC is determined by the maximum sine wave amplitude that can be put at the input of the ADC, without overdriving it, and the ADC's integrated noise in the signal bandwidth. The signals going through a receiver hardly resemble a sine wave. Advanced modulation and coding schemes are used for the data transmission, and the resulting signals spectrally look like noise. The crest factor defines the ratio between the peak amplitude of the signal divided by its (time-averaged) RMS value (signal power). Equation 4.11 gives the worst case crest factor, in case the signal is built up of x signals with different crest factors.

$$G_{crest}^{dB} = 20 \cdot \log_{10}\left(\frac{|\hat{V}_1| + |\hat{V}_2| + \cdots + |\hat{V}_x|}{\sqrt{V_{1,rms}^2 + V_{2,rms}^2 + \cdots + V_{x,rms}^2}}\right) \quad [\text{dB}] \qquad (4.11)$$

In Table 4.2 crest factors for a few modulation schemes are summarized.

In the design of the receiver, the crest factor should be accounted for in the calculation of the maximum receiver gain to be sure the circuits used in the receiver are not overdriven by the peaks of the modulated signals. This means additional top-end dynamic range has to be accounted for.

Waveform	Crest factor	
Square wave	0	dB
sine wave	3	dB
GMSK (BT=0.3)	6.2	dB
QPSK	3.5–4	dB
64-QAM	7.7	dB
128-QAM	8.2	dB

Table 4.2: *Crest factor for different digital modulation schemes*

4.3 Receiver Gain

To reduce the noise requirement of the ADC, the gain in front of the ADC should be as large as possible. To be able to calculate the maximum gain in front of the ADC, a model of the receiver is introduced in Fig. 4.17. The model splits the receiver in three parts being the antenna, the RF front-end and the ADC. In the model the input and output impedances of the different blocks are indicated. The

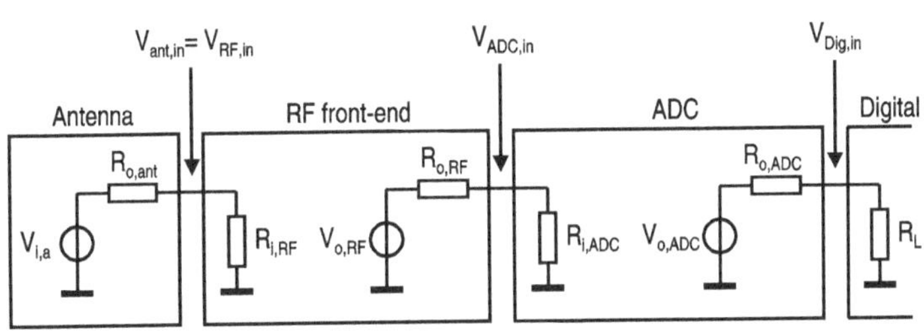

Figure 4.17: *Receiver model used for max. signal level and loaded voltage gain calculations*

loaded voltage gain G_{Vl} of a block is its output voltage divided by the input voltage, when driven with the output impedance of the preceding block and loaded with the input impedance of the following block. The accumulated voltage gain of the RF front-end $G_{Vl,RF}$ is determined by the gain of the LNA, the IF amplifier, the duplexer and the mixer.

The upper limit for the maximum ADC input signal is the maximum signal swing that is possible within the ADC power supply voltage. If a differential sine wave ADC input is assumed the maximum rms differential input signal of a single ADC is:

$$V_{ADC,in,max,rms}^{dBV} = 20\log_{10}\left(\frac{x \cdot V_{DDA}}{\sqrt{2}}\right) \quad [\text{dBV}] \qquad (4.12)$$

where x is the ratio between the single ended, peak-to-peak ADC input signal swing and the ADC supply voltage. The maximum loaded voltage gain becomes:

$$G^{dB}_{Vl,RF,max} = V^{dBV}_{ADC,in,max,rms} - V^{dBV}_{int,ADC,in,max,rms} - G^{dB}_{crest} \text{ [dB]} \quad (4.13)$$

4.3.1 Narrow vs. Broad Band AGC

For the implementation of the AGC loop two choices can be made: narrow band or broad band AGC. Narrow band AGC acts on the wanted signal only and has no awareness of interferer presence. The receiver always has to reckon for the most demanding interferer to be expected at the antenna, independent on the wanted signal level. When a broad band AGC loop is used, interferer absence information can be translated in a power saving for the analog blocks. In case the wanted signal is very strong, gain reduction is allowed, as the noise and distortion floor of the different blocks have become less relevant.

4.4 Bottom-End of the ADC DR

The receiver noise figure is determined in a receiver sensitivity test. In this test, the wanted signal power is brought at a level where the receiver output BER specified in the communication standard is just met. In this section the digital modulation scheme, receiver sensitivity and RF front-end and ADC noise requirements will be discussed.

4.4.1 Receiver SNR Requirement

In modern cellular and connectivity receivers digital modulation schemes are used for data transmission. The data is modulated onto one or more carriers which are transmitted by the transmitter. The required receiver SNR is determined by the SNR needed to detect the bits at a certain BER at the receiver demodulator, and is dependent on the digital modulation scheme used. For a given digital modulation scheme and for transmission at bit-error-rate BER, the required SNR per bit is E_b/N_0 [Hz/(bits/s)], which is called energy per bit to noise spectral density ratio. Multiplied by the link spectral efficiency, which is the ratio of the physical channel bit-rate R_b and the utilized channel bandwidth B, the channel's required SNR can be calculated which is needed for reliable transmission at bit-error-rate BER. Equation 4.14 relates the maximum channel capacity according to Shannon's theorem [10], to the energy per bit to noise spectral density ratio and the link spectral efficiency.

$$R_{b,max} < B \cdot \log_2(1 + \text{SNR}) = B \cdot \log_2\left(1 + \frac{E_b}{N_0} \cdot \frac{R_b}{B}\right) \quad \text{[bits/s]} \tag{4.14}$$

Figure 4.18 shows the Shannon bound on E_b/N_0 for reliable transmission of bits along a transceiver path with a normalized rate of R_b/B. The figure shows that there is a limit $10 \cdot \log_{10}(\ln(2))$ to the required E_b/N_0 even for very small R_b/B. In the same figure several digital modulation [11] schemes are compared for their

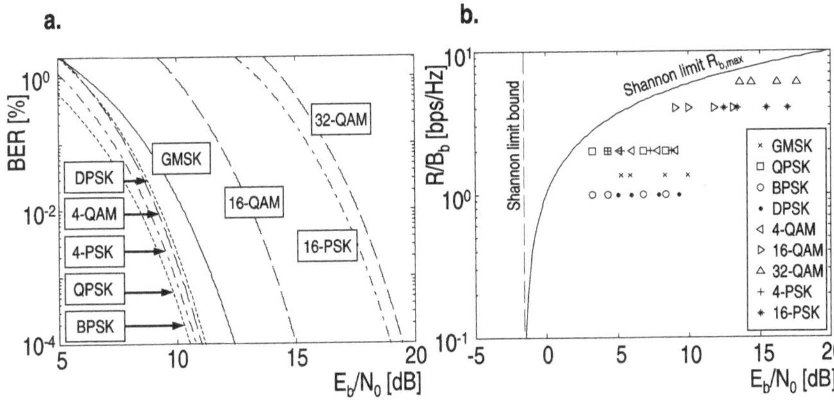

Figure 4.18: *(a) BER as a function of the energy per bit to noise spectral density ratio for different types of digital modulation schemes. (b) Energy per bit to noise spectral density ratio for different modulation schemes and compared to Shannon limit (BER = 2%)*

E_b/N_0 required to achieve a BER of 2%, which is the specified BER at sensitivity for a GSM receiver. As can be seen from the figure the digital modulation schemes have an E_b/N_0 which is quite far for the Shannon bound.

The channel capacity of an information channel can be further improved by using advanced techniques, like error correcting or spread spectrum coding. This way the effective data rate over the channel can be increased, without increasing the physical data rate R_b.

4.4.2 Receiver Noise Figure and ADC Noise Floor

From the SNR required by the digital modulation scheme, for reliable transmission at a certain BER, the receiver noise figure (NF) can be calculated. The noise analysis of the receiver starts with the available noise power, which is kTB. For each block along the receiver chain, a noise figure defines how much noise the block adds compared to kTB. In addition a gain is defined for each block, also shown in Fig. 4.19. In the figure it is illustrated that the gain amplifies the sig-

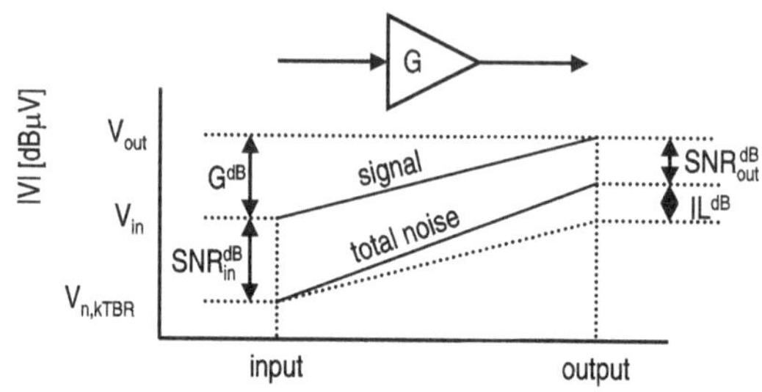

Figure 4.19: *Noise figure, implementation loss and SNR*

nal and noise with G to the output of the block, and additional noise is added by the block itself. The implementation loss (IL) is the difference between the input SNR and the output SNR, and in this case is equal to the NF. The implementation loss of a block in the receiver is equal to the difference between the cumulative receiver NF at the output and input of that block.

From the minimum receiver SNR requirement, the signal level defined by the sensitivity test and $kTBR$, the maximum receiver noise figure can be calculated. To calculate the ADC noise floor, the RF front-end (including Duplexer, LNA, mixer and IF amplifier) is seen as a black box with a cumulative noise figure and gain. The noise contribution of the digital part is assumed to be negligible. This is shown in Fig. 4.20. At the input of the receiver, the wanted signal is at sensitivity level of $V_{w,ant,in}$. Being at sensitivity, the receiver just has enough SNR, SNR_{Rx}, to be able to detect the incoming bits at the required BER. The required receiver noise floor at the antenna $V^{dBV}_{n,Rx,in,max,rms}$ is $V^{dBV}_{w,ant,in,rms} - \mathrm{SNR}^{dB}_{Rx,min}$. The difference between the receiver noise floor $V_{n,Rx,in,max,rms}$ and the $kTBR$ noise $V_{n,ant,kTBR,rms}$ is the maximum receiver noise figure, with which the receiver will pass the sensitivity test. The noise will be distributed over the RF front-end (duplexer, LNA, mixer and IF amplifier) and the ADC. This is indicated in Fig. 4.20. In the figure the ADC implementation loss is indicated. The ADC has 0 dB gain, and should have an input referred noise floor which is:

$$V^2_{n,ADC,in,max,rms} \leq V^2_{n,Rx,out,max,rms} - V^2_{n,RF,out,rms} \ [\mathrm{V}] \tag{4.15}$$

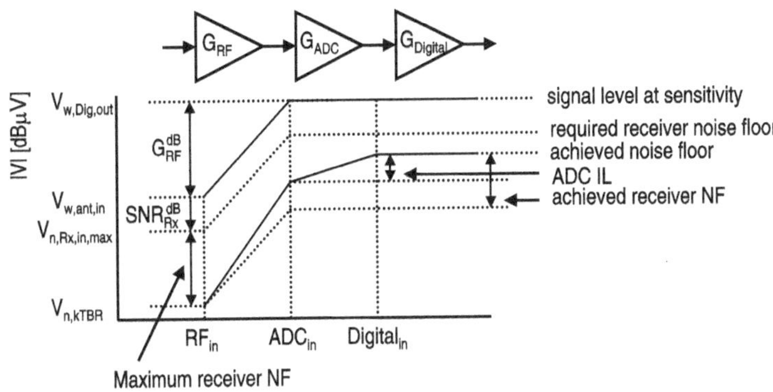

Figure 4.20: *Cumulative noise figure of a cascade of RF front-end, ADC and digital part*

under the condition of:

$$V^{dBV}_{n,RF,out,rms} = V^{dBV}_{n,ant,kTBR,rms} + G^{dB}_{Vl,RF,max} + \mathrm{NF}_{RF} \quad [\mathrm{dBV}] \tag{4.16}$$

$$V^{dBV}_{n,Rx,out,max,rms} \leq V^{dBV}_{n,ant,kTBR,rms} + G^{dB}_{Vl,RF,max} + \mathrm{NF}_{Rx,max} \quad [\mathrm{dBV}] \tag{4.17}$$

4.5 DR of the ADC

The combination of the outcome of Eqs. 4.12 and 4.18 leads to an ADC DR of:

$$\mathrm{DR}^{dB}_{ADC} \geq V^{dBV}_{int,ADC,in,max,rms}|_{G_{Vl,RF}=G_{Vl,RF,max}} - V^{dBV}_{n,ADC,in,max,rms} \quad [\mathrm{dB}] \tag{4.18}$$

4.5.1 DR of a Quadrature ADC

A quadrature system requires two ADCs to convert the quadrature signals into the digital domain. In case the input signals to the ADCs are quadrature sine waves, the required peak SNR of the individual ADCs is relaxed with 3 dB as the I&Q signal components are correlated but the noise contributions of the I and Q ADCs are not correlated. It is clear that this advantage comes at the cost of two times the ADC power consumption. Furthermore, this 3 dB advantage is independent of the IF choice.

4.6 RF Front-End and ADC Linearity Requirements

The linearity requirements of a receiver are normally defined in an IIP2 and IIP3 test. In these tests large out-of-band interferers are defined which should be handled by the receiver, without getting too far into compression, yielding too high intermodulation products.

A/D converter linearity is often qualified by measuring its harmonic- or intermodulation distortion with a full-scale sine wave signal at its input. In receivers the linearity of the ADC is most relevant at high power out-of-band interferers. The interferers inter-modulate which can yield in-band intermodulation products. In this section the receiver IP2 and IP3 and ADC harmonic and intermodulation distortion will be related. In Sect. 4.7.4 these relations will be used to analyse the distortion requirement interchange between RF front-end and ADC.

4.6.1 Second and Third Order Harmonic Distortion

A general transfer function which models the second and third order distortion of a circuit block is:

$$y = x_{in} + a \cdot x_{in}^2 + b \cdot x_{in}^3 \tag{4.19}$$

The HD2 and HD3 is calculated with an input sine-wave of frequency ω_w and amplitude A:

$$x_{in}(t) = A \cdot \sin(\omega_w \cdot t) \ \ [\text{V}] \tag{4.20}$$

Combining Eqs. 4.19 and 4.20 yields:

$$Y_w \approx A \ \ [\text{V}] \tag{4.21}$$

$$Y_{DC} = \frac{a \cdot A^2}{2} \ \ [\text{V}] \tag{4.22}$$

$$Y_{HD2} = \frac{a \cdot A^2}{2} \ \ [\text{V}] \tag{4.23}$$

$$Y_{HD3} = \frac{b \cdot A^3}{4} \ \ [\text{V}] \tag{4.24}$$

Next to the harmonic components, the HD2 causes a DC component at $\frac{a \cdot A}{2}$ compared to full scale, and the HD3 causes a negligible (if $b \cdot A^2 \ll 1$) signal component ω_w, which has been omitted from Eq. 4.21. The spectrum of y is shown in Fig. 4.21. The input signal related HD2 and HD3 distances are:

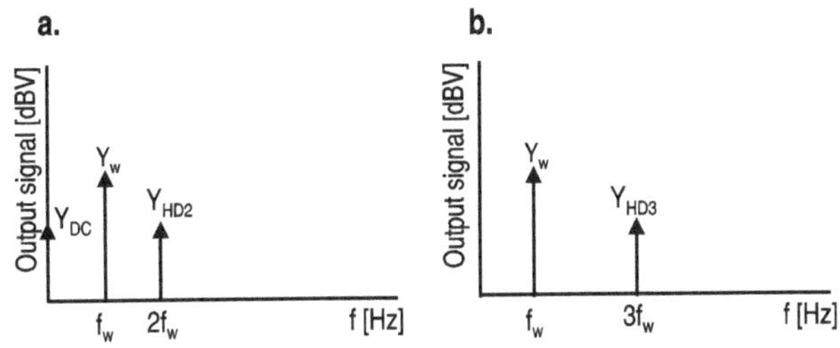

Figure 4.21: *Second (a) and third (b) order harmonic distortion spectra*

$$\mathrm{HD2D}^{dB} = 20\log_{10}\left(\frac{2}{a \cdot A}\right) \quad [\mathrm{dB}] \tag{4.25}$$

$$\mathrm{HD3D}^{dB} = 20\log_{10}\left(\frac{4}{b \cdot A^2}\right) \quad [\mathrm{dB}] \tag{4.26}$$

4.6.2 Second and Third Order Intermodulation and IP2 and IP3

The intermodulation distortion is determined with two test tones at frequency ω_1 and ω_2, and amplitudes A_1 and A_2. The amplitudes are related by $A_1 + A_2 = A$ and $R_{A_1/A_2} = A_1/A_2$. The input signal becomes:

$$\begin{aligned} x_{in}(t) &= A_1 \cdot \sin(\omega_1 \cdot t) + A_2 \cdot \sin(\omega_2 \cdot t) \quad [\mathrm{V}] \\ &= A \cdot \left(\frac{\sin(\omega_1 \cdot t)}{1 + \frac{1}{R_{A_1/A_2}}} + \frac{\sin(\omega_2 \cdot t)}{1 + R_{A_1/A_2}}\right) \quad [\mathrm{V}] \end{aligned} \tag{4.27}$$

The intermodulation products are calculated by combining Eqs. 4.19 and 4.27. The IM products which might appear in the signal bandwidth are:

$$Y_1 \approx A_1 \tag{4.28}$$

$$Y_2 \approx A_2 \tag{4.29}$$

$$Y_{DC} = \frac{a \cdot (A_1^2 + A_2^2)}{2} = \frac{a \cdot A^2 \cdot (1 + R_{A_1/A_2}^2)}{2(1 + R_{A_1/A_2})^2} \quad [\mathrm{V}] \tag{4.30}$$

$$Y_{IM2} = a \cdot A_1 \cdot A_2 = \frac{a \cdot A^2 \cdot R_{A_1/A_2}}{(1 + R_{A_1/A_2})^2} \quad [\mathrm{V}] \tag{4.31}$$

$$Y_{IM3} = \frac{3 \cdot b \cdot A_1^2 \cdot A_2}{4} = \frac{3 \cdot b \cdot A^3 \cdot R_{A_1/A_2}}{4(1 + R_{A_1/A_2})^3} \quad [\mathrm{V}] \tag{4.32}$$

The IM2 distortion produces a DC component, and a distortion component at $f_2 - f_1$. The IM3 distortion causes a distortion component at frequency $2f_1 - f_2$

and $2f_2 - f_1$, the latter being neglected being out of band. The IM3 distortion also causes additional input signal components f_1 and f_2 which are neglected. The relevant IM2 and IM3 products are displayed in Fig. 4.22a, b. In Fig. 4.23, the DC

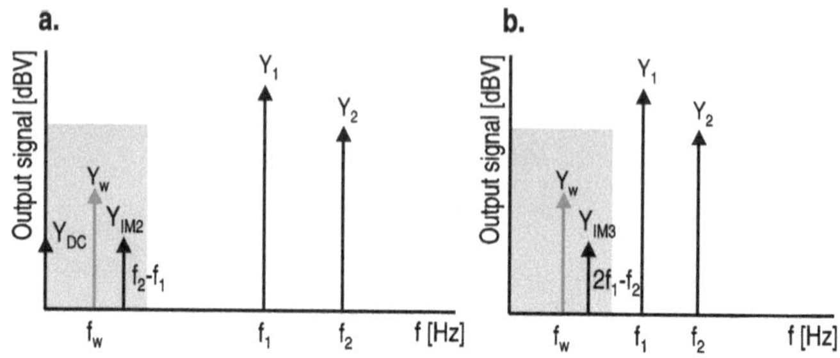

Figure 4.22: *Second (a) and third (b) order intermodulation distortion*

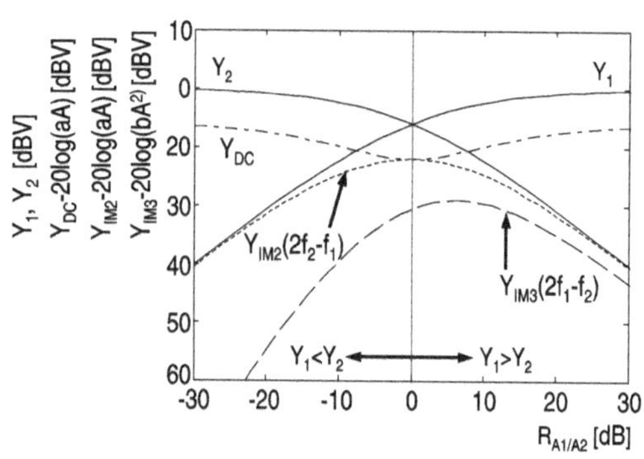

Figure 4.23: *DC offset and IM2/IM3 products as function of* R_{A_1/A_2}

offset, IM2 and IM3 products are displayed as a function of R_{A_1/A_2}. From the figure it can be seen that the Y_{IM2} component at $f_2 - f_1$ is symmetrical around $R_{A_1/A_2} = 1$ or 0 dB. The Y_{IM3} components gets smaller when $A_1 < A_2$, which unfortunately is in contradiction with filtering, because due to filtering, Y_1 will be attenuated more than Y_2.

The ADC intermodulation is normally specified with 2 equal amplitude sine waves at the receiver input. If $A_1 = A_2 = \frac{1}{2}A$ and thus $R_{A_1/A_2} = 1$, the IM2 and IM3 distances become:

$$\mathrm{IM2D}^{dB} = 20\log_{10}\left(\frac{2}{a \cdot A}\right) \;\; [\mathrm{dB}] \tag{4.33}$$

$$\mathrm{IM3D}^{dB} = 20\log_{10}\left(\frac{16}{3 \cdot b \cdot A^2}\right) \;\; [\mathrm{dB}] \tag{4.34}$$

Y_{DC}^{dB} will be at $20\log_{10}(a \cdot A^2/4)$. In case $A_1 = A_2 = \frac{1}{2}A$, the combined equations 4.23/4.24, 4.33/4.34 predict that the $\mathrm{HD2D}^{dB}$ is equal to the $\mathrm{IM2D}^{dB}$, and the $\mathrm{IM3D}^{dB}$ is 2.5 dB (16/12x) bigger than the $\mathrm{HD3D}^{dB}$, if the IM and HD products are related to the input signal amplitude of a single tone ($\frac{1}{2}A$). When the IM products are also related to full scale (A), the $\mathrm{IM2D}^{dB}$ and $\mathrm{IM3D}^{dB}$ figures are 6 dB and 8.5 dB better than the $\mathrm{HD2D}^{dB}$ and $\mathrm{HD3D}^{dB}$ figures respectively.

From the equations above it can be derived that the input output power relation between the input signal power and its IM2 and IM3 intermodulation products is 2 dB/dB and 3 dB/dB respectively. The IP2 and IP3 are defined by the extrapolated point where the IM2 and IM3 components are equally big as the input signal. The input and output IP2 and IP3 are displayed in Fig. 4.24. From the figure the IIP2 and IIP3 can be calculated.

$$Y_{IM2}^{dBV} = 2 \cdot V_{in}^{dBV} - \mathrm{IIP2}^{dBV} \quad \text{or} \tag{4.35}$$

$$\mathrm{IIP2}^{dBV} = V_{in}^{dBV} + \mathrm{IM2D}^{dB} \;\; [\mathrm{dBV}] \tag{4.36}$$

$$Y_{IM3}^{dBV} = 3 \cdot V_{in}^{dBV} - 2 \cdot \mathrm{IIP3}^{dBV} \quad \text{or} \tag{4.37}$$

$$\mathrm{IIP3}^{dBV} = V_{in}^{dBV} + \frac{\mathrm{IM3D}^{dB}}{2} \;\; [\mathrm{dBV}] \tag{4.38}$$

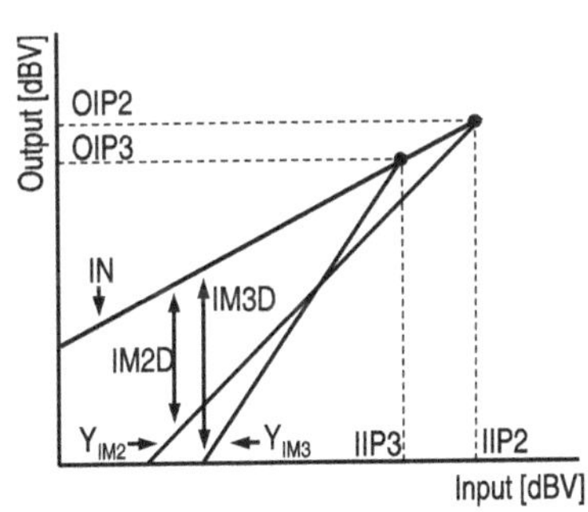

Figure 4.24: *Second and third order intercept point*

Using Eqs. 4.33 and 4.34 the IIP2 and IIP3 becomes:

$$\text{IIP2} = -6 + A^{dBV} + \text{IM2D}^{dB} \tag{4.39}$$

$$= 20\log_{10}\left(\frac{1}{a}\right) \quad [\text{dBV}] \tag{4.40}$$

$$\text{IIP3} = -6 + A^{dBV} + \frac{1}{2}\text{IM3D}^{dB} \tag{4.41}$$

$$= 20\log_{10}\left(\sqrt{\frac{4}{3\cdot b}}\right) \quad [\text{dBV}] \tag{4.42}$$

In a receiver the IM requirements of the individual blocks is subject to the selectivity that was added to the receiver. This can cause the IM tones to have different amplitude $A_1 \neq A_2$ (or $R_{A_1/A_2} \neq 1$). From Eq. 4.40 it can be seen that the IM2 level rises 1 dB/dB for both A_1 and A_2. Equation 4.42 predicts the IM3 level to rise with 2 dB/dB with A_1 and 1 dB/dB with A_2. The signal to distortion ratio can be calculated and yields:

$$\begin{aligned}\text{SDR}^{dB}_{IM2} &= A_w^{dBV} - A_1^{dBV} - A_2^{dBV} + \text{IIP2}^{dBV} \\ &= A_w^{dBV} - 2\cdot A^{dBV} + R^{dB}_{1/R} \\ &\quad + R^{dB}_{R} + \text{IIP2}^{dBV} \quad [\text{dB}]\end{aligned} \tag{4.43}$$

with

$$R_{1/R} = 1 + \frac{1}{R_{A_1/A_2}} \quad \text{and} \quad R_R = 1 + R_{A_1/A_2} \quad [\text{-}]$$

$$\begin{aligned}\text{SDR}^{dB}_{IM3} &= A_w^{dBV} - 2\cdot A_1^{dBV} - A_2^{dBV} \\ &\quad + 2\cdot \text{IIP3}^{dBV} \\ &= A_w^{dBV} - 3\cdot A^{dBV} + 2R^{dB}_{1/R} + R^{dB}_{R} \\ &\quad + 2\cdot \text{IIP3}^{dBV} \quad [\text{dB}]\end{aligned} \tag{4.44}$$

4.6.3 Third Order Cross-Modulation

When an out-of-band IM test is combined with an input signal, cross-modulation products appear around the input signal. The input signal of the receiver $x(t)_{in}$ contains 4 signals: the IM test tones f_1, f_2, the wanted signal f_w and a neighboring channel f_n.

$$\begin{aligned}x_{in}(t) &= A_w \cdot \sin(\omega_w \cdot t) + A_n \cdot \sin(\omega_n \cdot t) \\ &\quad + \frac{A}{1 + \frac{1}{R_{A_1/A_2}}} \cdot \sin(\omega_1 \cdot t) + \frac{A}{1 + R_{A_1/A_2}} \cdot \sin(\omega_2 \cdot t) \quad [\text{V}]\end{aligned} \tag{4.45}$$

The wanted and neighboring channel are modulated by the IM tones due to 3rd order distortion. Figure 4.25 shows the relevant resulting IM distortion products. The intermodulation products of the wanted signal are not shown, because they are related to the wanted signal amplitude and in this case are very small. Combining

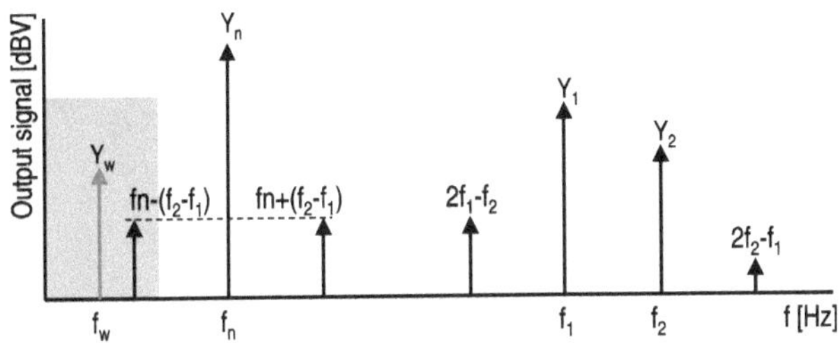

Figure 4.25: *Third order cross-modulation*

Eqs. 4.42 and 4.19 yields the level of the CM products, which have an equal amplitude.

$$Y_{CM3} = \frac{3 \cdot b \cdot A_1 \cdot A_2 \cdot A_n}{8} \text{ [V]} \tag{4.46}$$

The modulated cross modulation products of f_n might appear in the wanted signal bandwidth, degrading the SNDR ratio. The SDR can be calculated by combining Eqs. 4.42 and 4.46 which yields:

$$\text{SDR}_{CM3}^{dB} = A_w^{dBV} - 2 \cdot A^{dBV} - A_n^{dBV} + 2 \cdot \text{IIP3}^{dBV} + 6 \text{ dB} \text{ [dB]} \tag{4.47}$$

4.6.4 Distortion in a Quadrature ADC

The ADCs used in the I&Q receiver produce distortion which might interfere with the wanted signal. If it is assumed that this distortion is caused by the distortion of the I&Q ADC input differential pairs (which is often the case in a $\Sigma\Delta$ modulator), and the both differential pairs match perfectly, the distortion is correlated.

The different IM2 and IM3 product amplitudes at the output of the I&Q ADC are given by Table 4.3, when it is assumed that the input signal consists of two sine waves of amplitude $\frac{1}{2}A$.

The different signal and distortion components are displayed in Fig. 4.26. The figure shows, that if an NZIF system is taken, the IM2 product $f_1 - f_2$ falls out of the signal bandwidth, and Table 4.3 predicts an IM2 relaxation of 3 dB compared to Eq. 4.33.

Second order distortion components		Third order distortion components	
Frequency	Amplitude	Frequency	Amplitude
$2f_2$	$\sqrt{2}aA^2/16$	$2f_2 - f_1$	$3bA^3/32$
$f_1 + f_2$	$\sqrt{2}aA^2/8$	f_2	$9bA^3/32$
f_1	$\sqrt{2}aA^2/16$	f_1	$9bA^3/32$
$f_2 - f_1$	$\sqrt{2}aA^2/8$	$2f_1 - f_2$	$3bA^3/32$
0	$\sqrt{2}aA^2/4$	$-3f_1$	$bA^3/32$
$f_1 - f_2$	$\sqrt{2}aA^2/8$	$-2f_1 - f_2$	$3bA^3/32$
$-2f_1$	$\sqrt{2}aA^2/16$	$-2f_2 - f_1$	$3bA^3/32$
$-f_1 - f_2$	$\sqrt{2}aA^2/8$	$-3f_2$	$bA^3/32$
$-2f_2$	$\sqrt{2}aA^2/16$		

Table 4.3: *Second and third order distortion components in an I&Q system*

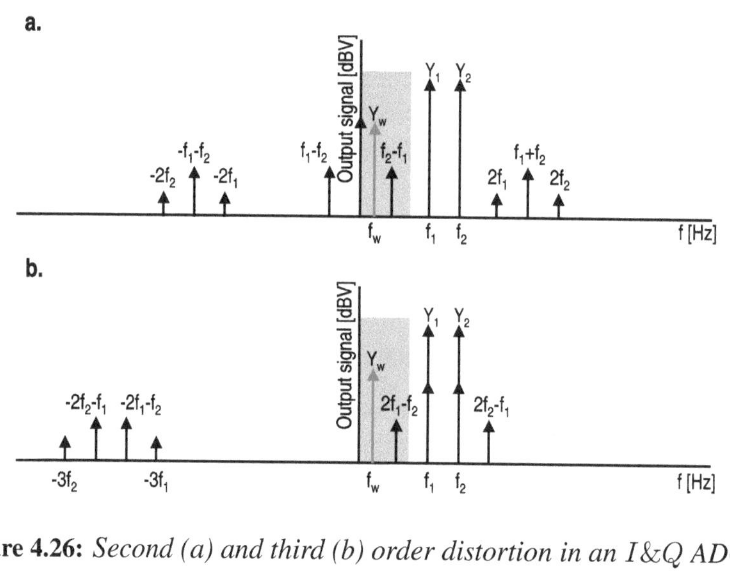

Figure 4.26: *Second (a) and third (b) order distortion in an I&Q ADC*

In Appendix A it is calculated that when the input signal components have a positive frequency, the HD3 components are only present in the negative frequency band. In the case of finite image rejection of the I&Q ADC, or offset differences in the I&Q input pairs, a third harmonic distortion components will appear in the positive frequency band as well, as the cancellation of the distortion products is not perfect.

4.7 Example Receiver Partitioning: Receiver for a GSM Mobile Phone

This section elaborates on the system choices for, and the partitioning of an NZIF direct conversion receiver for GSM, and the influence on the ADC requirements.

Table 4.4 gives a summary of the GSM system in relation to the receiver. In

Communication standard	GSM
System bit-rate	270.8333 kbps
System bandwidth	200 kHz
Receiver SNR requirement (BER = 2%) SNR_{Rx}^{dB}	6.5 dB
Digital modulation scheme	GMSK

Table 4.4: *GSM receiver system summary*

GSM the modulation scheme used is GMSK. The BER as a function of E_b/N_0 for GMSK modulated signal plotted in Fig. 4.18 shows that for a BER of 2% the required E_b/N_0 is 5.2 dB. In GSM the data rate R_b is 270.833 kbps, and the channel bandwidth is 200 kHz, which yields and R_b/B of 1.3 dB (Eq. 4.14). This means that the SNR requirement for the complete GSM receiver $SNR_{Rx,min}^{dB}$ is $5.2 + 1.3 = 6.5$ dB.

The in- and out-of-band signal levels for GSM are defined in the communication standard specification. Figure 4.27 shows the interferer/blocking/wanted signals specification for a GSM system. The x-axis displays the frequency distance to

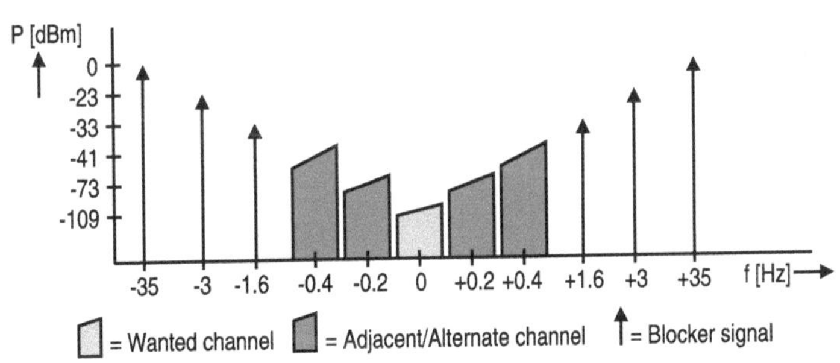

Figure 4.27: *Signal definitions for a GSM receiver*

the wanted signal, the y-axis displays the signal power. The out-of-band signals defined in the specification can either be modulated or un-modulated.

From the $\mathrm{SNR}_{Rx,min}$ and signal definitions for a GSM receiver, the RF front-end and ADC noise and distortion requirements will be calculated. The wanted signal level at sensitivity in dBV minus the $\mathrm{SNR}^{dB}_{Rx,min}$ and the margin in the noise floor reserved for the distortion, yields the maximum receiver noise floor $V^{dBV}_{n,Rx,max}$ at the antenna. The maximum noise figure for the receiver NF^{dB}_{Rx} is determined by the difference of $V^{dBV}_{n,Rx,max}$ and the V^{dBV}_{kTBR} noise, which is shown in Fig. 4.28. In the partitioning of the receiver, a margin is held in the receiver noise

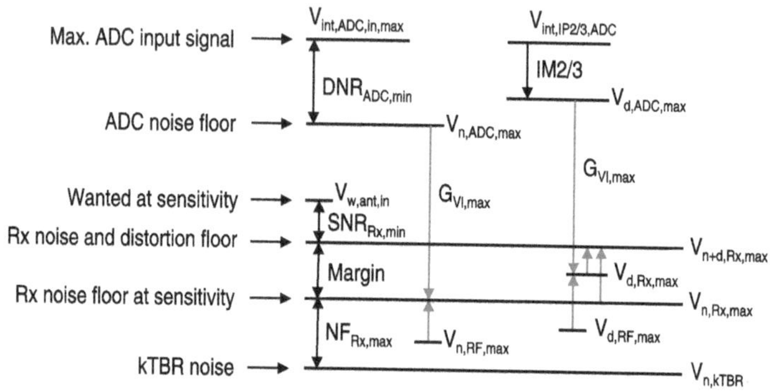

Figure 4.28: *Signal level diagram of the GSM receiver*

floor to account for the distortion contributions of the RF front-end and ADC. The $V_{d,Rx}$, $V_{d,RF}$, $V_{d,ADC}$ represents the maximum distortion product level of the receiver, the RF front-end and the ADC respectively. The accumulated noise and distortion of the RF front-end and ADC are chosen such that the receiver will just pass the sensitivity test in the IP2/3 tests.

Before diving into the ADC and RF front-end noise and distortion requirements, the IF is related to the image rejection requirement.

4.7.1 IF Choice and Image Rejection

As discussed in Sect. 4.1.4, the IF choice is a trade-off between the DC offset and 1/f noise requirements, and image rejection. As GSM is a narrow band system, NZIF is the preferred choice.

In an NZIF system, the image interference is not only dependent on the IRR but also on the level of adjacent channels. The co-channel interference ratio $\mathrm{IR}_{cochint}$ requirement for a GSM receiver is 9 dB. In the GSM co-channel interference test the wanted signal is at -82 dBm. The maximum level of the co-channel interference is:

$$P^{dBm}_{cochint,max} = P^{dBm}_{w} - \text{IR}^{dB}_{cochint} = -82 - 9 = -91 \text{ [dBm]} \tag{4.48}$$

Depending on the IF frequency the image appearing in the wanted signal bandwidth will come from a different adjacent channel. The IRR requirement for the in band signal can be calculated by:

$$\text{IRR}^{dB}_{in-band} = P^{dBm}_{-f_{IF}} - P^{dBm}_{cochint,max} \text{ [dB]} \tag{4.49}$$

In which $P^{dBm}_{-f_{IF}}$ is the power of the adjacent channel which will become the image of the wanted.

The image appearing in adjacent channels also puts a requirement on the receiver IRR. The adjacent channels have an image at $f_{image} = -(2 \cdot f_{IF} + f_{int})$ which is limited by the GSM spectral mask definitions (Fig. 4.27).

The in- and out-of-band image rejection as a function of the IF frequency is shown in Table 4.5:

f_{IF}	100	200	300	kHz
IRR^{dB} in-band image	$-73-(-91)=18$	$-41-(-91)=50$	$-33-(-91)=66$	dB
IRR^{dB} image adj. of +200 kHz	$-41-(-73)=32$	$-33-(-41)=8$		dB
IRR^{dB} image adj. of +400 kHz	$-33-(-41)=8$			dB

Table 4.5: *Receiver image rejection requirements*

In some cases the IRR requirement indeed is determined by an out-of-band (adjacent) image. Furthermore, the higher the IF is chosen, the higher the IRR requirement will be. This asks for a trade-off between 1/f noise and IF frequency, to come to an optimal IRR and 1/f noise specification. In this receiver the IF is set to 100 kHz, which leads to a moderate receiver IRR^{dB}_{Rx} specification of 32 dB.

In the calculation above, the IF filtering is neglected as the IRR in a GSM receiver is determined by close-by adjacent channels. The IRR_{Rx} specification can be divided in an IRR specification for the I&Q mixers, LO signals and quadrature ADC. The actual image requirement of the ADC will be subject to the IF filtering in front of the ADC. The IRR^{dB}_{ADC} is set to > 40 dB, to have sufficient margin.

4.7.2 Top-End of the ADC Dynamic Range

The top-end of the ADC dynamic range is determined by the GSM spectral mask, the filtering in front of the ADC, and the crest factor of the different signals to be expected at the ADC input. In Table 4.6 the most relevant signal definitions out of the GSM system specification [12] are shown. It concerns the signal definition for the sensitivity test, the spectral mask of the interferers/blockers/adjacent channels and the IP2 and IP3 test signals. The signals are specified at the antenna and are calculated to the input of the ADC in Table 4.6, under the assumption that:

- all receiver blocks have a loaded voltage gain of 0 dB
- the duplexer reduces the input power at the antenna outside the 35 MHz wide GSM band (Table 4.6)
- a first order filter $F(f)$ with a cut-off frequency f_c of 400 kHz is added to the IF amplifier stage

The result is displayed in Table 4.6. In the table it is assumed that the wanted signal is present under all tests specified, which can increase the effective crest factor (Eq. 4.11). With the chosen selectivity, the maximum (peak) signal to be

	Freq. offset [MHz]	Antenna [dBm]	$V_{ant,in,rms}$ [dBμV]	Duplexer [dB]	$F(f)$ [dB]	$V_{RF,out,rms}$ [dBμV]	Crest [dB]	Eff. Crest [dB]	$V_{int,ADC,in,max,peak}$ [dBμV]
Wanted at sensitivity	0	−109	−2.01	0	0.00	8	6.2	6.2	14.2
Interferers from spectral mask									
Interferer 1	0.2	−73	34.0	0	−1	33.0	6.2	6.7	39.7
Interferer 2	0.4	−41	66.0	0	−3.0	63.0	6.2	6.2	69.2
Interferer 3	1.6	−33	74.0	0	−12.3	61.7	3.0	3.0	64.7
Interferer 4	30	−23	84.0	0	−17.6	66.4	3.0	3.0	69.4
Interferer 5	35	0	107.0	−23	−38.8	45.2	3.0	3.2	48.3
Wanted signal, and IP2/3 interferers									
Wanted signal	0.0	−99	8.0	0	0	8.0	6.2	6.2	
IP3 interferers CW (modulated)	0.8	−49	58.0	0	−7.0	51.0	3.0	7.0	59.1
	1.6	−49	58.0	0	−12.3	45.7	6.2		
IP2 interferer	6.0	−31	76.0	0	−23.5	52.5	6.2	6.2	58.7
ADC maximum input signal									69.4

Table 4.6: *Antenna signal level definitions according to the GSM specification and signal level impact throughout the receiver chain*

expected at the ADC input $V^{dBV}_{int,ADC,in,max,peak}|_{G^{dB}_{Vl,RF}=0}$ is: 69.4 dBμV, when the total loaded voltage gain up to the ADC is 0 dB. To reduce the noise contribution of the ADC to the noise figure of the receiver, the gain in front of the ADC is set to maximum. Equation 4.12 predicts a maximum rms ADC input signal of 113.1 dBμV or a differential rms sine wave input signal of 0.45 V when the supply voltage of the ADC is 1.2 V and $x = 0.53$. If Eqs. 4.12 and 4.13 are combined the input signal of a single ADC becomes:

$$G^{dB}_{Vl,RF,max} = 20 \cdot \log_{10}\left(\frac{x \cdot V_{DDA}}{\sqrt{2}}\right) - V^{dBV}_{int,ADC,in,max,rms}|_{G^{dB}_{Vl,RF}=0} \quad [\text{dB}] \tag{4.50}$$

which yields a $G^{dB}_{Vl,RF,max}$ of 46.6 dB.

The GSM specification defines that the power level of the wanted signal can be −15 dBm at maximum. In this case, the gain in the receiver has to be reduced, to prevent the ADC to be overdriven, which calls for AGC circuitry preceding the ADC.

4.7.3 Receiver Sensitivity Requirement and the Bottom-End of the ADC Dynamic Range

In the GSM standard the sensitivity test is defined with a wanted input signal at $P^{dBm}_{w,ant,in} = -102$ dBm. In the product definition of a radio, the NF_{Rx} is often chosen lower than the minimum required by the radio specification standard, which increases the sensitivity of the radio and gives a competitive advantage. In the receiver described in this book the reference sensitivity level is set to −109 dBm or $V^{dBV}_{w,ant,in,rms} = -2.0$ dBμV as shown in Table 4.6. At the input of the receiver the noise is $kTBR = -13.83$ dBμV which yields receiver SNR of 11.82 dB at the antenna. With minimum required SNR in a GSM receiver of 6.5 dB (Table 4.4) and a margin of 0.32 dB, leaves a maximum $\text{NF}^{dB}_{Rx,max}$ of 5 dB. The $\text{NF}_{Rx,max}$ can be distributed over the RF front-end and ADC, if it is assumed that the IL_{Dig} is negligible.

The interchange between NF_{RF} and ADC DR is shown in Fig. 4.29 as a function of the ratio between the integrated noise contributions of RF front-end and ADC, $R_{n,RF/ADC}$. At high $R_{n,RF/ADC}$ the RF dominates the noise contributions and the NF_{RF} approximates NF_{Rx}, and the ADC DR increases. At low $R_{n,RF/ADC}$ the ADC dominates the noise contributions, and the required ADC DR reduces. It can be seen from Fig. 4.29, that dependent on the noise contribution ratio between RF and ADC, the ADC dynamic range varies roughly between ∼80–100 dB.

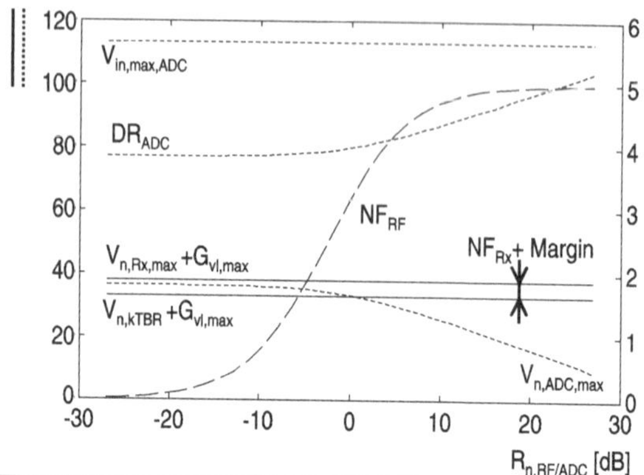

$V^{dBV}_{in,ADC,max,rms}$	dBμV	DR_{ADC}	dB	NF_{RF}	dB
$V^{dBV}_{n,kTBR,rms}$	dBμV	$V^{dBV}_{n,Rx,max,rms}$	dBμV	NF_{Rx}	dB
$V^{dBV}_{n,ADC,max,rms}$	dBμV	Margin	dB	G^{dB}	dB

Figure 4.29: *RF front-end NF and ADC DR requirement as a function of* $R_{n,RF/ADC}$

4.7.4 Receiver Linearity Requirement and ADC Linearity

In the GSM standard the receiver linearity is defined in the IP2 and IP3 tests. In the IP2 and IP3 test the wanted signal in both cases is defined at −99 dBm. In the IP2 test an interferer is defined at a level of −31 dBm, and at a frequency offset of 6 MHz to the wanted signal. In the IP3 test, a CW and modulated interferer is defined both at a level of −49 dBm, and at frequency distance of 800 and 1600 kHz to the wanted signal respectively. In both tests the receiver is at maximum gain. To achieve the BER of 2% the required SNR at the output should be 6.5 dB.

The reference sensitivity to be taken in the linearity requirement calculations is subject to the choice of narrow or broad-band AGC. The GSM specification allows for receiver de-sensitization to −99 dBm, at high out-of-band interference power. In case of narrow band AGC, the receiver can not act on out-of-band interference as it is not aware of interference. Therefore, the reference sensitivity of −109 dBm must be maintained always, as the AGC can only act on the wanted signal itself. In case of broad-band AGC, the receiver is aware of interference, and the sensitivity requirement can be relaxed to −99 dBm. In the IP2/3 calculations the IP2/3 products are allowed to reduce the margin of 0.32 dB taken in the noise figure calculations to 0 dB, independent on the AGC scheme.

	Broad band AGC	Narrow band AGC	
Reference sensitivity at antenna	–99	–109	dBm
Rx SNR	6.5	6.5	dB
Required Rx noise floor at antenna	–105.5	–115.5	dBm
Rx noise floor at antenna at max. gain	–115.8	–115.8	dBm
Max. IP2/3 level at the antenna $P_{d,Rx}^{dBm}$	–105.9	–127.0	dBm
Max. IP2/3 level at the ADC input $V_{d,Rx,rms}^{dBV}$	47.7	26.7	dBμV

Table 4.7: *Max. IP2/3 product level as function of reference sensitivity*

The influence of AGC bandwidth on the IP2/3 requirement is summarized in Table 4.7 This has a major impact on the linearity requirements of the receiver. Figure 4.30 shows the linearity requirements of the RF front-end and ADC in both cases, as a function of $R_{dist,RF/ADC}$. $R_{dist,RF/ADC}$ represents the ratio of the RF and ADC distortion products contribution at the antenna, $V_{d,RF}^{dBV}$ and $V_{d,ADC}^{dBV} - G_{Vl,RF,max}^{dB}$. Therefore:

$$R_{dist,RF/ADC}^{dB} = V_{d,RF}^{dBV} - (V_{d,ADC}^{dBV} - G_{Vl,RF,max}^{dB}) \text{ [dB]} \qquad (4.51)$$

From the figure it can be seen that the ADC linearity requirements from system level perspective are limited. This is because the IP2 and IP3 interferers are filtered by the IF amplifier. This allows for a relaxed RF front-end IP2/3, at the cost of only a slightly higher ADC linearity. In case of narrow band AGC, the ADC IM2D and IM3D vary between ~40–70 and ~50–80 respectively, depending on the distortion contribution ratio $R_{dist,RF/ADC}$ of RF front-end and ADC. For broad band AGC, the ADC IM2D and IM3D vary between ~20–50 and ~30–60 respectively, depending on the distortion contribution ratio $R_{dist,RF/ADC}$ of RF front-end and ADC. This means that broad band AGC relaxes the distortion requirements of the receiver by 20 dB, compared to narrow band AGC.

4.8 ADC Requirements, the System Quality Indicators and ΣΔ Modulators as the ADC Architecture

In this section the ADC requirements for a direct conversion receiver with first order filtering are gathered and evaluated from the system quality indicators perspective of Chap. 2.

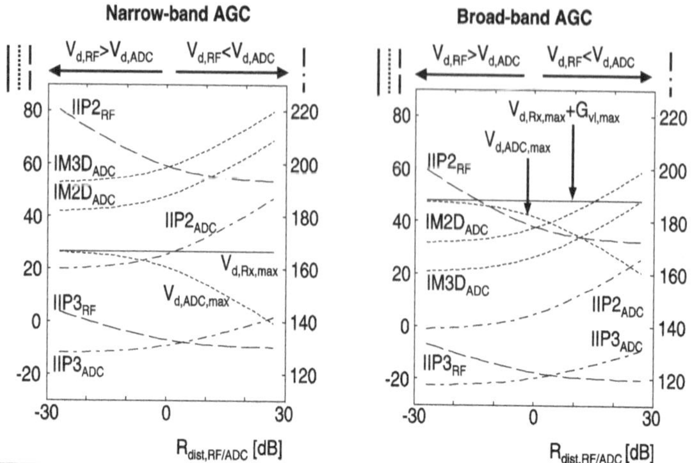

$IIP2_{RF}$	dBμV	$IM2D_{ADC}$	dB	$IIP2_{ADC}$	dBμV
$IIP3_{RF}$	dBμV	$IM3D_{ADC}$	dB	$IIP3_{ADC}$	dBμV
$V^{dBV}_{d,Rx,max,rms}$	dBμV	$V^{dBV}_{d,ADC,max,rms}$	dBμV		

Figure 4.30: *RF front-end and ADC linearity requirement as a function of distortion ratio* $R^{dB}_{dist,RF/ADC}$

1. Accuracy:

   - ADC DR: is a trade-off between analog filtering, RF front-end NF and receiver SNR requirement. For the GSM example this trade-off led to a ADC DR between ~80–100 dB.
   - ADC linearity: is a trade-off between analog filtering, RF front-end IIP2/3, type of AGC and receiver SNR requirement. For the GSM example this trade-off led to a ADC IM2/3 specification between ~20–80 dB.
   - ADC bandwidth: system bandwidth requirements are in the range of 200 kHz–28 MHz. The exact ADC bandwidth is dependent on the ZIF/NZIF choice. If the digital modulation scheme allows for it, for the higher bandwidths ZIF will be chosen, as 1/f noise will have little influence. For the low bandwidths NZIF is the preferred choice.
   - The IRR^{dB}_{ADC} should be larger than 40 dB.

The ADC requirements revealed, match with the properties of a $\Sigma\Delta$ modulator. $\Sigma\Delta$ modulators use over-sampling and noise shaping, pushing the in-band quantization noise out-of-band, which makes them extremely suitable for high DR, narrow/medium bandwidth applications. The DR of $\Sigma\Delta$

modulators will be further analyzed in Chap. 5. In this chapter a number of ΣΔ modulator architectures will be reviewed on their algorithmic accuracy.

2. Robustness to secondary inputs: The highly digitized receiver architecture chosen, has limited analog filtering in front of the ADC. This means that an anti-alias filter can be required. To be able to use broad band AGC, the receiver blocks, and thus the ADC, should provide out-of-band signal information. Furthermore, the ADC will be subject to technology impairments/changes and supply voltage- and temperature variations. A CT ΣΔ modulator uses an analog noise shaping loop filter, which also acts as a built-in anti-alias filter, which increases receiver power efficiency and robustness to interference. A ΣΔ modulator can supply additional out-of-band interference information, as it uses over-sampling. Furthermore, a 1-bit modulator is insensitive to technology impairments like matching. These and more subjects are presented in Chap. 6. Relations between ΣΔ modulator performance indicators (like DR, THD and aliasing) and their costs (like power and area) will be derived.

3. Flexibility: As said, the ADC will be used in a multi-standard receiver. This means bandwidth (200 kHz–28 MHz), resolution and clock frequency programmability. The flexibility of ΣΔ modulators can be implemented by changing the loop filter coefficients, and clock frequency, to accommodate for the different DR and bandwidth requirements. This will be explained in Chaps. 5 and 7. Particularly in Chap. 7 it will be investigated how a ΣΔ modulator can be made flexible.

4. Efficiency: ΣΔ modulators are widely known for their power efficiency. A thorough ΣΔ modulator efficiency analysis will be presented in Chap. 8.

5. Emission of secondary outputs: In the multi-pipe receiver of the modern mobile phone, different receivers should co-exist and co-habit, and the receiver should not interfere with other circuits on the same chip. Therefore, an ADC architecture should be chosen which emits as little as possible interference. As a 1-bit ΣΔ modulator only has a relatively simple 1-bit quantizer and 1-bit DAC which are switching, the expected emitted interference is negligible. As said in the introduction, emission is not further investigated in this book.

4.9 Conclusions

For narrow band systems $(B \approx f_{c,1/f})$, NZIF is the preferred IF because of the 1/f noise. For larger bandwidth systems $(B \gg f_{c,1/f})$, ZIF is the best choice, because of the limited impact of 1/f noise on the ADC accuracy and the high ADC bandwidth requirement. It depends on the digital modulation scheme used, whether the receiver will be robust against DC offset in a ZIF configuration.

Several relations between the performance parameters of the RF front-end and ADC have been derived. Choosing a high RF front-end NF or a low dynamic range for the ADC, might lead to an unrealistic ADC DR or RF front-end NF respectively, with the consequence of an unrealistic power consumption of the RF front-end or ADC. Furthermore, it was shown in this chapter that only limited filtering in front of the ADC, can alleviate the ADC linearity requirements dramatically. The relations presented in this chapter will be used in Chap. 6 to link the system specifications to the ADC circuit design trade-offs, to come to the most optimal ADC implementation, considering the quality indicators. Of course the same should be done for the RF front-end, but this is out of the scope of this book.

The specifications and characteristics derived for an ADC for a GSM receiver, very well match the properties of $\Sigma\Delta$ modulators. To validate the choice to use the $\Sigma\Delta$ modulator ADC architecture as basis for the receiver ADC, its properties are judged on the quality indicators presented in Chap. 2. In particular, continuous-time $\Sigma\Delta$ modulators are investigated.

Chapter 5

$\Sigma\Delta$ Modulator Algorithmic Accuracy

$\Sigma\Delta$ modulators are a well appreciated A/D converter choice for implementing the A/D conversion in receiver architectures for quite some time, as they can achieve a high resolution in low to moderate bandwidths at low power consumption. This chapter elaborates about the algorithmic accuracy of $\Sigma\Delta$ modulators. The algorithmic accuracy of a $\Sigma\Delta$ modulator is defined as the theoretical modulator performance measured in the Signal to Quantization Noise Ratio (SQNR). Circuit imperfections like noise, distortion, aliasing, etcetera are not taken into account in the algorithmic accuracy, and will be the subject of Chap. 6.

Sigma Delta modulators trade amplitude resolution with time resolution. Only a limited number of bits b are used for the quantization process. Instead, the modulators' sample frequency is taken much larger than the nyquist criterion dictates, which defines the over-sampling ratio (OSR) by $f_s/f_{s,nyquist} \gg 1$. In combination with an Lth order filter in the loop, the modulator succeeds in shaping part of the quantization noise out of the signal bandwidth, yielding a higher in band SQNR.

In this chapter, $\Sigma\Delta$ modulator analysis is limited to $\Sigma\Delta$ modulators which are used as an ADC. Furthermore, only $\Sigma\Delta$ modulators without input sampler are considered (the reason for this is explained in Chap. 6). The loop filter of the modulator is either continuous-time (CT) or partly CT and partly digital. Furthermore, in this chapter only feed-forward loop filters will be analyzed.

R.H.M. van Veldhoven, A.H.M. van Roermund, *Robust Sigma Delta Converters*,
Analog Circuits and Signal Processing, DOI 10.1007/978-94-007-0644-6_5,

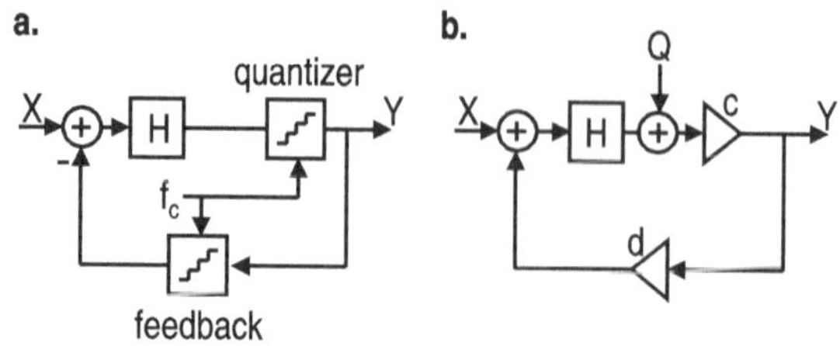

Figure 5.1: *Block diagram of a ΣΔ modulator with Lth order, CT, feed-forward loop filter*

An example ΣΔ modulator used as an ADC is shown in Fig. 5.1a. It consists of an analog input X, a CT loop filter H, a quantizer (or ADC), a digital output Y and a feedback DAC. In Fig. 5.1b, a model of the modulator is presented. The quantizer is replaced by a quantization noise source Q with Gaussian distribution and a gain c leading to a linear model[1], and the DAC is replaced by a gain d. From the model in Fig. 5.1b, it can be written that:

$$Y = \frac{cH}{1 + cdH} X + \frac{c}{1 + cdH} Q \tag{5.1}$$

From Eq. 5.1, it can be seen that in case H is a low pass filter which at low frequencies has a high gain ($\frac{cH}{1+cdH} \approx 1/d$), the input signal appears at the output unfiltered. The quantization noise however, is suppressed by the loop filter gain, and is pushed to higher frequencies where the gain of H is low. Therefore, at low frequencies a high resolution can be achieved, if the out-of-band quantization noise is removed by a digital filter.

Equation 5.2 shows a first order approximation of the relation between the modulators loop filter order L, number of bits b in the quantizer and DAC, and the over-sampling ratio OSR for single loop ΣΔ modulators [13]. In the derivation of Eq. 5.2 it is assumed that the performance of the modulator is not bounded by the stability criteria of the loop. Therefore, the equation will give very optimistic SQNR figures for higher order modulators.

$$\mathrm{SQNR}^{dB} = 10 \cdot \log_{10}\left(\frac{3}{2}\left(\frac{2L+1}{\pi^{2L}}\right)\left(2^b - 1\right)^2 \mathrm{OSR}^{2L+1}\right) \quad [\mathrm{dB}] \tag{5.2}$$

In this chapter the algorithmic accuracy will be explored for five different ΣΔ modulator architectures:

[1]The linearized model presented here is only valid when there is very limited or no correlation between input signal and quantization noise.

1. $\Sigma\Delta$ modulators with 1-bit quantizer and 1-bit DAC.
2. $\Sigma\Delta$ modulators with b-bit quantizer and b-bit DAC.
3. $\Sigma\Delta$ modulators with 1.5-bit quantizer and 1.5-bit DAC.
4. $\Sigma\Delta$ modulators with multiple quantizers and 1-bit DAC.
5. $\Sigma\Delta$ modulators with 1 or b-bit quantizer and DAC, and with additive error feedback loop.
6. Cascaded $\Sigma\Delta$ modulators.

In the later implementation Chap. 9, modulator examples of items 1, 3 and 4 will be shown. The $\Sigma\Delta$ modulator architecture of item 5 actually is being implemented at the time of writing this book. By using an additional error feedback loop, the need for the cascaded modulator stage of item 6, which was presented in [14], is avoided to reduce complexity. The modulator architectures of item 2, 5 and 6 will not be represented in the implementation chapters.

In the exploration of the algorithmic accuracy of the different $\Sigma\Delta$ modulator architectures, no distinction will be made between a feed-forward or feedback loop filter, as generally the achievable algorithmic accuracy of both feed-forward and feedback modulators is the same. The same holds for the choice of a discrete or continuous-time (CT) implementation of the loop filters and DACs.

5.1 $\Sigma\Delta$ Modulators with 1-bit Quantizer and 1-bit DAC

Equation 5.2 of the introduction shows that a high SQNR is possible by increasing L and OSR, without using an excessively number of bits in the quantizer and DAC. In this section 1-bit highly over-sampled $\Sigma\Delta$ modulators will be subject of investigation.

The design of the $\Sigma\Delta$ modulator starts with the design of the loop filter. A generalized $\Sigma\Delta$ modulator loop filter is shown in Fig. 5.2. The integrators are unscaled ideal integrators and coefficients m_n $(n = [1..L])$ are pre-calculated values, determined from for example a loop filter of order L with maximally flat amplitude (Butterworth) or phase (Chebychev) response. The ratios between the m_n coefficients determine the stability of the modulator. At high frequencies, the loop filter is forced to first order by the feed-forward coefficients m_n to guarantee loop stability. At the same time, the m_n ratios also determine the noise shaping efficiency

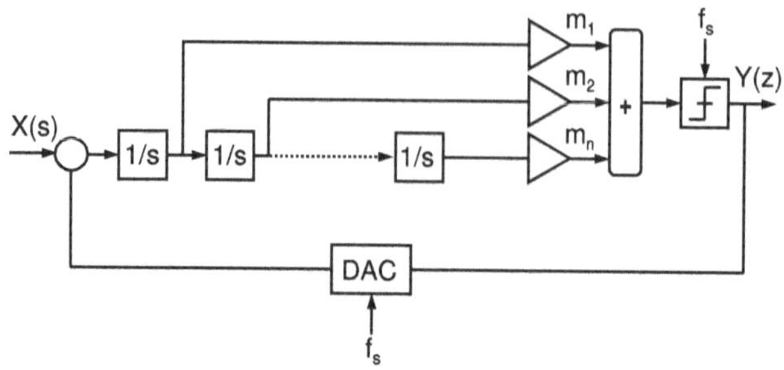

Figure 5.2: *Block diagram of a 1-bit ΣΔ modulator with Lth order, CT feed-forward loop filter*

of the modulator, and a trade-of between noise shaping and stability is required. As this ΣΔ modulator is going to be implemented in an IC technology with a maximum supply voltage, the m_n coefficients need to be re-distributed over the loop filter such, that the internal signal swings on the integrator outputs are within the supply. Of course a margin in the signal swing is required, as the integrator circuits need to stay within their linear region of operation. To be able to scale the swing on the outputs, the m_n coefficients are split in integrator coefficients i_n and feed-forward coefficients j_n in such a way, that the signal swings on the integrator are made equal, without changing the pole zero locations of the loop filter transfer function. The coefficients i_n are normalized to a peak value of 1 at a sample frequency of 1 Hz.

The calculation of the loop filter coefficients is demonstrated for a modulator on which most of the implemented ΣΔ modulators presented in this book are based. Its block diagram is shown in Fig. 5.3. It uses a fifth order feed-forward loop filter with two resonators. The loop filter is built up out of 5 integrator stages with unity gain ω_n, feed-forward coefficients a_n and resonator coefficients b_n. The resonator coefficients are used to spread the quantization noise of the modulator more evenly over the signal bandwidth to increase the SQNR in a certain bandwidth. The output signal of the loop filter is quantized by a 1-bit quantizer and fed back to the input by the feedback DAC. Both the quantizer and DAC are sampled at f_s. The architecture of Fig. 5.3 can be used for lower or higher order ΣΔ modulators, by adding or leaving out integrator stages and feed-forward coefficients. To create a resonator, at least two integrator stages are required.

As the linearity requirements of the later integrator stages are lower than the first integrator stage (Sect. 6.6.1), higher signal swings can be tolerated on the later

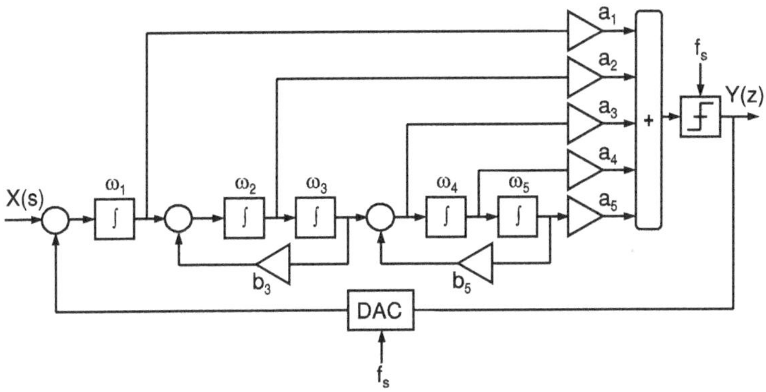

Figure 5.3: *Block diagram of a 1-bit, 5th order, CT feed-forward ΣΔ modulator*

integrator stages. As this will lead to higher integrator unity gain frequencies and thus smaller capacitors in the loop filter, the area of the loop filter will be reduced. To be able to do the scaling of the i_n coefficients, an integrator output clip level l_n is defined. These coefficients l_n define the maximum signal swing at the output of the integrator n.

For a fixed maximum integrator output signal level l_1, the unity gain of the first integrator can now be calculated with:

$$\omega_1 = i_1 \cdot \frac{l_1}{A_{X,in,max}} \cdot f_s \text{ [rad/s]} \tag{5.3}$$

in which $A_{X,in,max}$ is the maximum input signal of the modulator. The unity gain frequencies for integrator $n = [2 - L]$ can be calculated by:

$$\omega_n = i_n \cdot \frac{l_n}{l_{n-1}} \cdot f_s \text{ [rad/s]} \tag{5.4}$$

Note that the requirements on the integrator unity gain frequencies are dependent on the clip level after and in front of the integrator.

Next to the unity gain frequencies, the feed-forward coefficients are also dependent on the clip levels chosen for the integrator stages. If the allowed signal swing on the output of integrator n is made larger, its feed-forward coefficient a_n should be made smaller. In a general form, the feed forward coefficients can be calculated by:

$$a_n = \frac{j_n}{l_n} \text{ [-]} \tag{5.5}$$

The coefficients j_n thus relate the maximum integrator output swing l_n to the feed-forward coefficients a_n.

If the set of coefficients m_n (or the combination of coefficients sets i_n and j_n) define a stable modulator, the feed-forward coefficients and unity gain frequencies can be generated for a stable modulator sampled at any given f_s, and with any input signal amplitude $A_{X,in,max}$. When implementing the modulator in an IC technology, a choice has to be made for the maximum integrator swings l_n, which is determined by integrator circuit biasing, the supply voltage of the modulator and the required linearity of each integrator stage.

As stated earlier, local feedback coefficients b_n can be used to decrease the in-band quantization noise. The b_n coefficients that make two integrator stages a resonator can be calculated by:

$$b_3 = \frac{\omega_B^2 \cdot k_3^2}{\omega_2 \cdot \omega_3} \quad \text{and} \quad b_5 = \frac{\omega_B^2 \cdot k_5^2}{\omega_4 \cdot \omega_5} \quad [\text{-}] \tag{5.6}$$

The k coefficients determine the resonator frequencies related to the signal bandwidth ω_B.

From the above coefficients only, the SQNR of a modulator with loop filter order L can not be calculated. To be able to calculate the modulators SQNR, a model for the quantizer is needed. From the theory presented in [15] and by doing several assumptions, a reasonably accurate model of the quantizer can be derived. The assumptions are:

- The quantizer gain is calculated with no quantizer input signal. Therefore, the modulator is assumed to be idling with 1010 and 1100 patterns.
- Because of the high frequency patterns (compared to the input signal), the contribution of the first integrator and its feed-forward coefficient is dominant in the quantizer input signal.
- The quantizer output switches between $+1$ and -1.
- The quantization noise is assumed to have a Gaussian distribution, and is completely de-correlated from the modulator input signal by the loop filter. Therefore, its spectrum is assumed to be white.

From the quantizer input and output power the quantizer gain can be calculated. From the quantizer gain, the quantizer input power and the assumption that the quantization noise is white and that the output switches between +1 and −1, the quantization noise power can be calculated. The resulting quantizer model consists of a linear gain and a Gaussian noise source, representing the quantizer gain and quantization noise respectively. Once the quantizer model is known, the SQNR of the modulator can be calculated by integrating the quantization noise in the signal bandwidth using the quantizer model and the loop filter coefficients presented in this chapter. The SQNR of the modulator presented in Fig. 5.3 is calculated to be:

$$\mathrm{SQNR} = \sqrt{\left(\frac{(1-\frac{2}{\pi})^{-1} j_5^2 i_2^2 i_3^2 i_4^2 i_5^2 \mathrm{OSR}^{11}}{\frac{1}{11}\omega_B^{11} - \frac{2}{9}\omega_B^9\alpha + \frac{1}{7}\omega_B^7(2\beta+\alpha^2) - \frac{2}{5}\omega_B^5\alpha\beta + \frac{1}{3}\omega_B^3\beta^2}\right)} \quad [-] \tag{5.7}$$

with $\alpha = \omega_2\omega_3 b_3 + \omega_4\omega_5 b_5$ and $\beta = \omega_2\omega_3 b_3 \omega_4\omega_5 b_5$.

Figure 5.4 presents the modulators SQNR as a function of the k_3 and k_5 for an OSR of 40. From the figure the optimal value of the k coefficients can be read. The maximum SQNR is achieved for $b_3 = 0.54$ and $b_5 = 0.91$, or $b_3 = 0.91$ and

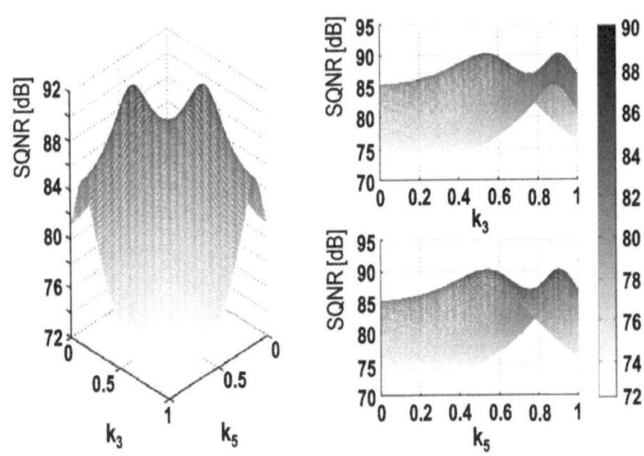

Figure 5.4: *SQNR as a function of k_3 and k_5 coefficient values*

$b_5 = 0.54$. These values can also be found by equating the derivative of Eq. 5.7 to zero.

In Fig. 5.5 the result of Eq. 5.7 is shown for a ΣΔ modulator without the resonators ($k_3 = k_5 = 0$) and for a ΣΔ modulator with optimal k_3 and k_5 values (ratio between resonator frequencies and bandwidth is constant). In the same figure transient simulated results of both the modulators are shown (discrete points). The difference between the simulations, and the difference between calculations

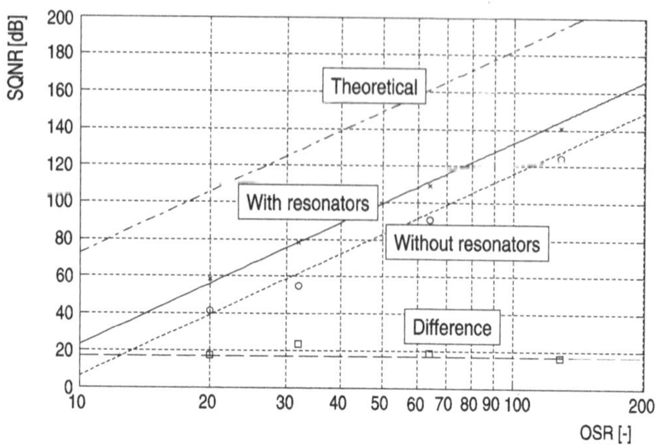

Figure 5.5: *Simulated and calculated SQNR for a 5th order, 1-bit ΣΔ modulator with or without resonators*

of the two different modulators are plotted in the same figure. As a reference, the stability-unbounded theoretical performance predicted by Eq. 5.2 is also plotted in the figure. Introducing the additional feedback paths b_3 and b_5 results in a fixed SQNR improvement of 17 dB independent of OSR, for the given loop filter architecture and coefficients, as long as the resonator frequencies are sufficiently below the frequency where the loop filter is forced back to first order. The fixed improvement of 17 dB can be explained by the fact that the ratio between bandwidth and resonator frequency is fixed.

As a feed-forward loop filter can be directly transformed in a feedback loop filter, the same coefficients and equations presented in this chapter can be used for a feedback modulators. Using the same coefficients, the performance of both modulators will be exactly the same [16].

5.2 ΣΔ Modulators with b-bit Quantizer and b-bit DAC

A special class of ΣΔ modulators are b-bit modulators. The 1-bit quantizer and DAC are replaced a b-bit quantizer and DAC, and loop filter coefficients are changed accordingly.

The reason to choose for a multi-bit modulator can be twofold. Firstly, increasing the loop filter order of a 1-bit converter to high orders gives limited performance increase above loop filter orders of about 5 as the loop stability criteria more and more limits the amount of noise shaping [17]. Secondly, over-sampling of the 1-bit modulator can not be increased further because of speed limitations (e.g. due to technology).

The merits of a transition from a 1 to a b-bit modulator are:

1. The multi-bit modulator performance increases with approximately 6 dB per bit [17].
2. The input range of the modulator increases compared to a 1-bit modulator by [18]:
$$\frac{V_{in,max,N=x}}{V_{in,max,N=2}} = \frac{0.7 + x - 2}{0.7(x-1)} \text{ [-]} \tag{5.8}$$
if it is assumed that the maximum input signal of the modulator is 0.7 times (or approximately -3 dB) below its feedback DAC output level (which is true for the modulators presented in this book).
3. The lower quantization noise compared to 1-bit modulators allows for a lower cost (area, power) decimation filter. For μm instead of nm-technologies, this might be a valid consideration, as digital cells in μm technologies are bulky. In nm-technologies this area advantage is questionable, as digital cell area scales with s_T^2 [4] and Sect. 6.1.
4. The quantizer gain is better defined, and stability is achieved more easily [6, 17, 19, 20].

The b-bit modulator performance disadvantages are:

1. Both multi-bit quantizer and DAC are required which are complex circuits.
2. The integral linearity requirement of the feedback DAC should be at least equal to the overall ΣΔ modulator's integral linearity requirement (Sect. 6.6.4.2).
3. Absolute loop gain accuracy is required (Sect. 6.4.2).

5.3 ΣΔ Modulators with 1.5-bit Quantizer and DAC

A special class of a b-bit modulator is a 1.5-bit ΣΔ modulator. It uses three levels in the feedback DAC. Going from two levels (1-bit) to three levels (1.5-bit) the quantization error reduces by a factor of two. And according to Eq. 5.8 the input range of the modulator is increased by 1.6 dB. This gives a total improvement in

SQNR of 7.6 dB. A big advantage of a 1.5-bit modulator compared to a modulator with more than 3 levels is, that a 1.5-bit DAC can be kept linear when implemented in a differential mode as will be shown in Sect. 6.6.4.3.

5.4 ΣΔ Modulators with Multiple Quantizers and 1-bit DAC

As seen in the previous section, a multi-bit ΣΔ modulator can achieve a higher performance than a 1-bit modulator sampled at the same frequency. However, the linearity of the DAC has to be conform the intended performance of the modulator. This section shows a new class of modulators, in which the advantages of the higher performance multi-bit modulator and the inherently linear 1-bit modulator are combined. The modulator uses a multi-bit quantizer to increase its SQNR, but preserves the 1-bit inherently linear feedback DAC. Figure 5.6a shows the

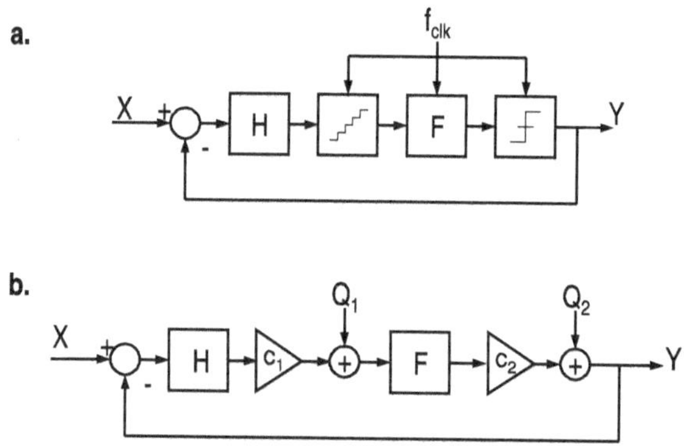

Figure 5.6: *Modulator architecture (a) and model (b)*

proposed ΣΔ modulator architecture. It comprises an analog filter H, a b-bit quantizer, a digital filter F, a 1-bit quantizer and a 1-bit feedback path. The 1-bit quantizer is implemented digitally by taking the MSB of the output word of the digital filter F, which is fed back to the input via the 1-bit DAC. Therefore the 1-bit quantizer costs no additional hardware.

If it is assumed that the quantization noise of the b-bit and 1-bit quantizer is completely de-correlated by the digital filter F, the b-bit and 1-bit quantizers can be replaced by quantization noise sources Q_1 and Q_2 and with gains of c_1 and c_2 respectively. The model is redrawn in Fig. 5.6b, which has the frequency domain transfer function:

$$Y = \frac{c_1 c_2 H F}{1 + c_1 c_2 H F} X + \frac{c_2 F}{1 + c_1 c_2 H F} Q_1 + \frac{1}{1 + c_1 c_2 H F} Q_2 \tag{5.9}$$

If it is assumed that H and F have high gain at low frequencies, Eq. 5.9 changes into:

$$Y \approx X + \frac{1}{c_1 H} Q_1 + \frac{1}{c_1 c_2 H F} Q_2 \tag{5.10}$$

$$= X + \frac{1}{c_1 H} \left(Q_1 + \frac{1}{c_2 F} Q_2 \right) \tag{5.11}$$

From Eq. 5.11 it can be seen that Q_1 is shaped by H, while Q_2 is shaped by the product of H and F. If the gain of $c_2 F$ is higher than the number of bits times 6 dB, Q_2 will be below Q_1 at the output of the modulator.

Next to the analog filter, loop stability now is dependent on the digital filter as well. This means that the delay of the digital filter has to be made very small at frequencies close to f_s to avoid instability. With an inverted IIR low-pass filter this can be done, as it has a direct feed-forward path from filter input to output. This will be shown for an implementation example in Chap. 9.

One drawback of this architecture compared to its conventional multi-bit modulator (with multi-bit feedback DAC) is that due to the additional 1-bit quantizer, more quantization noise is injected in the ΣΔ modulator loop. This results in a larger output signal of the first integrator. Therefore, the first modulator unity gain has to be chosen lower compared to the original modulator only containing the b-bit quantizer to avoid overload of the b-bit quantizer. This will result in a lower SQNR for the modulator with multiple quantizers, but the SQNR will still be higher than for a 1-bit quantizer only modulator. The exact calculation of the coefficients of this type of modulator is out of the scope of this book.

It should be noted that every combination of first and second quantizer resolution, loop filter order and OSR is possible.

The general principle of using more quantizers and 1 feedback DAC, is patented in [21].

5.5 ΣΔ Modulators with Additive Error-Feedback Loops

Another way to increase the SQNR of a ΣΔ modulator is by adding error feedback to the quantizer. In [17] a fully digital implementation of quantizer error feedback is introduced stand-alone. In this section an analog equivalent of the error feedback loop is added to the main loop of a ΣΔ modulator. The architecture is presented in Fig. 5.7a. The architecture consists of the outer or main loop: in-

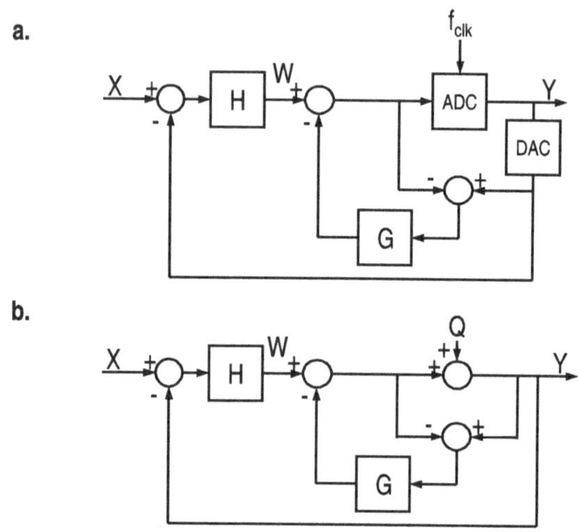

Figure 5.7: *Modulator architecture (a) and model (b)*

put X, loop filter H, A/D output Y and the feedback DAC. The input of the outer loop is X, its output Y. An additional error feedback path (inner loop) is included in the main loop, consisting of a summation node and filter G, which feeds the quantization error back into the loop via an additional summing node. The input of the inner loop is W, its output Y. From the model presented in Fig. 5.7b, the transfer function of the input signal X and the quantization error Q to the output Y can be calculated if the quantizer is modeled by a quantization noise source Q and a gain of 1. The feedback DACs are modeled by a gain of 1. The modulators input signal and quantization noise frequency domain transfer functions become:

$$Y = \frac{H}{1+H}X_1 + \frac{1-G}{1+H}Q \tag{5.12}$$

From Eq. 5.12 it can be seen that the original shaping behavior of the loop filter H is unchanged, and still reduces the quantization error Q in the part of the frequency spectrum where H has gain. Furthermore, the quantization error Q is multiplied by a factor $(1 - G)$. If G has a gain of 1, the quantization error is fully canceled at the output Y. In reality this is not implementable because the additional loop will always have implementation delays like for example a parasitic pole in the summation nodes.

In practical implementations the DAC will be split into two DACs. One for the outer loop and one for the inner loop. This is because the gain requirement of the DACs is different, or the summing on one of the summing nodes is done in the current domain. In this case an additional DAC is required to implement the error feedback loop. Fortunately, the noise and linearity requirements are limited, as the errors introduced in this part of the loop will be shaped by the loop filter H, similarly to the shaping of the quantization noise Q. This will be shown in Sect. 6.4.4.

In Fig. 5.8 the NTF of the inner loop only is shown for different loop filter orders of G ($L = [1..5]$) for two different number of quantization levels ($N = 2$ or $N = 9$). When the number of quantization levels is decreased, the loop filter coefficients have to be chosen less aggressive as the quantization noise power in the loop becomes too large overdriving the quantizer. A detailed stability analysis is out of the scope of this book. In the NTFs in Fig. 5.8 the smaller quantization error due to the increase in quantization levels is not included. An additional

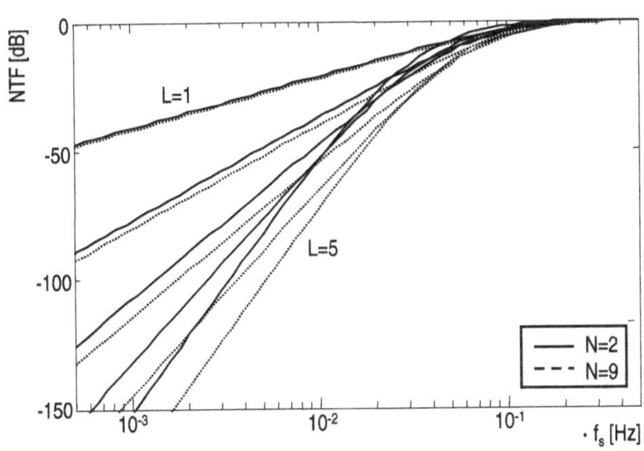

Figure 5.8: *NTF amplitude and phase response of error feedback loop for* $L = [1..5]$ *and* $N = 2$ *or* $N = 9$

advantage of the inner error feedback loop is that Filter G has no influence on the transfer function of W to the output Y, and does not affect the criteria which define stability of the outer loop when the linear model of Fig. 5.7 is considered.

This can be seen if the transfer function of W/Y is considered:

$$Y = W + (1 - G)Q \tag{5.13}$$

Figure 5.9 shows the SQNR of the additional loop only, as a function of the OSR for $N = 2$ and $N = 9$, and for loop orders two to five. The improvement in SQNR between the 2 level and 9 level quantizer is partly explained by the smaller quantization step (less quantization noise), and is partly explained by the more aggressive noise shaping in G allowed in the case of a smaller quantization step quantizer.

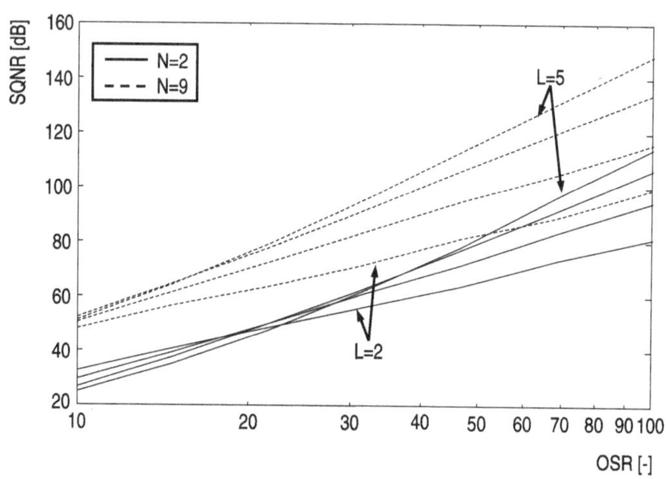

Figure 5.9: *SQNR of the error feedback loop for* $L = [1..5]$ *and* $N = 2$ *and* $N = 9$

Figure 5.10 shows an example spectrum of a 5th order 9-level ΣΔ modulator with and without an additional second-order error-feedback loop. The coefficients of the main loop are the same in both cases. Using the additional error feedback loop in the ΣΔ modulator main loop, the SQNR increase in this case is 30 dB at an oversampling of 36.

An example with an even more aggressive noise shaping is given in Fig. 5.11. In the ΣΔ modulator presented of Fig. 5.7, for loop filter H a second order filter with one resonator is chosen, and filter G is a sixth order filter. The quantizer and DAC are of 3-bit resolution. The aggressive noise shaping is possible due to the nature of the inner loop; the quantization error only is fed back into the loop, unlike in the outer loop, where signal and quantization noise are fed back. The

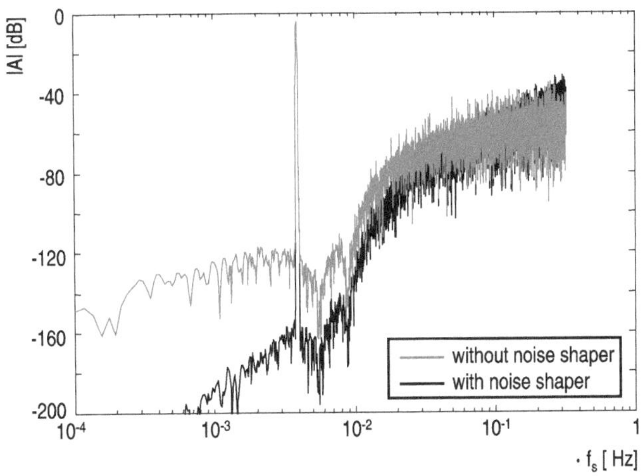

Figure 5.10: *Example 5th order 9-level ΣΔ modulator with and without an additional second order error feedback loop*

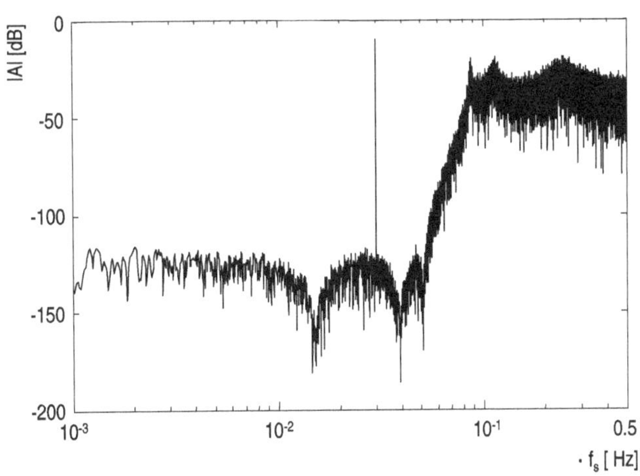

Figure 5.11: *Example 3-bit ΣΔ modulator with a second order loop filter with one resonator H, with an additional error feedback loop with a sixth order loop filter G*

additive error-feedback loop's noise shaping is not bounded by the control theory loop stability criteria in the same way as the outer loop. The low-pass modulator achieves 87 dB at 10 times oversampling. According to Fig. 5.5, a fifth order single loop modulator with two resonators at least requires an over-sampling ratio of 40 to achieve 87 dB.

Although the application of these error feedback loops in a $\Sigma\Delta$ modulator promise large SQNR improvements with low requirement hardware, in reality the performance increase will be limited by filter coefficient inaccuracies (Sect. 6.4.4).

In the examples of this section so far, the error feedback loop has been applied to a quantizer in a $\Sigma\Delta$ modulator, making its noise shaping more aggressive, without increasing the stability constraints of the outer loop. These error feedback loops can also be used in other types of A/D converters, to add noise shaping to its quantizer(s). Furthermore, in digital noise shapers an additional error feedback loop can also be used to get a more aggressive noise shaping. The advantage of digital loops are, that the filter coefficients can be designed as accurate as required, whereas in analog (to digital) noise shaping loop the filter coefficients are subject to process variations.

A similar error feedback loop can also be used to shape the (linearity) errors of a DAC. An example is displayed in Fig. 5.12a. Figure 5.12b shows a model of the

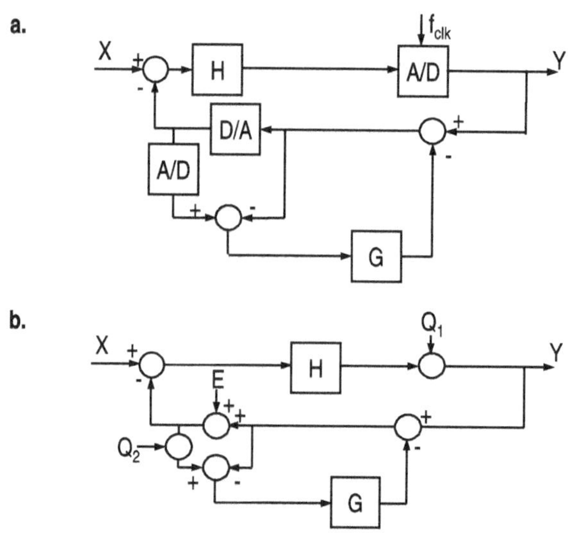

Figure 5.12: *Error shaping loop used to shape the linearity errors in the feedback DAC of a $\Sigma\Delta$ modulator (a) and its model (b)*

modulator. The errors made in the DAC are modeled by E, the quantization errors made by the main quantizer are modeled by Q_1 and the quantization errors made by the auxiliary ADC are modeled by Q_2. The transfer function derived from the model is presented in Eq. 5.14, in which it is assumed that the gain of the main feedback DAC is 1. The DAC of subject in this case is part of a $\Sigma\Delta$ modulator loop. The DAC input and output signals are subtracted from each other and fed through filter G. Consequently, the DAC errors E are shaped by the filter G. For frequencies where $G = 1$, the errors of the DAC will not be visible in the output spectrum of the $\Sigma\Delta$ modulator.

$$Y = \frac{H}{1+H}X + \frac{1}{1+H}Q_1 + \frac{GH}{1+H}Q_2 + \frac{H(G-1)}{1+H}E \tag{5.14}$$

A big disadvantage is the extra required A/D converter in the error feedback loop. The requirements on this ADC are as high as the requirements on the overall loop. The (quantization) error Q_2 of the extra ADC is introduced at the input of the modulator. This makes this type of DAC linearity error shaping unattractive.

The general principle of using additional error feedback loops in a $\Sigma\Delta$ modulator, is patented in [22].

5.6 Cascaded ΣΔ Modulators

Another way to increase the performance of a $\Sigma\Delta$ modulator is to cascade 2 or more $\Sigma\Delta$ modulators. In a cascaded $\Sigma\Delta$ modulator, the quantization error of the previous $\Sigma\Delta$ modulator stage is digitized by its consecutive $\Sigma\Delta$ modulator stage. In the digital domain, the quantization error is subtracted, yielding a higher order noise shaping [17]. The advantage of a cascaded $\Sigma\Delta$ modulators is that low order $\Sigma\Delta$ modulator stages can be used, which have lower stability requirements than high order $\Sigma\Delta$ modulator stages, and noise shaping can be chosen more aggressive. This means that a 2 times second-order cascaded $\Sigma\Delta$ modulator will have a better SQNR than a 1 stage 4th order $\Sigma\Delta$ modulator at the same over-sampling, at the cost of one extra quantizer, two extra DACs and the digital compensation filter.

For accurate quantization noise cancellation of the first stage the modulator coefficients and digital compensation filter coefficients should match. Therefore, DT modulators are preferred but these are known to be power hungry. Efforts have been done to use a CT modulator as the basis of a cascaded modulator ([23, 24]), but this approach requires additional calibration algorithms. Therefore, this type of modulator is not investigated further in this book.

5.7 Conclusions

This chapter has shown the algorithmic accuracy of different types of modulators.

The most straightforward way to achieve a certain SQNR is to take a ΣΔ modulator with an Lth order loop filter (either feed-forward or feedback), an inherently linear 1-bit quantizer and DAC, and a sufficiently high sampling frequency. As the loop filter order is bounded by stability criteria and the OSR is limited by the speed of technology, alternative loop filter architectures have been proposed to further increase the SQNR of ΣΔ modulators.

Multi-bit ΣΔ modulators give an increase of the SQNR of approximately 6 dB per bit, but are limited in performance by the feedback DAC linearity due to unit cell mismatch. A promising direction might be using a 1.5-bit modulator as it decreases the quantization noise by about 6 dB, and has a larger stable input range of 1.6 dB compared to a 1-bit modulator, and can be implemented with perfect linearity.

The linearity issue of multi-bit ΣΔ modulators can also be avoided by multi-quantizer ΣΔ modulators, which have multi-bit quantization but use 1-bit feedback.

ΣΔ modulators with additive error feedback loops seem promising as their performance can be increased with hardware that requires only limited performance. As this is a relatively new modulator topology, no implementations are available at the time of writing this book.

Cascaded modulators use multiple ΣΔ modulator stages to achieve a high order noise shaping, at the cost of an interstage gain matching requirement and quite some additional hardware. Therefore, this architecture is not further explored in this book.

In summary, the architecture of the modulator implementations presented in this book is limited to:

1. Single loop, Lth order, 1-bit feed-forward modulators
2. Single loop, Lth order, 1.5-bit feed-forward modulators
3. Single loop, Lth order, 1-bit feed-back modulators
4. Single loop, Lth order, multi-quantizer modulators with 1-bit feedback

and of each architecture one or more implementations will be presented in this book.

$\Sigma\Delta$ modulators with additive error feedback loops will be further analyzed on circuit imperfections in the next chapter, but no implementations will be shown in this book.

Chapter 6

ΣΔ Modulator Robustness

In the analysis of the ΣΔ modulator's algorithmic accuracy in Chap. 5, the circuits are assumed to have no imperfections. In reality, the performance of the circuits will come at the cost of resources (area and power). Trade-offs have to be made to come to a good performance-to-resource ratio which can be verified in a benchmark with other ΣΔ modulators. The benchmarking of modulators will be done in Chap. 8. In this chapter, the relation between architecture choices and circuit imperfections and their impact on the ΣΔ modulator performance and cost will be discussed. The subjects of discussion are:

- Technology
- Continuous-time vs. discrete time loop filter
- Feed-forward vs. feedback loop filter
- Gain accuracy
- Circuit noise
- Linearity
- Aliasing
- Excess loop delay
- Clock jitter

A schematic overview of the above is presented in Fig. 6.1. The block diagram on the bottom of Fig. 6.1 illustrates all relevant imperfections for a continuous-time ΣΔ modulator which will be discussed in this chapter. In this chapter, first it will

R.H.M. van Veldhoven, A.H.M. van Roermund, *Robust Sigma Delta Converters*,
Analog Circuits and Signal Processing, DOI 10.1007/978-94-007-0644-6_6,

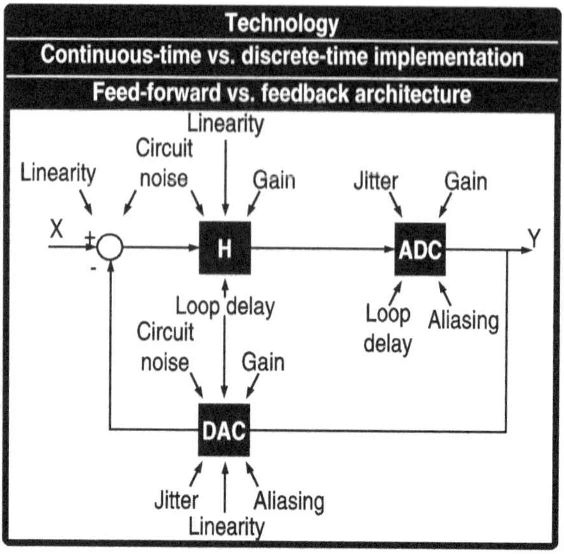

Figure 6.1: *Chapter overview. The block diagram holds for continuous-time $\Sigma\Delta$ modulators*

be explained why it is advantageous, to replace analog circuits by digital circuits where possible. As analog IP will never be completely digital, the most robust architecture for remaining analog functions should be searched for. In this chapter, the latter is done for $\Sigma\Delta$ modulators. The focus will be mainly on $\Sigma\Delta$ modulators with a 1-bit feedback DAC, as most of the implementations presented in this book have a 1-bit feedback DAC (one implementation has a 1.5-bit feedback DAC). Where relevant, modulators with multi-bit DACs will be discussed.

6.1 Portable, Technology Robust Analog IP and Time-to-Market

The quest to increase digital processing per unit area has led to scaling of CMOS technologies, yielding faster and smaller transistors. In [4] the technology scaling factor s_T is introduced to investigate the impact of technology scaling on digital circuits. The technology scaling factor s_T is defined as the ratio between the minimum L of a transistor in the current technology node and the minimum L of the previous IC technology node or $s_T = L_{min,\ current\ node}/L_{min,\ previous\ node}$ and is approximately 0.7.

For digital circuits the advantages going to a smaller feature size technology are evident. Its area decreases with s_T^2 and transistor speed increases with $1/s_T$ for the constant field scaling period of the years 1990 to 2000 [4]. This makes the amount of processing power per area increase with $1/s_T^3$. The power-delay product ($V \cdot I \cdot \Delta T_{min}$) of digital circuits decreases with $1/s_T^3$. For the constant voltage scaling period from the year 2001 to date, speed and efficiency increases with $s_T^{-\frac{1}{2}}$ and s_T^{-1} respectively. Clear area, speed and efficiency advantages are seen for digital circuits in newer technologies. Furthermore, digital circuits are much more easy to port to the next technology node, as the noise margin in digital circuits allows for a higher abstraction description level of digital circuits which enables a very high degree of automation of the port. The porting of analog IP to the next technology node normally requires lots of handcraft designing. Furthermore, the analog IP is very sensitive to changes of supply voltage and transistor parameters in newer technologies, which makes the port more difficult, and very often the analog IP is on the critical time path of a SoC tape-out. The challenge is to port the analog IP to the next technology node in the same pace as is done for the digital IP. This enables to port complex SoCs containing both analog and digital circuits to the next technology node, without an increase in time-to-market. Another challenge is to increase the robustness of analog circuits such that a SoC becomes more robust to technology changes.

This section shows the impact of the digital-processing-per-area-driven scaling of CMOS technologies on a limited number of analog design parameters. Furthermore, this section will present a top-down and bottom-up design methodology for the design of analog IP blocks in order to increase the portability and thus technology-change robustness of these analog IP blocks.

6.1.1 Technology Scaling and Its Impact on Analog Design Parameters

This section shows a flavor of the impact of technology scaling on analog design parameters, when L_{min} scales with s_T. The presented scaling factors are a combination of the ones presented of the technology scaling parameters presented in [4] and the ones in [3].

1. The ratio V_{supply}/σ_{V_T} is more or less constant. This means that analog blocks that require amplitude resolution (for example required in Nyquist and multi-bit $\Sigma\Delta$ modulator ADCs), at best will have the same performance when ported to the next technology node.

2. The gm/I slightly increases (which for example helps to increase parasitic pole frequencies for the same power).

3. The supply voltage decreases which means that the biasing of telescopic amplifiers will become more difficult, and cascaded stages (or folded cascode stages) will be needed to supply sufficient gain. Due to the parallel current paths, this can lead to an increase in power consumption, as less current is re-used.

The list above is far from complete and shows just a flavor of the changes in analog design parameters. Conclusion is, that careful re-simulation of all analog circuit blocks is required to do a successful port to the next technology node. In some cases even a change in circuit topology might be required to achieve the same analog performance in the next technology.

6.1.2 A Design Methodology to Increase the Portability of Analog IP

The design methodology presented in this section prescribes the replacement of analog circuits by digital circuits where possible, as far as the ratio between system performance and system implementation cost optimization allows. The remaining analog circuits should be designed like it is done in the digital design flow. The quality of a limited amount of circuits with poor fundamental analog performance is boosted by digital processing at limited cost (area and power). This trend is not a luxury but a requirement, as most of the parameters determining analog circuit performance get worse in newer technologies, as shown in the previous section. Analog functions can be assisted, calibrated or even be replaced by digital functions to improve the performance of the analog IP. To take the most out of the advantages the scaling of technology brings (more signal processing per area and faster transistors, and increasing efficiency of digital circuits), the digitization should be carried through four different levels, as shown in Fig. 6.2 to come to portable, high quality mixed-signal IP. These levels are:

- System/application level
- Analog (sub) IP block level
- Circuit level
- Layout level

An excellent example of digitization at system level is the highly digitized receiver of Chap. 3. Analog filtering in front of the ADC is reduced as much as possible,

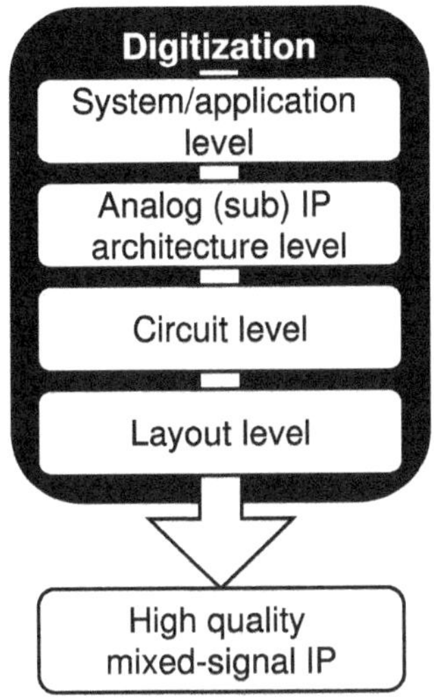

Figure 6.2: *Digitization of a system at various levels*

and replaced by more flexible digital filtering after the ADC. Analog functionality is reduced and a flexible receive pipe is created, which can be used for different cellular and wireless standards. The burden is put on the ADC and although the amount of extreme performance analog IP should be reduced, as this kind of IP will be the most difficult to port, the amount of analog hardware is greatly reduced. Section 9.1 will show some example ADC implementations to prove that such a high performance ADC can still be built at competitive power consumption and that it can be scaled to newer technologies.

At sub analog IP block architecture level, the analog functionality required to implement the function of the analog IP should be kept as simple as possible, or should be replaced by digital functionality. An example of an analog IP block with a digitized architecture is the multi-quantizer $\Sigma\Delta$ modulator of Sect. 5.4. By reducing the order of the loop filter of the modulator and introducing a multi-bit quantizer with limited performance requirements and a digital filter instead, the analog complexity of the modulator is greatly reduced. Section 9.2 will show an implementation example of this $\Sigma\Delta$ modulator architecture.

At circuit level the analog blocks should be designed the digital way. The normal analog design procedure is to optimize each circuit independently by transistor sizing. This requires a lot of custom design and handcraft layout which is very time consuming. The digital design methodology is much more robust to technology porting. Digital circuits are defined by a descriptive language like VHDL. The VHDL code is synthesized using a library with standard digital cells. Each cell has a p-cell layout suitable for automatic routing. If analog IP can also be designed this way, time-to-market can be reduced considerably. To enable this methodology for analog circuits, top-down and bottom-up iterations are required to come to good circuit architectures. Commonalities in circuit blocks should be searched for, to break-up the analog circuits in unit cells, and to reduce the number of analog functions that need attention when ported to the next technology node. This reduces the number of analog sub blocks that need maintenance, and enables automation of unit cell design. On the bottom side, simple unit cell circuits containing only a few transistors are put in a library. From top-level IP block architecture side, the top-level is split in to sub-blocks which are built out of the unit cells out of this library. A few iterations of top-level splitting and unit cell design will be necessary before the unit cells exactly fit the sub-blocks of the top-level architecture. In Sect. 9.3 (and to some extend in Sect. 9.2) some ΣΔ modulator examples will be shown, which were designed following this approach.

At layout level the unit cells should be transformed in p-cell layouts, which enables the usage of digital routing tools to generate the layout of the analog IP. The layout of the p-cells can be generated automatically using a layout programming language like skill. An example circuit and layout of a digital and analog p-cell is shown in Fig. 6.3. In the example the analog inverter layout has all the required properties to enable the use of the digital layout tool to (hierarchically) layout the

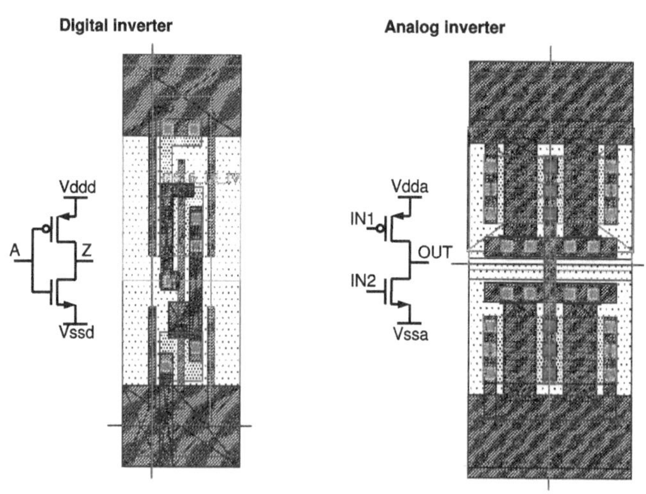

Figure 6.3: *Analog inverter circuit and its example p-cell layout*

analog IP block. On each analog sub IP block, extractions are done to see if the analog performance of the sub block is not decreased by the layout parasitics introduced during the automated layout procedure. A typical flow diagram to design the p-cells and the analog IP is presented in Fig. 6.4. In Sect. 9.3 example $\Sigma\Delta$

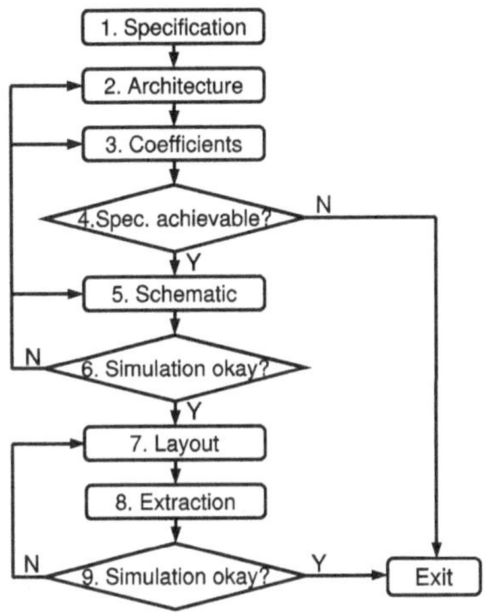

Figure 6.4: *Design flow of analog p-cells and IP*

modulators built up out of analog p-cells will be shown.

For the unavoidable analog parts in an analog IP that can not be replaced by digital circuits, robust implementations should be searched for, which is the done in this chapter for $\Sigma\Delta$ modulators.

6.2 Continuous Time vs. Discrete Time Loop Filter

Discrete time (DT) $\Sigma\Delta$ modulator are implemented with switched capacitor (SC) circuits. Continuous-time (CT) $\Sigma\Delta$ modulators are implemented with gm-C or RC circuitry. The main differences between DT and CT $\Sigma\Delta$ modulators are summarized below.

- Because a DT modulator has a sampler at the input, sampling non-linearities at the input of the modulator will be directly visible at the output. In CT modulators sampling non-linearities do not occur in the loop filter, as there is no sampling operation in the loop filter.
- Because a DT modulator loop filter has a sampler at the input, aliasing can occur at the input of the modulator. A CT modulator however has no

sampler in the loop filter. The aliasing caused by the sampling in the quantizer of a CT modulator is suppressed by the gain in front of the sampler (Sect. 6.7.1).

- DT modulators use a SC feedback DAC, which can cause sampling errors and aliasing at the input as well. The feedback DAC of a CT modulator can be implemented with either SC, switched resistor (SR), or switched current (SI) circuitry. Sections 6.7.2, 6.7.3 and 6.7.4 respectively will introduce equations to calculate the expected aliasing for these three DAC topologies.
- Because of the SC nature of the circuits in DT modulators, high bandwidth (compared to the signal bandwidth) circuits are required because of settling requirements [25]. This gives CT modulators a power consumption advantage, unless special circuit techniques are used in the SC circuits of DT modulators.
- The coefficients of DT modulators can be derived from a digital (numerical) modulator in a straightforward way. For CT modulators this procedure is slightly more complex.
- As the coefficients in DT modulators are all related to capacitor ratios, a DT modulator is not very sensitive to process spread. CT modulator coefficients are determined either by gm-C or RC filters, and therefore are more sensitive to process spread.
- The quantization noise shaping in a DT modulator scales with its clock frequency f_s, as all coefficients scale with f_s. In CT modulators special measures have to be taken, to enable clock frequency scaling.
- DT modulators are generally known to be clock jitter insensitive, while CT modulators are said to be more prone to clock jitter. In this book it will be shown that a CT modulator with switched capacitor DAC is less sensitive to clock jitter compared to a fully discrete time $\Sigma\Delta$ modulator as it eliminates the need for an input sampler (Sect. 6.9).

In radio receivers for mobile communication, which is the main application area for the modulators presented in this book, power consumption is the key parameter, which favors CT $\Sigma\Delta$ modulators. Next to that, CT $\Sigma\Delta$ modulators have implicit anti-alias filtering for aliases occurring due to quantizer sampling, which can be of benefit in the interferer-rich receiver environment. If a SC DAC is used together with a CT loop filter, the modulator gets more robust to high frequency jitter (Sect. 6.9), but sacrifices alias suppression if not properly designed

(Sect. 6.7.4). The loop filter coefficients calculation is not difficult once the way of calculating them is known, and the coefficient spread in a CT modulators can be solved by calibration. Although the quantization noise shaping scaling advantages of DT modulators seems evident, the equivalent noise impedance of the modulator scales with f_s as well ($R_{noise,eq} = 1/(f_s C)$. In CT modulators the input impedance is constant over clock frequency scaling. Coefficient scaling can be done by changing the integrator capacitors, as will be shown in Chap. 7. In summary, a CT $\Sigma\Delta$ modulator is the best choice, if combined with a SC feedback DAC, and mainly because of its lower power consumption.

6.3 Feed-Forward vs. Feedback Loop Filter

The loop filter of the $\Sigma\Delta$ modulator can either be feed-forward or feedback [16, 17]. Below, a $\Sigma\Delta$ modulator with feed-forward loop filter is compared with a $\Sigma\Delta$ modulator with feedback filter.

- Generally, the achievable algorithmic accuracy (SQNR) of both feed-forward and feedback modulators is the same, as the quantization noise transfer function can be made exact the same for both modulators (with the same amount of hardware).
- The out-of-band filtering of a feed-forward and a feedback architecture is first and Lth order respectively.
- Due to the steeper out-of-band STF, feedback modulators have a built in Lth order anti-alias filter for the aliasing occurring due to the sampling by the quantizer (Sect. 6.7).
- The aliasing occurring in the $\Sigma\Delta$ modulator's input feedback DAC is the same for both architectures (Sect. 6.7).
- The output signal of the integrators used in feedback modulators have more correlation to the input signal compared to feed-forward modulators. This leads to higher linearity requirements for the integrator circuits.
- Due to the higher linearity requirements, the power consumption of a feedback modulator is expected to be higher.
- Unlike feed-forward modulators, feedback modulators do not recover from overload (except for a first order modulator).

- Feed-forward modulators require feed-forward coefficients and an additional summing node to sum the feed-forward coefficients, which adds an additional pole in the loop. A feedback modulator needs L feedback DACs where the feed-forward modulator needs only 1, but does not require the feed-forward coefficients nor the additional summing node.

As a $\Sigma\Delta$ modulator in a receiver application requires high linearity, a large signal component at the integrator output is unwanted. Furthermore, as large interference is present at the input of the receiver application, overload recovery is very important. These two arguments both plead for a feed-forward filter. Unfortunately, an additional summing node is required for the feed-forward filter, which introduces additional excess phase.

The feedback filter does not have such an additional summing node (and thus no extra parasitic pole that comes with the summing node), which makes it more suitable for high sample frequencies. Furthermore, the feedback loop filter has inherently Lth order out-of-band filtering, a key-feature in a receiver application. Unfortunately, the Lth order filtering can not be used to reduce the requirements on the $\Sigma\Delta$ modulator in terms of dynamic range, as the wanted signal can be much smaller than the out-of-band interferer. The interferer together with the wanted signal has to be delivered to the ADC within the power supply range to avoid distortion in the circuit preceding the ADC. As the input gm or resistor of the gm/C or RC integrator modulator input stage is normally the dominant noise contributor, this means that the noise requirement of the ADC does not change. The additional filtering in the feedback modulator compared to a feed-forward filter might alleviate the requirements on the decimation filter.

Although feedback modulators have better suppression of aliases due to quantizer sampling, the mechanism which causes aliases in the feedback DAC, when an RC integrator is used as an input stage, is the same for a feed-forward and feedback filter. In case the aliasing in the DAC is dominant over the aliasing in the quantizer, there is no difference for both loop filter topologies.

As in digitized receivers a high ADC linearity is required, a feed-forward modulator is preferred as it uses less power than a feedback modulator, to meet linearity requirements.

As power consumption, linearity and overload recovery are seen as the most important advantages, and the additional out-of-band filtering of a feedback modulator does not help to reduce the requirements of the ADC, a feed-forward loop filter is in favor, unless the summing node becomes a too large contributor to delay in the loop, and a feedback filter is the only way out. In Sect. 6.8.2 a compensation method is introduced, which fully compensates the excess phase of the additional summing node required in a feed-forward loop filter, eliminating the final hurdle for a feed-forward $\Sigma\Delta$ modulator.

6.4 Gain Accuracy

To reduce the in-band quantization noise to the required level, the loop filter gain of a $\Sigma\Delta$ modulator should be sufficiently large. There can be also a requirement on the absolute accuracy of the gain. These issues will be discussed in the next paragraphs for the different modulator architectures.

In the calculations on the gain accuracy of different modulator architectures, the linear quantizer model presented in Sect. 5.4 used.

6.4.1 $\Sigma\Delta$ Modulator with 1-bit Quantizer and 1-bit DAC

In a 1-bit $\Sigma\Delta$ modulator the absolute loop gain accuracy required is very low. The quantizer input signal amplitude is not so important, as the quantizer only has to decide whether its input signal is smaller or larger than its threshold.

6.4.2 $\Sigma\Delta$ Modulator with b-bit Quantizer and b-bit DAC

The absolute gain requirements in a multi-bit ($b > 1$) converter are high. The output signal of the loop filter should exactly match the input range of the quantizer. If the input signal to the quantizer is too small, comparators at the top and bottom of the quantizer input range are not used and the modulator performance will drop by the number of levels used divided by the total number of levels. If the input signal is too large, the quantizer will be overdriven and the $\Sigma\Delta$ modulator will dramatically loose performance, as high frequency quantization noise will fold back into the signal bandwidth. Furthermore the input signal will be distorted.

6.4.3 $\Sigma\Delta$ Modulator with Multiple Quantizers and 1-bit DAC

In $\Sigma\Delta$ modulators with multiple quantizers, the requirements on the loop gain is dependent on the number of bits used in the quantizers. For quantizers with $b = 1$ the loop gain requirements described in Sect. 6.4.1 hold, and for the quantizers with $b > 1$ the loop gain requirements described in Sect. 6.4.2 hold.

6.4.4 ΣΔ Modulator with Additive Error Feedback Loops

ΣΔ modulators with additive error feedback loops have requirements on the gain to the quantizer equal to single-loop modulators. Additionally they have gain requirements on the feedback paths in the additional error feedback loop. Starting point is the model of the modulator with additional error feedback loop of Fig. 5.7b. In Fig. 6.5, the gain mismatch d between the two feedback paths and the gain of the main feedback path e are modeled additionally. Furthermore, E models the errors of the additional DAC in the error feedback loop. First the transfer

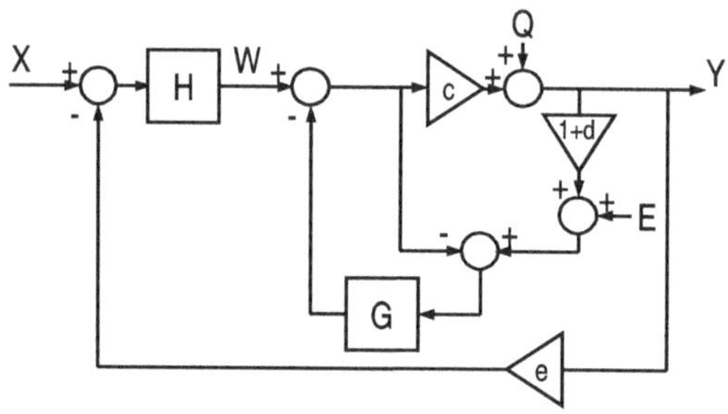

Figure 6.5: *Model of a ΣΔ modulator with b-bit quantizer and DAC and with additive error feedback loop with feedback gain error*

function between W and Y is calculated, which yields:

$$Y = \frac{c}{1 + G(c(1+d) - 1)} W + \frac{1 - G}{1 + G(c(1+d) - 1)} Q + \frac{-cG}{1 + G(c(1+d) - 1)} E \tag{6.1}$$

Equation 6.1 shows that if $c = 1$ and $d = 0$, W directly appears at the output Y without being filtered or attenuated, independent on the poles and/or zeros in G, as $(c(1+d) - 1) = 0$. This means that under these conditions, the stability of the main loop is unchanged.

The relation between inputs X, Q and E and output Y of the ΣΔ modulator is given by:

$$Y = \frac{cH}{1 + ceH - G(c(1+d) - 1)} X + \frac{1 - G}{1 + ceH - G(c(1+d) - 1)} Q + \frac{-cG}{1 + ceH - G(c(1+d) - 1)} E \tag{6.2}$$

Equation 6.2 shows that the errors E of the additional DAC are shaped by the loop filter H. This does not mean that if H has extremely high gain, extremely large linearity errors are allowed in this DAC, as linearity errors can be transformed into signal dependent DAC gain errors.

If G has a value different than 1, the inner additional noise shaping loop will add limited or no noise shaping to the outer loop. To illustrate the effect, a first order example of G is given by:

$$G = \frac{1+g}{s+1} \tag{6.3}$$

If $g = 0$, the value of G well below the -3 dB cut off frequency is close to 1. This means that in this portion of the frequency spectrum Q is suppressed. At higher frequencies the value of G becomes smaller and smaller than 1, and the quantization noise in the output spectrum will rise. If G is put into Eq. 6.1, and it is assumed that $c = e = 1$ and $d = 0$ this results in:

$$Y = \frac{s+1}{s+1+(1+g)d}W + \frac{s-g}{s+1+(1+g)d}Q \tag{6.4}$$

Equation 6.4 shows that if $d = 0$, the input signal W of the error feedback loop is not affected by the loop filter G, as the quantization error is fed back perfectly. This means that the stability criteria of the outer loop are not affected. If $d \neq 0$, loop filter G will be partly visible in the transfer function Y/W, and might compromise the stability of the $\Sigma\Delta$ modulator. In Fig. 6.6 a simulation of the transfer function Y/W for different values of d is shown. Even with a mismatch between the two feedback paths of 0.1, the main loop's gain will change less than 1 dB (Fig. 6.6a). If the non-linearity of the additional DAC leads to signal dependent gain changes in d, the effects on the noise shaping as well as the signal transfer function Y/W are negligible. Furthermore, at values of d higher than 0.1, the additional phase shift in the main loop due to the noise shaper will be less than 3 degrees (Fig. 6.6b). At high frequencies where stability of the modulator is defined, this is even less. If a lower pole frequency is chosen in G, the phase shift at high frequencies can be reduced further, at the cost of noise shaping as will be shown in the next simulation.

In Fig. 6.7 a simulation of the NTF (Q/Y) of the additive error feedback loop only (without the outer loop attached) is shown for different values of g. If $g = 0$ the loop filter G gives perfect first order shaping to the quantization noise. If $g \neq 0$ the shaping will be first order for only a limited part of the spectrum. This is shown in Fig. 6.7. As g deviates more and more from zero, the term $1 - G$ will

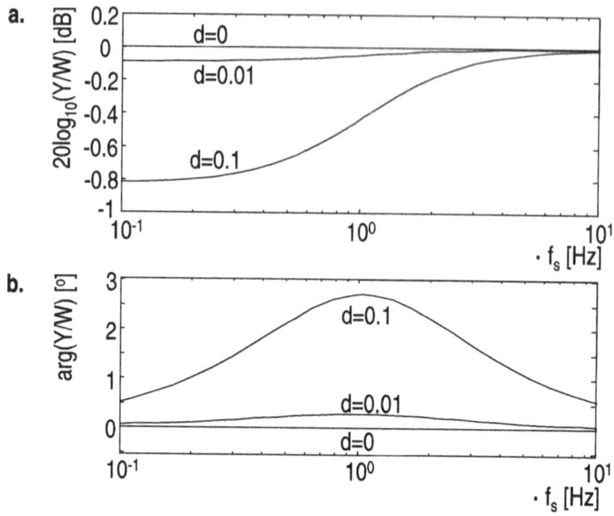

Figure 6.6: *Simulation of a ΣΔ modulator with additive error feedback loop with a gain error d in the additional feedback DAC*

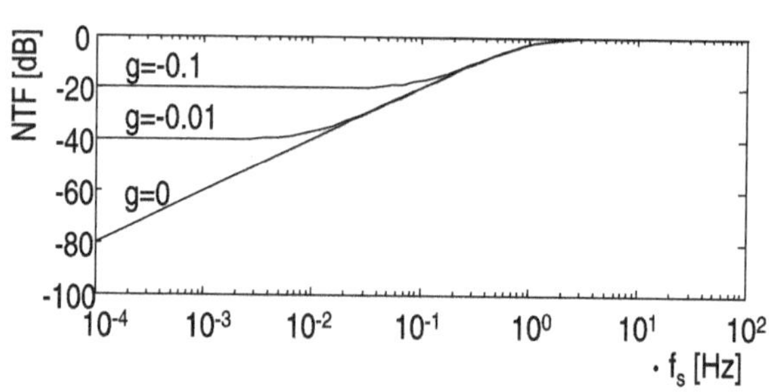

Figure 6.7: *Simulation of the NTF of the additive error feedback loop with a gain error g in filter G*

also deviate more and more from 0. This means that the additional error feedback loop will add less noise shaping to the main loop at low frequencies.

6.4.5 Cascaded ΣΔ Modulators

Cascaded ΣΔ modulators do not only require well defined gain to the quantizer when multi-bit (Sect. 6.4.2), but also have gain requirements between the two stages. If a cascaded modulator architecture, built-up out of two CT modulators is chosen, accurate matching is required between the analog loop filters and the digital noise cancellation filter [23, 26]. Due to the complexity of such a modulator, it is seen as an unattractive alternative. Therefore, this architecture is not further analyzed nor represented in the implementation chapters.

6.5 Circuit Noise of the Modulator's Input Stage and DAC

In terms of circuit noise, the most critical blocks are the input stage and the feedback DAC, as an error (and thus noise) introduced by these blocks will be directly visible in the output of the ΣΔ modulator. The noise requirements on the blocks later in the loop are limited as their noise and distortion is shaped by the preceding integrator stages.

In all modulators presented in this book, an RC integrator is used for the input stage. This is because it can handle large input signal swings at excellent linearity (Sect. 6.6.1.2). Therefore, noise analysis of other input stages is not part of this book. The implemented DACs presented in this book are all 1-bit and therefore the noise analysis of multi-bit DACs will be excluded here. The feedback DAC can either be implemented with switched current (SI), switched resistor (SR) or switched capacitor (SC) circuits. The three different feedback DAC topologies are shown in Fig. 6.8. The input referred noise of the modulator will be calculated in

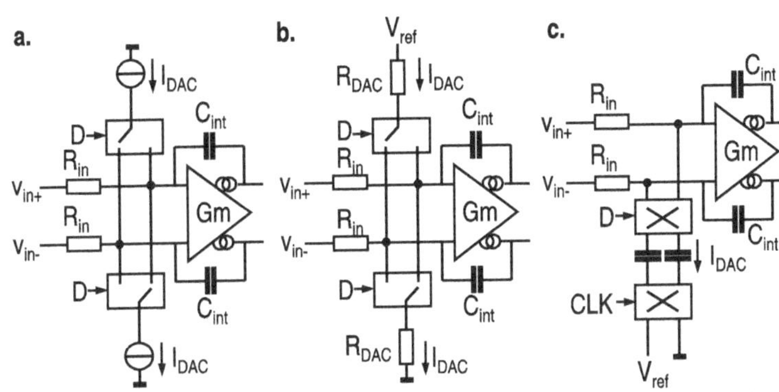

Figure 6.8: *RC integrator input stage with SI (a), SR (b) and SC (c) feedback DAC*

the next few paragraphs assuming a modulator with an RC integrator input stage with either an SI, SR or SC 1-bit feedback DAC.

6.5.1 RC Integrator Input Stage and SI Feedback DAC

If the noise introduced by the later integrators is neglected, because of the gain of the first integrator, the total input referred noise of the modulator is only determined by the input resistors, the feedback DAC current sources and the circuit noise of the OTA:

$$S_{\Sigma\Delta M,in,SI}(\omega) \approx 8kTR_{in} + S_{OTA}(\omega)\left((\omega R_{in}C_{int})^2 + 1\right) + 2i_{n,dac}^2 R_{in}^2 \ [\text{V}^2/\text{Hz}] \tag{6.5}$$

The current source noise $i_{n,dac}$ also includes 1/f-noise, which requires large area current sources which can limit the maximum switching speed. As the modulators presented in this book are clocked over 100 MHz, the use of a current source DAC is unattractive. Therefore the SI DAC is not used in the implementations presented in this book. In SR and SC DAC implementations the 1/f noise problem is decoupled from the DAC unit element size, because the 1/f noise originates from the reference voltage instead of the unit elements themselves. This gives an additional degree of freedom.

6.5.2 RC Integrator Input Stage and SR Feedback DAC

Following the same reasoning as in the previous paragraph, the input referred noise of a modulator with SR DAC can be calculated by expanding the result in [16] with the OTA and reference voltage noise leading to:

$$S_{\Sigma\Delta M,in,SR} \approx 8kTR_{in}\left(1 + \frac{R_{in}}{R_{DAC}}\right) + 2S_{Vref}\left(\frac{R_{in}}{R_{DAC}}\right)^2 + S_{OTA}\left((\omega R_{in}C_{int})^2 + \left(1 + \frac{R_{in}}{R_{DAC}}\right)^2\right) \ [\text{V}^2/\text{Hz}] \tag{6.6}$$

Like the SI DAC current sources, the DAC reference also introduces 1/f noise. This can either be improved by increasing area, filtering and/or using chopping techniques. Anyhow the amount of 1/f noise on the DAC reference and the maximum switching speed of the ΣΔ modulator are independent issues. This gives an extra degree of freedom.

6.5.3 RC Integrator Input Stage and SC Feedback DAC

The combination of Eq. 6.6 with the results of [27] and [28] leads to a modulator input referred noise of an RC integrator stage with SC feedback DAC [29]:

$$S_{\Sigma\Delta M,in,SC}(\omega) \approx 8kTR_{in}\left(1 + R_{in}f_sC_{DAC}\right) + 2S_{Vref}(\omega)(R_{in}f_sC_{DAC})^2 + S_{OTA}(\omega)\left((\omega R_{in}C_{int})^2 + \frac{2R_{sw} + R_{in}^2 f_s C_{DAC}}{2R_{sw}}\right) \ [\text{V}^2/\text{Hz}] \tag{6.7}$$

For a switched capacitor DAC the total modulator input referred noise density of Eq. 6.7 is only true if it is assumed that the OSR is large, and the sinc function in the SC DAC can be neglected [27]. Like in a $\Sigma\Delta$ modulator with SR DAC, the amount of 1/f noise on the DAC reference and the maximum switching speed of the $\Sigma\Delta$ modulator are independent issues, which gives an extra degree of freedom in the design.

6.5.4 Impact of Supply Voltage on the Circuit Noise Requirements

In a power-efficient design, the noise of the input resistors and the DAC equivalent resistors are dominant in the noise contributions over the OTA and DAC reference voltage noise. This will yield smaller OTA input currents and optimizes the ratio between modulator linearity and OTA bias current, as will be shown in Sect. 6.6.1.2.

For a modulator with SR or SC DAC, the total modulator input referred noise density now reduces to:

$$S_{\Sigma\Delta M,in,SR\ or\ SC} \approx 8kTR_{in}\left(1+\frac{R_{in}}{Z_{DAC}}\right)\ [\mathrm{V^2/Hz}] \tag{6.8}$$

in which Z_{DAC} is either R_{DAC} (SR DAC) or $1/(f_s C_{DAC})$ (SC DAC). In the calculation of the noise contributions of R_{in} and the DAC feedback impedance, it is assumed that the modulators input range is 0.7 times the DAC reference voltage [30]. For the circuit proposed in Fig. 6.8 this means that:

$$\frac{2V_{in,rms}}{V_{ref}} = \frac{R_{in}}{R_{DAC}}\ [\text{-}] \tag{6.9}$$

If Eqs. 6.8 and 6.9 are combined the most optimum SNR can be calculated.

$$\begin{aligned}\mathrm{SNR}_{circuit\ noise} &= \frac{V_{in,rms}}{\sqrt{8kTR_{in}\left(1+\frac{R_{in}}{Z_{DAC}}\right)}}\\ &= \frac{V_{in,rms}}{\sqrt{8kTR_{in}\left(1+\frac{2V_{in,rms}}{V_{ref}}\right)}}\ [\text{-}]\end{aligned} \tag{6.10}$$

Equation 6.10 shows that both $V_{in,rms}$ and V_{ref} should be as large as possible. However, both are limited by the supply. This leads to the extreme case were

$V_{in,rms,max} = V_{dda}/\sqrt{2}$ and $V_{ref} = V_{dda}$, where the input signal and reference voltage are rail-to-rail. This changes Eq. 6.9 into:

$$\text{SNR}_{circuit\ noise} = \frac{V_{dda}}{\sqrt{16kTR_{in}\left(1+\sqrt{2}\right)}} \quad [\text{-}] \tag{6.11}$$

From Eq. 6.10 it can be seen that a proportionally higher supply will allow for a quadratically higher noise impedance for the same $\text{SNR}_{circuit\ noise}$, as the input signal and reference voltage can be made bigger. This means the same SNR with quadratically less current, or a higher SNR for the same current.

In practical situations the modulator input signal will not be rail-to-rail, as it needs to be driven by the preceding circuit, which needs some headroom. The reference voltage is normally generated by a bandgap circuit followed by a reference buffer, to decouple the reference voltage from the supply. Therefore, the modulator input signal amplitude and reference voltage are not very likely to be rail-to-rail, which will lead to a lower $\text{SNR}_{circuit\ noise}$ for the same R_{in}, or will lead to a higher power consumption because of the lower R_{in} required.

6.6 Non-linearity

In this section the non-linearity of different modulator circuit blocks and their impact on the performance of the ΣΔ modulator are discussed. The Input stage, quantizer and feedback DAC are discussed consecutively.

6.6.1 Non-linearity in the Input Stage

Harmonic distortion in the input stage of a ΣΔ modulator causes input signal related harmonics at the output. Furthermore, inter-modulation of high frequency quantization noise and/or high frequency input signals causes noise and/or signal folding into the signal bandwidth. The relation between circuit distortion and the required resources to reduce the distortion will be investigated in this paragraph.

For the loop filter, the non-linearity analysis is limited to the non-linearity of the first integrator, as the non-linearity errors introduced in the later integrator stages are suppressed by the gain of the integrators in front of these stages.

First, the HD3 distance (HD3D) will be calculated for different differential input pair configurations of which the input transistors are biased in strong or weak inversion. Second, a selection will be made what the best configuration is for a ΣΔ modulator in terms of linearity.

In the linearity analysis, only the non-linearity of the $I_D - V_{gs}$ transfer function of MOS input stage transistors is considered. Furthermore, only HD3 will be considered as HD2 is (transistor) mismatch related, and in theory is absent in a fully differential design. In practice, the unbalance in the input stage will cause second order harmonic distortion.

6.6.1.1 Non-linearity of Differential Pairs

In Fig. 6.9a, an NMOS differential pair is shown. In [31] the third order distortion

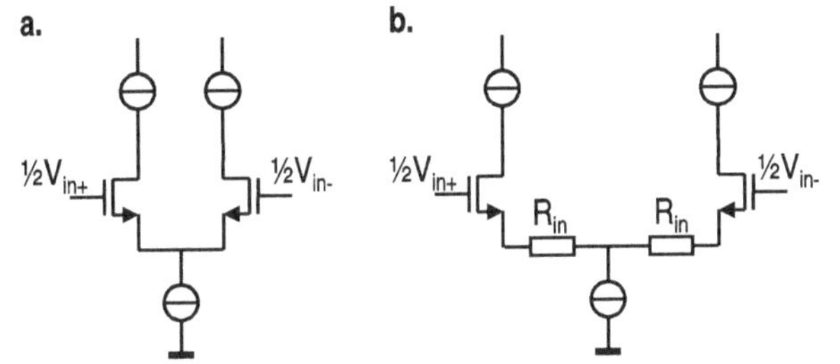

Figure 6.9: *(a) NMOS differential pair. (b) Degenerated NMOS differential pair*

of a MOS differential pair biased in strong inversion is calculated. The HD3D is given by:

$$\text{HD3D} = \frac{32 \cdot V_{GT}^2}{\hat{V}_{in}^2} = \frac{128 \cdot I_D^2}{\hat{V}_{in}^2 \cdot gm^2} = \frac{64 \cdot I_D^2}{V_{in,rms,max}^2 \cdot gm^2} \; [\text{-}] \tag{6.12}$$

In the degenerated differential pair of Fig. 6.9b, the input signal V_{in} is reduced by a factor of approximately $gm \cdot R_{in}$ to the input transistor's gate-source voltage. In combination with Eq. 6.12 the HD3D of a degenerated differential pair with transistors in strong inversion can be calculated:

$$\text{HD3D} \approx \frac{64 \cdot I_D^2 \cdot R_{in}^2}{V_{in,rms,max}^2} \; [\text{-}] \tag{6.13}$$

In Appendix B the HD3D of a MOS differential pair in weak inversion is calculated:

$$\text{HD3D} = \frac{96 \cdot I_D^2}{\hat{V}_{in}^2 \cdot gm^2} = \frac{48 \cdot I_D^2}{V_{in,rms,max}^2 \cdot gm^2} \; [\text{-}] \tag{6.14}$$

HD3D for a degenerated differential pair biased in weak inversion now becomes:

$$\text{HD3D} \approx \frac{48 \cdot I_D^2 \cdot R_{in}^2}{V_{in,rms,max}^2} \; [\text{-}] \tag{6.15}$$

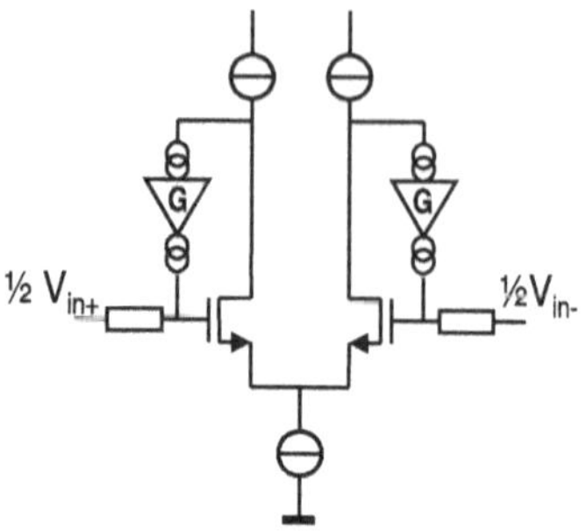

Figure 6.10: *NMOS differential pair in feedback configuration*

Figure 6.10 shows a differential pair in feedback configuration. If the input transistors are biased in strong inversion it can be calculated that:

$$\text{HD3D} = \frac{64 \cdot I_D^2 \cdot R_{in}^2 \cdot G^2}{V_{in,rms,max}^2} \quad [-] \tag{6.16}$$

In which G is the current amplification factor. If the differential pair in feedback configuration is biased in weak inversion, the HD3D becomes:

$$\text{HD3D} = \frac{48 \cdot I_D^2 \cdot R_{in}^2 \cdot G^2}{\hat{V}_{in,rms,max}^2} \quad [-] \tag{6.17}$$

6.6.1.2 Non-linearity of a ΣΔ Modulator Input Stage

There are basically two options for a linear input stage: either the input stage of Fig. 6.9, or the input stage in a feedback configuration of Fig. 6.10. In case of a degenerated input stage, the DAC output current is directly integrated on the integrator capacitor. This is shown in Fig. 6.11a. As stated earlier the input stages of the modulators presented in this book all use RC OTA integrator input stages like the one displayed in Fig. 6.11b. In Fig. 6.11b, the differential pair is part of the ΣΔ modulator loop, and overall feedback is provided by the feedback DAC. In [16], Eq. 6.12 is used to calculate the distortion of a ΣΔ modulator with RC OTA input stage, with the input transistors biased in strong inversion.

$$\text{HD3D} = \frac{32 \cdot gm \cdot I_D^2 R_{in}^3}{(1 + \frac{R_{in}}{R_{DAC}}) V_{in,rms,max}^2} \quad [-] \tag{6.18}$$

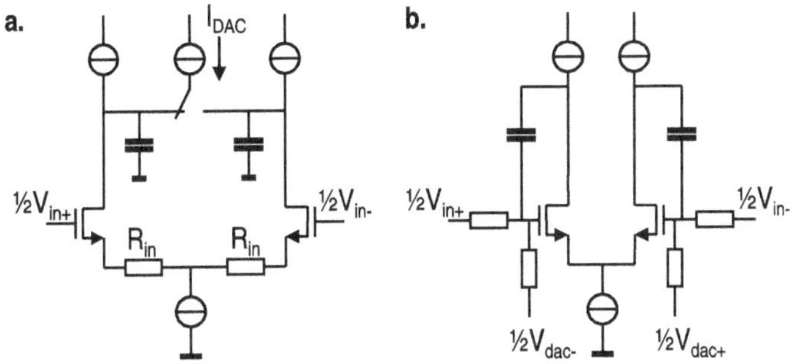

Figure 6.11: *Degenerated differential pair (a) or differential pair with feedback (b) as input stage of a* $\Sigma\Delta$ *modulator loop*

With the help of Eq. 6.14, the same calculation can be performed for an input pair biased in weak inversion, which yields:

$$\text{HD3D} = \frac{24 \cdot gm \cdot I_D^2 R_{in}^3}{\left(1 + \frac{R_{in}}{R_{DAC}}\right) V_{in,rms,max}^2} \quad [\text{-}] \tag{6.19}$$

At first, an input pair in strong inversion seems to out perform an input pair in weak inversion by a factor of $x = 4/3$. It should be noted though that $\frac{gm}{I_D}_{wi} > \frac{gm}{I_D}_{sat}$ which partly compensates x.

The advantage of the degenerated integrator input stage over the RC integrator input stage is that the quantization noise is not present at the input of the differential pair. This avoids quantization noise folding into the signal bandwidth due to the non-linearity of the input pair. In the RC integrator input stage the quantization noise is present at the input of the differential pair, and significant quantization noise folding might occur. However, when Eq. 6.13 is compared with Eq. 6.18, or Eq. 6.14 is compared with Eq. 6.19, conclusion is that the RC OTA integrator input stage outperforms a degenerated input pair by a factor of $(gm \cdot R_{in})/(2(1 + R_{in}/RDAC)$. From linearity perspective, this makes the RC integrator stage the preferred input stage topology, under the condition that the quantization noise folding is no issue.

6.6.2 Non-linearity in the Quantizer Decision Levels

Random offset in the decision levels of the quantizer causes distortion. The quantizer offset requirement in a 1-bit $\Sigma\Delta$ modulator is very low, as the offset introduced by the 1-bit comparator is suppressed by the loop filter gain. In a multi-bit quantizer the variance of the offset in the comparators is limited to a fraction of an LSB to at least ensure monotonicity. The non-linearity of the quantizer is shaped by the loop filter, and normally is no issue. For quantizers with a large number of bits, the monotonicity requirement requires large comparator input stages or offset cancellation techniques, which both are not very attractive in terms of area and complexity respectively. In this book quantizers with only a limited number of bits are used.

6.6.3 Inter-Symbol-Interference in the Feedback DAC

Inter-symbol-interference or ISI can cause non-linearity due to memory effects in the feedback DAC. The effect is illustrated in Fig. 6.12a. The area of one 1-symbol is not equal to half of the area of two consecutive 1-symbols, which makes the feedback data dependent and thus introduces a non-linearity. One solution to

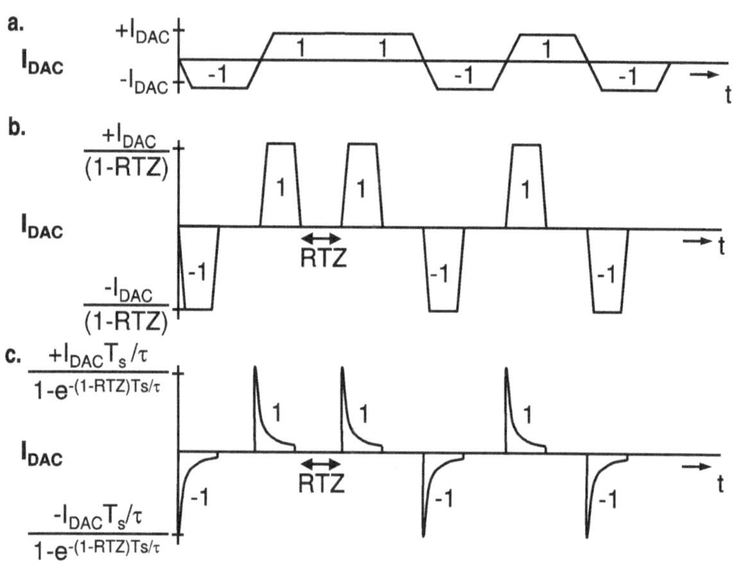

Figure 6.12: *Inter-symbol-interference, and return-to-zero for an SI/SR and SC DAC*

avoid ISI is the introduction of return-to-zero (RTZ) in the feedback pulses [30]. Figure 6.12b shows the result of introducing RTZ in the DAC feedback pulses. After each symbol the DAC returns to zero output. Each pulse now has the same amount of edges, and the area of each pulse is independent of the data pattern. To compensate for the shorter pulse period, the amplitude of the pulses have to be increased by a factor 1/(1-RTZ), where RTZ=[0..1].

Another way to reduce the sensitivity to ISI, is using SC feedback. When the available settling time is infinite, an SC pulse decays to zero, and in theory the ISI of consecutive pulses is zero. In practice due to bandwidth limitations in the circuits and the finite settling-time available due to the f_c sampled nature of the DAC, the switched capacitor will not be completely discharged to zero from symbol to symbol, and some ISI will remain. To get rid of ISI, an SC DAC can be combined with RTZ. An example waveform is shown in Fig. 6.12c.

To be able to always feedback a unity charge Q_{DAC} of $I_{DAC}T_s$ in one clock cycle, the SC feedback current is defined by:

$$\begin{aligned} I_{SC}(t) &= I_{peak}e^{-t/\tau} \ \text{[A]} \\ &\quad \text{for } t = [n \cdot T_s, n + (1 - \text{RTZ})T_s\rangle \\ I_{SC}(t) &= 0 \ \text{[A]} \\ &\quad \text{for } t = [n + (1 - \text{RTZ})T_s, (n+1)T_s\rangle \\ \text{and for } n &= [0..\infty] \end{aligned} \tag{6.20}$$

with a SC DAC peak output current of:

$$I_{peak} = \frac{I_{DAC}\frac{T_s}{\tau}}{1 - e^{-(1-\text{RTZ})T_s/\tau}} \ \text{[A]} \tag{6.21}$$

6.6.4 Non-linearity in the Output Levels of the Feedback DAC

The non-linearity of the feedback DAC will directly show up at the output of the modulator as its error is directly fed to the input of the modulator. The linearity error will cause harmonic distortion of the input signal, and high frequency quantization noise folding into baseband, reducing the algorithmic accuracy.

6.6.4.1 Non-linearity in the Output Levels of a 1-bit DAC

As 1-bit DACs only have two levels, they are inherently linear, and matching requirements in unit elements are avoided.

6.6.4.2 Non-linearity in the Output Levels of a b-bit DAC

The performance advantages of a multi-bit $\Sigma\Delta$ modulator described in Sect. 5.2, come with the disadvantage of the linearity requirements for the DAC, which requires accurate matching of unit cells. As matching is related to area, this makes multi-bit modulators unattractive, as large area limits the speed of operation. In literature several solutions to the DAC linearity problem have been proposed like

DEM, DWA and barrel-shifting. A good overview is given in [17]. The solutions proposed either increase cost, area, complexity, loop delay or a combination, at limited to moderate performance increase. Therefore, multi-bit DACs are avoided in the implementations presented in this book.

6.6.4.3 Non-linearity in the Output Levels of a 1.5-bit DAC

A special member of the b-bit DACs is the 1.5-bit DAC. It has three output levels which are 1, 0 and -1. In principle the 1.5-bit DAC has the same linearity issues as a b-bit DAC ($b > 1.5$-bit). The difference is, that by using smart switching for the unit cells, the non-linearity theoretically can be eliminated. The principle of operation is shown in Fig. 6.13 for a 1.5-bit current DAC. When both current

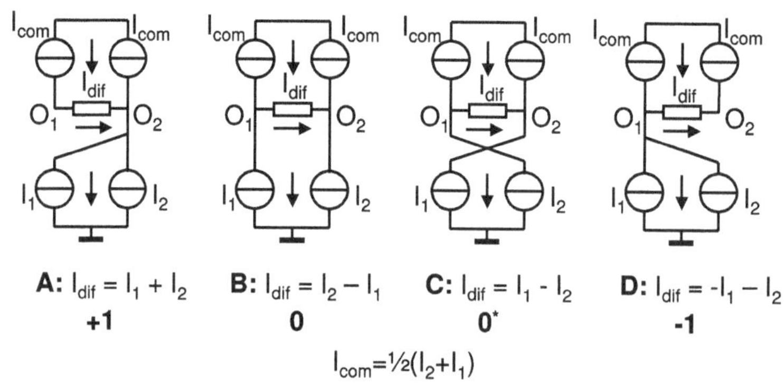

Figure 6.13: *1.5-bit current DAC switching diagram*

sources are connected to O_2 (A) the differential current I_{dif} is $I_1 + I_2$. When I_1 is connected to O_1 and I_2 is connected to O_2 (B) there is a differential current $(I_2 - I_1)$. When I_1 is connected to O_2 and I_2 is connected to O_1 $I_{dif} = (I_1 - I_2)$ (C). When both current sources are connected to O_1 the differential current I_{dif} is $-(I_1 + I_2)$ (D). When $I_1 < I_2$ the lower non-linear curve (dotted) of Fig. 6.14 describes the transfer function of the DAC. The non-linearity is clearly visible. When $I_1 < I_2$ and $I_1 = I$ and $I_2 = 1.01I$ due to a mismatch of 1% in the current sources there only is a gain error of $(2.01I)/2I = 1.005$ or 0.043 dB, which can be neglected. A more serious problem is the non-linearity of the transfer function. From Fig. 6.14 it can be seen that a non-linearity occurs for code 0 because for codes $+1$ and -1 the absolute value of the differential current is identical. To improve linearity 0-data chopping is proposed. This technique switches between circuit B and C dependent on 0-data. The idea is illustrated in Fig. 6.14. The output current at 0-data changes sign dependent on the data that is put on to the D/A converter, by chopping I_1 and I_2 between O_1 and O_2. Each 0-data or RTZ period,

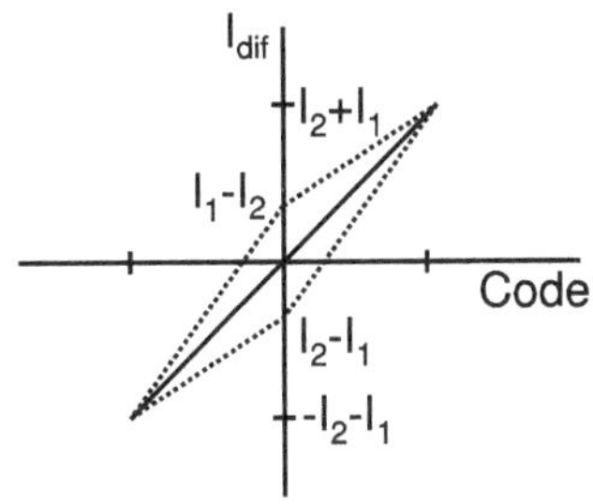

Figure 6.14: *Transfer function of a 1.5-bit current DAC with unit cell mismatch*

the sign of the current in the 0-data is inverted. The transfer function of the D/A converter is chopped between the two dashed lines, resulting in a perfectly linear straight line. The data pattern and DAC output current are shown if Fig. 6.15. An

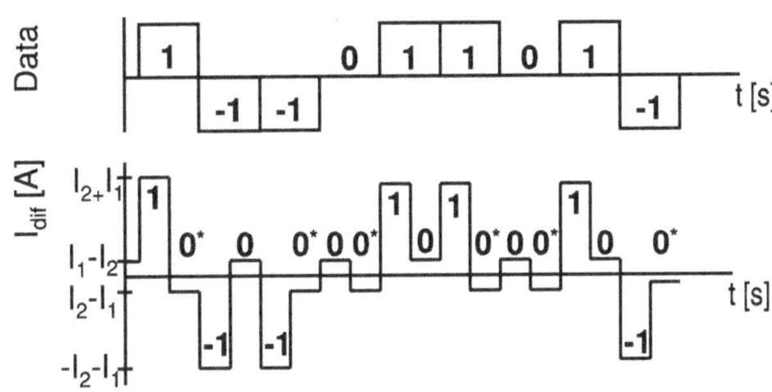

Figure 6.15: *DAC input data and output current using the zero-chopping technique*

example simulation is shown in Fig. 6.16. The 1.5-bit DAC is employed in a 4th order $\Sigma\Delta$ modulator with 1 resonator stage. The dashed line gives the ideal modulator output spectrum. If there is 1% mismatch between I_1 and I_2, the output spectrum transforms into the straight black line. Due to the non-linearity there is a lot of noise folding and harmonic distortion can be seen from the spectrum. If the proposed technique is used, the D/A converter is linearized and the theoretical algorithmic accuracy of the modulator is restored.

Note that the two current sources I_{com} in Fig. 6.13 are providing an equal amount of current as an example. If these current sources are not equal, chopping can be used to make them equal.

The technique is patented in [32, 33].

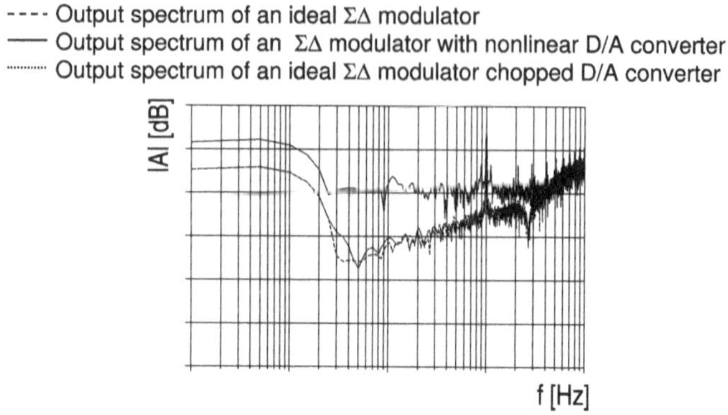

Figure 6.16: *Simulation of a 4th order 1.5-bit ΣΔ modulator, with and without the chopping technique proposed with 1% current source mismatch*

6.7 Aliasing in ΣΔ Modulators

If in the sampling process in an ADC, which is sampled at f_s, an input signal of frequency f_{in} does not fulfill the Nyquist criterion, input signals beyond $\frac{1}{2}f_s$ fold back into the bandwidth $0 - \frac{1}{2}f_s$ Hz causing an in-band alias at f_a according to $f_a = f_s - f_{in}$ [34]. This alias might interfere with the wanted signal to be converted by the ADC. Therefore, if high frequency signals are expected at the input of the ADC (like in highly digitized receivers in which analog signal conditioning is minimized) an anti-alias filter will be required.

A continuous-time ΣΔ modulators have a built in anti-alias filter [30] which effectively reduce the in-band aliases introduced by the quantizer sampling.

Another source of aliases often forgotten is the ΣΔ modulator's feedback DAC. In this section, theory is presented with which the aliasing in the DAC can be calculated. The DAC is in a RC integrator input stage configuration, and the calculation will be done for three different DAC types, which are an SI, an SR and an SC DAC.

6.7.1 Aliasing in the Quantizer

In CT ΣΔ modulators, the loop filter acts as an anti-aliasing filter for the aliasing which occurs due to sampling operation in the quantizer. The amplitude of the aliased frequency components can be calculated by using the model presented in Fig. 6.17 in which the quantizer has been replaced by a linear gain c. A signal X

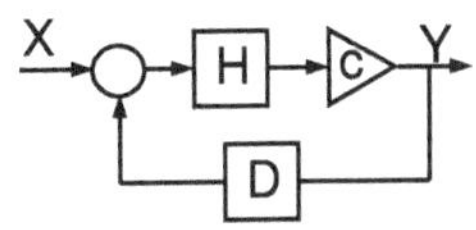

Figure 6.17: $\Sigma\Delta$ *modulator model*

with frequency f_{in} close to f_s will only have limited feedback from the DAC due to the sinc function of the DAC:

$$D(f) = d \cdot \frac{\sin(\pi \cdot f/f_s)}{\pi \cdot f/f_s} \tag{6.22}$$

This means that the input signal will be amplified by the loop filter with approximately $H(f_{in})$. Arrived at the quantizer, the signal will be sampled and an alias signal component will appear at $f_s - f_{in}$. Like the quantization error, the alias component will be shaped by the loop filter gain at $f_s - f_{in}$ which is $H(f_s - f_{in})$. The theoretical amplitude of the alias product at the output of the $\Sigma\Delta$ modulator is:

$$Y(f_s - f_{in}) = \frac{cH(f_{in})}{(1 + cD(f_{in})H(f_{in}))\,(1 + cD(f_s - f_{in})H(f_s - f_{in}))} \cdot X(f_{in}) \tag{6.23}$$

As $\Sigma\Delta$ modulators use over-sampling, the highest frequency alias product to be expected in-band is at $f_s(1 - 1/2/\text{OSR})$. This means that even at an OSR as low as 10, $D(f_{in}) \approx 0$ and $D(f_s - f_{in}) \approx d$. This simplifies Eq. 6.23 to:

$$Y(f_s - f_{in}) = \frac{H(f_{in})}{H(f_s - f_{in})} \cdot \frac{X(f_{in})}{d} \tag{6.24}$$

6.7.2 $\Sigma\Delta$ Modulator with an SI Feedback DAC

A $\Sigma\Delta$ modulator input stage including the first integrator and an SI DAC is shown in Fig. 6.18. The current sources I_{DAC} are switched to the virtual ground nodes depending on the data, or are dumped via the RTZ switch. When the current sources are connected to the virtual ground this node is loaded with the R_{out} of the current sources. In RTZ mode, the currents I_{DAC} are dumped and the R_{out} of the current sources is disconnected from the virtual ground nodes. Due to the RTZ period in the data signals, the signal on the virtual ground node is modulated due to the time varying impedance at that node because of the DAC current source switching. This can cause in-band aliases of input signal frequencies close to f_s.

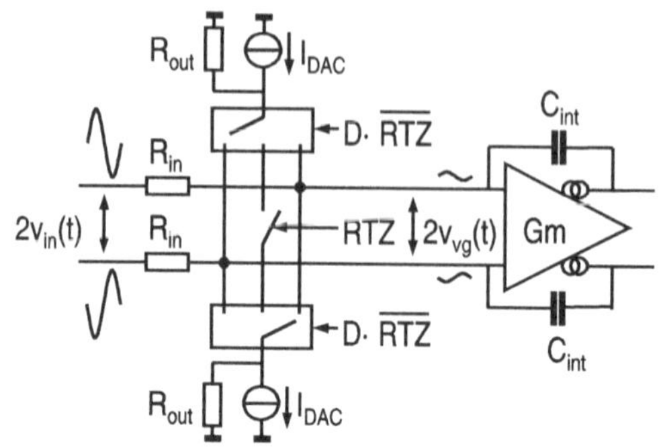

Figure 6.18: *Model to calculate the aliasing in a ΣΔ modulator with switched current DAC*

The signal on the virtual ground node due to the input signal can be calculated to be:

$$\frac{v_{vg}}{v_{in}} \approx \frac{1}{1 + 2Gm \cdot R_{in}} \quad [\text{-}] \tag{6.25}$$

in which the DAC output impedance is neglected. The signal on the virtual ground nodes is modulated by the output impedances of the current sources, resulting in a current i_{vg}. The current drawn from the virtual ground node by the feedback path switches between v_{vg}/R_{out} (Data period) and zero (RTZ period). To calculate the alias amplitude, only the fundamental frequency is taken ($n = 1$) as the fundamental frequency is the one causing the dominant contribution to the aliasing due to down modulation. The higher order harmonics are neglected. Using Eq. C.1 of Appendix C with $n = 1$, the amplitude of the fundamental frequency component can be calculated to be:

$$|i_{vg,n=1}| = \frac{2v_{vg}}{R_{out}\pi} \sin(\pi(1 - \text{RTZ})) \quad [\text{A}] \tag{6.26}$$

The anti-alias distance (AAD) is defined as the ratio between the high frequency input signal amplitude at the input and the amplitude of the aliased signal calculated back to the input of the modulator. To calculate the AAD, i_{vg} current is compared to the input current v_{in}/R_{in}. If the input signal is a sine wave with frequency f_{in}, an alias component is expected at $f_s - f_{in}$ with an AAD of:

$$\text{AAD}^{dB} = 20 \log_{10} \left(\frac{R_{out}\pi(1 + 2Gm \cdot R_{in})}{R_{in} \sin(\pi(1 - \text{RTZ}))} \right) \quad [\text{dB}] \tag{6.27}$$

An important observation is the fact that if there is no RTZ (RTZ=0), there is no aliasing of input signals close to f_s. Obviously, if RTZ=1 there is no aliasing due to the fact that there is no feedback at all. If it is assumed that $2Gm \cdot R_{in} \gg 1$ which is normally the case, the AAD increases with the $Gm \cdot R_{out}$.

6.7.2.1 Simulations

Figure 6.19 shows simulations of a fifth order, 1-bit ΣΔ modulator with SI DAC sampled at 25 MHz. In the simulation, the modulator has input resistors R_{in} of 10 kΩ, an input stage with a Gm of 2 mA/V, an RTZ of 0.5 and an R_{out} of 100 kΩ unless indicated otherwise. Figure 6.19a shows the AAD for different RTZ values. At high RTZ, the feedback current amplitude has to be higher because the pulse gets shorter. A current source with a higher output current means a lower output impedance. In this simulation R_{out} is kept constant to clearly see the effect of the RTZ period, in reality however the lower R_{out} will give a worse AAD. However, the output impedance can be increased by circuit techniques like cascoding [35] and gain-boosting [36, 37]. The calculation and simulation of the AAD for different RTZ periods is within ±1 dB.

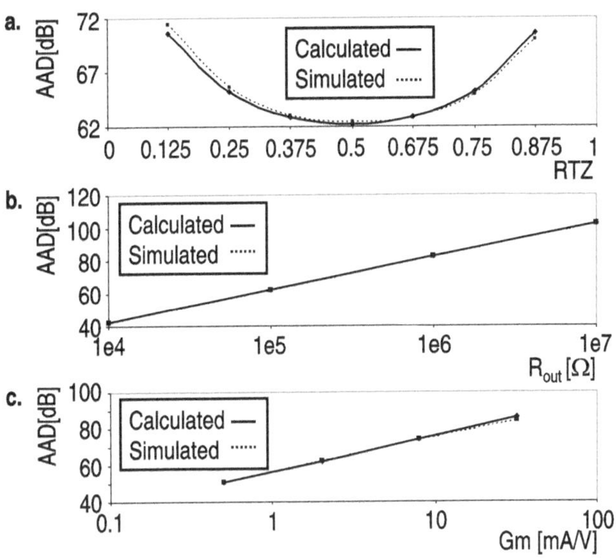

Figure 6.19: *Simulations on aliasing in a ΣΔ modulator with switched current DAC*

Figure 6.19b shows the AAD for different values of R_{out}. Higher R_{out} means higher AAD, as the i_{vg} gets smaller. Calculation and simulation are within ±1 dB.

Figure 6.19c shows the AAD for different Gm values. At larger Gm values the signals on the virtual ground node get smaller and the AAD will be larger. Calculation and simulation are within ±1 dB.

6.7.3 ΣΔ Modulator with an SR Feedback DAC

In a ΣΔ modulator with SR DAC, aliasing can occur due to output impedance differences of the DAC. Figure 6.20 shows a ΣΔ modulator input stage with SR DAC. Due to impedance mismatch between the switches, the virtual ground is

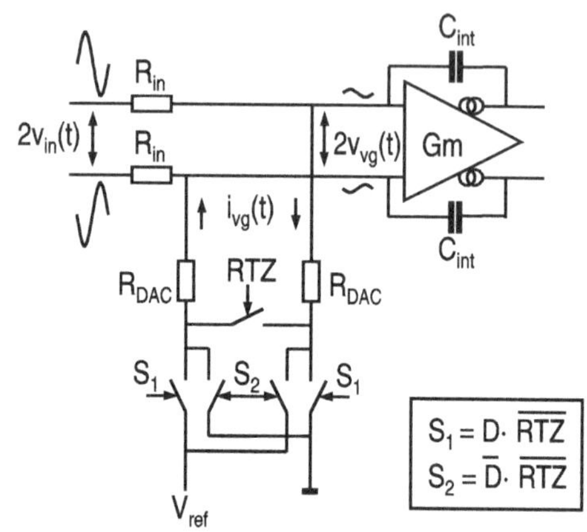

Figure 6.20: *Model to calculate the aliasing in a ΣΔ modulator with switched resistor DAC*

loaded by a different impedance during the different switching phases. This can cause aliasing in two ways. The first is a mismatch between the RTZ switch and the DATA switches. The second is an inter data switch mismatch. Both will be described in this paragraph. The aliasing due to the combination of both mismatches can be calculated by combining the equations of the next two paragraphs, and will not be shown here. A mismatch between the two DAC resistors R_{DAC} does not give aliasing as these resistors are always in use.

6.7.3.1 Mismatch Between Data Switches and RTZ Switch

If the RTZ switch impedance is not equal to two times the data switch impedance, aliasing can occur due to the non-constant loading on the virtual ground node. In the data period the load is $R_{DAC,eq} = 2R_{DAC} + 2R_{sw,D}$. In the RTZ period the load is $2R_{DAC} + R_{sw,\mathrm{RTZ}} = 2R_{DAC} + 2R_{sw,D} = R_{DAC,eq}$. If $R_{sw,\mathrm{RTZ}}$ is $2R_{sw,D} + \Delta R$, the signal on the virtual ground node will be modulated by the time varying DAC impedance, yielding a current i_{vg}. The amplitude of the fundamental component of i_{vg} can be derived in a similar way as explained in

Sect. 6.7.2 which yields:

$$|i_{vg,n=1}| = \frac{4v_{vg}\sin(\pi(1-\text{RTZ}))}{\pi}\left(\frac{1}{2R_{DAC,eq}} - \frac{1}{2R_{DAC,eq}+\Delta R}\right)\ \text{[A]} \tag{6.28}$$

The AAD then becomes:

$$\text{AAD}^{dB} = 20\log_{10}\left(\frac{\pi(1+2Gm\cdot R_{in})}{2R_{in}\sin(\pi(1-\text{RTZ}))\left(\frac{1}{R_{DAC,eq}} - \frac{1}{R_{DAC,eq}+\Delta R}\right)}\right)\ \text{[dB]} \tag{6.29}$$

6.7.3.2 Inter Data Switch Mismatch

In case of inter data switch mismatch the aliasing distance is dependent on the data pattern. If it is assumed that only one data switch deviates from the other data switches, the impedance on the virtual ground node switches between three conditions: $\overline{\text{RTZ}}\cdot D$, $\overline{\text{RTZ}}\cdot\overline{D}$ and RTZ. The corresponding impedances are $2R_{DAC}+2R_{sw,D}+\Delta R = R_{DAC,eq}+\Delta R$, $2R_{DAC}+2R_{sw,D} = R_{DAC,eq}$ and $2R_{DAC}+R_{sw,\text{RTZ}} = 2R_{DAC}+2R_{sw,D} = R_{DAC,eq}$. In the part where data is "1", the virtual ground node is loaded by a different impedance compared to the other two states. If the input signal to the modulator is dc free, the modulator will generate patterns like "0101" and "0011" with an equal amount of zeros and ones. Figure 6.21a shows the example patterns with RTZ. As the number of ones

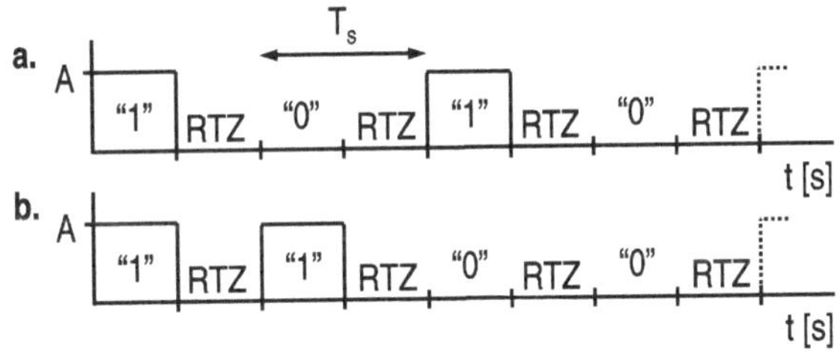

Figure 6.21: *"1010"- and "1100"-pattern data signal with RTZ*

and zeros in each pattern are equal, and there is an RTZ period, the amplitude of the frequency component $f_s = 1/T_s$ (Fig. 6.21) can be shown to be:

$$A_{f_s} = \frac{A}{\pi}\sin(\pi(1-\text{RTZ}))\ \text{[V]} \tag{6.30}$$

independent of the pattern. Therefore the AAD becomes:

$$\mathrm{AAD}^{dB} = 20\log_{10}\left(\frac{\pi(1+2Gm\cdot R_{in})}{R_{in}\sin(\pi(1-\mathrm{RTZ}))\left(\frac{1}{R_{DAC,eq}}-\frac{1}{R_{DAC,eq}+\Delta R}\right)}\right) \quad [\mathrm{dB}] \tag{6.31}$$

If due to a DC modulator input signal more zeros than ones appear at the output of the modulator, or vice versa, the AAD gets better or worse. In the example where one of the D switches has mismatch to the others is used, and the modulator only outputs zeros, aliasing will disappear, as the switch impedance will always be the same, and there is no impedance mismatch. In the case of only ones, the worst case aliasing occurs which converges to the AAD of Eq. 6.29.

6.7.3.3 Simulations

Figure 6.22 shows simulations on a fifth order, 1-bit ΣΔ modulator with SR DAC sampled at 25 MHz. The modulator has input resistors (R_{in}) of 10 kΩ, an input stage with a Gm of 2 mA/V, a ΔR of 200 Ω and an RTZ of 0.5 unless indicated otherwise. In this simulation it is assumed that the reference voltage is chosen as high as possible within the modulator supply, to optimize noise performance of the ΣΔ modulator (Eq. 6.6), as at higher reference voltages the feedback DAC resistors can be made larger for the same feedback current. This means that the feedback resistor R_{DAC} has to be scaled according to $R_{DAC} = \mathrm{RTZ}\cdot R_{DAC,\mathrm{RTZ=0}}$. In this simulation it is assumed $R_{DAC,\mathrm{RTZ=0}} = 10$ kΩ. Figure 6.19a shows the AAD for different RTZ values. At high RTZ, the feedback resistor has to become smaller as the same charge needs to be fed back in a smaller feedback pulse width. This leads to a relatively bigger error due to the switch mismatch. Calculation and simulation are within ±1 dB. Figure 6.22b shows the AAD for different values of ΔR for the RTZ-data switch and the inter data switch mismatch. At higher ΔR the AAD becomes lower due to the larger mismatch. The RTZ-data switch mismatch gives 6 dB larger aliasing compared to the data-data switch mismatch as expected. Calculation and simulation are within ±1 dB. Figure 6.22c shows the AAD for different Gm values. At larger Gm values the signals on the virtual ground node get smaller and the AAD will be larger. Calculation and simulation are within ±1 dB except for the simulation for $Gm = 32$ mA/V, which deviates 5 dB from calculation due to simulation inaccuracy.

6.7.4 ΣΔ Modulator with an SC Feedback DAC

If an RC integrator input stage is used together with an SC DAC, aliasing can occur due to the combination of a non-zero ohm input impedance of the virtual

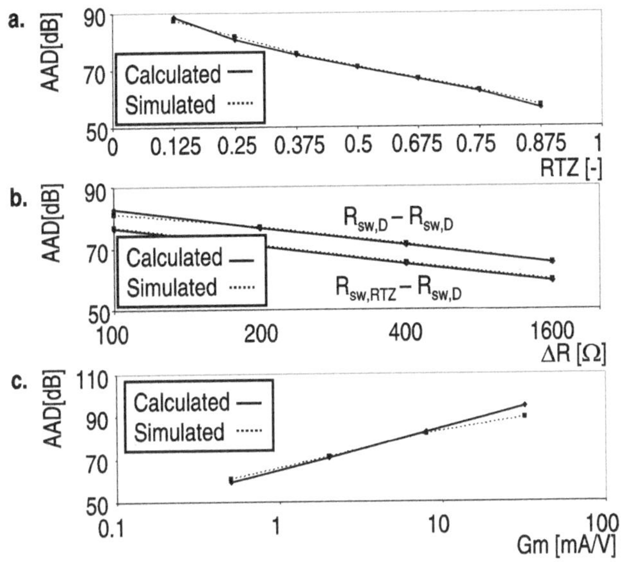

Figure 6.22: *Simulations on aliasing in a ΣΔ modulator with switched resistor DAC*

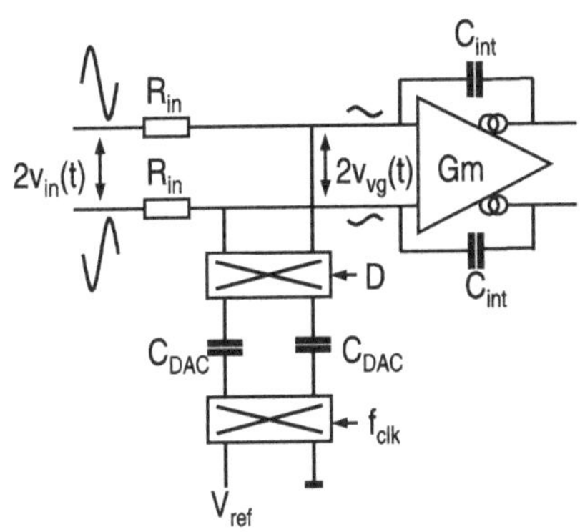

Figure 6.23: *Model to calculate the aliasing in a ΣΔ modulator with switched capacitor DAC*

ground node and the sampling operation of the DAC. The effect is illustrated in Fig. 6.23. Due to the non-zero input impedance of the virtual ground node, the input signal present at the input of the OTA can be calculated using Eq. 6.25. The signal at the virtual ground node is sampled on the DAC capacitors during the capacitor discharge phase. This causes a sub-sampled current through the equivalent switched capacitor resistor $1/(f_s C)$ into the virtual ground node, which means that aliases will appear in the output spectrum of the ΣΔ modulator.

The single-ended equivalent circuit used for the calculation of the aliasing occurring in the switched capacitor DAC is a sample-and-reset circuit shown in Fig. 6.24. The voltage on the virtual ground node is replaced by a voltage source

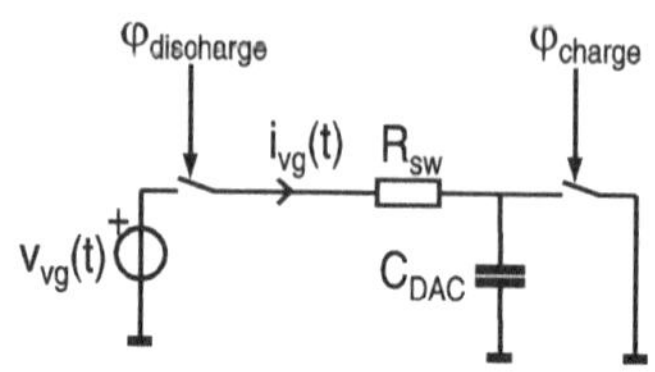

Figure 6.24: *Model to calculate the aliasing in the switched capacitor DAC*

which only contains the attenuated input signal with frequency f_{in} and amplitude V_{vg}. It is assumed that the DAC capacitor C_{DAC} is completely reset by the charge switch and thus the reference voltage can be replaced by a short.

The signal $v_{vg}(t)$ and the switched capacitor current are shown in Fig. 6.25a. The parameter of interest, is the current going into the discharge switch. This current can contain an aliased component of frequency $f_a = f_s - f_{in}$, which can fall into the signal bandwidth of the modulator. The current shape is displayed in Fig. 6.25b. The aliased current component is periodic with period time $mT_s =$

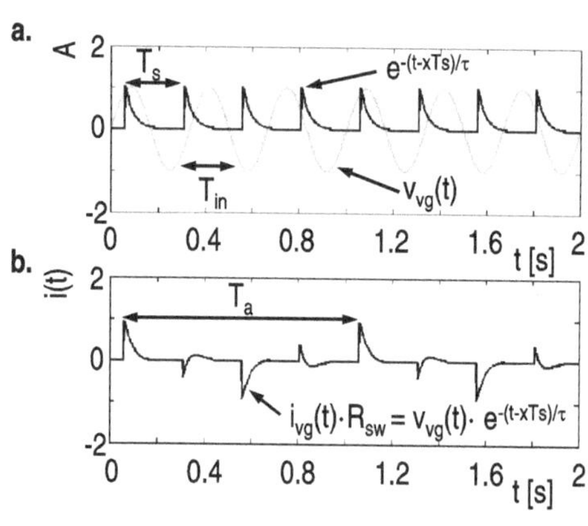

Figure 6.25: *Shape of the switch input current*

$f_s/(f_s - f_{in})T_s$, and has frequency f_a. To calculate amplitude of the aliased current component, the Fourier integral is used. As the Fourier integral can only be used for signals which are periodic over T, the integral has to be used on the aliased signal period T_a.

$$\alpha_n = \frac{1}{mT_s}\int_{T_a} i(t)e^{-jn\omega_a t}dt \ (n \geq 0) \tag{6.32}$$

in which n is the nth order Fourier coefficient. The spectrum of the current then is given by the Fourier series:

$$I(\omega) = \sum_{n=-\infty}^{\infty} \alpha_n e^{jn\omega_a t} \ \text{[A]} \tag{6.33}$$

The time signal of Fig. 6.25 can be written as the sum of time shifted switched capacitor currents:

$$i(t) = \sum_{x=0}^{m} \frac{V_{vg}}{R_s}\sin(\omega_{in}t)e^{\frac{-(t-xT_s)}{\tau}} \ \text{[A]} \tag{6.34}$$

in which V_{vg} is the signal amplitude on the virtual ground node, and R_s is the total series resistance during the discharge phase. If $1/gm$ approximates the total switch resistance R_{sw}, the gm should be taken into account, which yields $R_s = R_{sw} + 1/gm$. Furthermore $\tau = R_s C_{DAC}$.

Using the Fourier integral to calculate the nth order Fourier coefficient yields:

$$\alpha_n = \frac{1}{mT_s}\frac{V_{vg}}{R_s}\int_{xT_s}^{(x+a)T_s}\sum_{x=0}^{m}\sin(\omega_{in}t)e^{\frac{-(t-xT)}{\tau}}e^{-jn\omega_a t}dt \tag{6.35}$$

This can be rewritten to:

$$\alpha_n = \frac{1}{2jmT_s}\frac{V_{vg}}{R_s}\int_{xT_s}^{(x+a)T_s}\sum_{x=0}^{m} e^{\frac{xT_s}{\tau}}\left(e^{(j(\omega_{in}-n\omega_a)-\frac{1}{\tau})t} - e^{(-j(\omega_{in}+n\omega_a)+\frac{1}{\tau})t}\right)dt \tag{6.36}$$

which leads to:

$$\begin{aligned}\alpha_n = {} & \frac{1}{2jmT_s}\frac{V_{vg}}{R_s}\sum_{x=0}^{m}\left[\frac{1}{j(\omega_{in}-n\omega_a)-\frac{1}{\tau}}e^{j(\omega_{in}-n\omega_a)t-\frac{t-xT_s}{\tau}}\right]_{xT_s}^{(x+a)T_s} \\ & - \frac{1}{2jmT_s}\frac{V_{vg}}{R_s}\sum_{x=0}^{m}\left[\frac{-1}{j(\omega_{in}+n\omega_a)+\frac{1}{\tau}}e^{-j(\omega_{in}+n\omega_a)t+\frac{t-xT_s}{\tau}t}\right]_{xT_s}^{(x+a)T_s}\end{aligned} \tag{6.37}$$

Substituting the intervals $(x + a)T_s$ and xT_s yields:

$$\alpha_n = \frac{e^{j(\omega_{in}-n\omega_a)aT_s-\frac{aT_s}{\tau}} - 1}{2jT_s(j(\omega_{in} - n\omega_a) - \frac{1}{\tau})} \frac{V_{vg}}{R_s} \sum_{x=0}^{m} \frac{1}{m} e^{j(\omega_{in}-n\omega_a)xT_s} + \frac{e^{-j(\omega_{in}+n\omega_a)aT_s-\frac{aT_s}{\tau}} - 1}{2jT_s(j(\omega_{in} + n\omega_a) + \frac{1}{\tau})} \frac{V_{vg}}{R_s} \sum_{x=0}^{m} \frac{1}{m} e^{-j(\omega_{in}+n\omega_a)xT_s} \qquad (6.38)$$

α_n only has values unequal to zero for $n = \pm 1$ or $n = \pm\frac{\omega_{in}}{\omega_a}$. For all other n, the summation yields zero. This means that the output spectrum only contains the frequency components $\pm\omega_a$ and $\pm\omega_{in}$ as expected.

The single sided frequency spectrum amplitudes of the frequency components can be calculated by $A_n = 2|\alpha_n|$ and are summarized in Table 6.1.

Frequency	n	α_n	$\alpha_n(\tau \ll T_s$ and $1/gm \ll R_{sw})$
f_a	1	$\frac{e^{-j2\pi a-\frac{aT_s}{\tau}}-1}{-2T_s(\omega_s+j\frac{1}{\tau})}\frac{V_{vg}}{R_s}$	$\frac{-jf_sC_{DAC}V_{vg}}{2}$
$-f_a$	-1	$\frac{e^{j2\pi a-\frac{aT_s}{\tau}}-1}{2T_s(\omega_s-j\frac{1}{\tau})}\frac{V_{vg}}{R_s}$	$\frac{jf_sC_{DAC}V_{vg}}{2}$
f_{in}	$\frac{f_{in}}{f_a}$	$\frac{e^{-\frac{aT_s}{\tau}}-1}{2T_s(-j\frac{1}{\tau})}\frac{V_{vg}}{R_s}$	$\frac{-jf_sC_{DAC}V_{vg}}{2}$
$-f_{in}$	$-\frac{f_{in}}{f_a}$	$\frac{e^{-\frac{aT_s}{\tau}}-1}{-2T_s(j\frac{1}{\tau})}\frac{V_{vg}}{R_s}$	$\frac{-jf_sC_{DAC}V_{vg}}{2}$

Table 6.1: *Frequency components*

The frequency component f_a can fall into the ΣΔ modulator signal bandwidth. The frequency component f_{in} will add to the signal component f_{in} already present at the modulator input, and will have negligible influence. Note that f_s is not in the switched capacitor current.

If the single sided spectrum is considered, the amplitude of the alias component is $V_{in}(f_a) = V_{in}(f_in) = f_sC_{DAC}R_{in}V_{vg}$ referred to the ΣΔ modulator input. The aliasing distance in this case can be approximated by:

$$\frac{V_{in}(f_{in})}{V_{in}(f_a)} \approx \frac{gm}{f_sC_{DAC}} \text{ [-]} \qquad (6.39)$$

The calculated, approximated and simulated alias distance is shown in Fig. 6.26. If the τ/T_s is small, the approximated value of Eq. 6.39 holds. At higher τ/T_s,

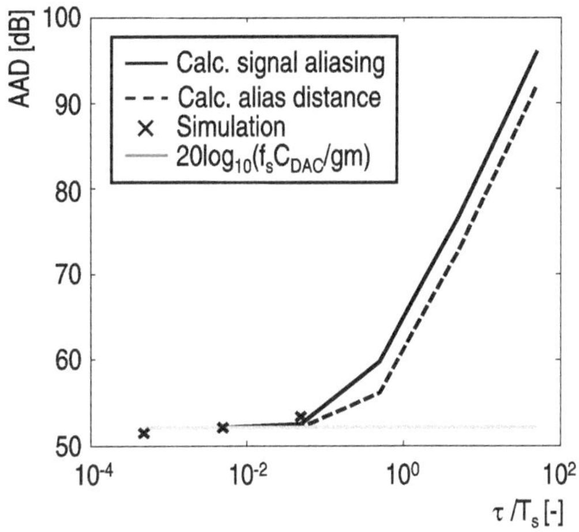

Figure 6.26: *Calculated, approximated and simulated AAD for a $\Sigma\Delta$ modulator with switched capacitor DAC*

the switch impedance becomes more relevant and the alias distance will become larger, however the feedback charge to the virtual ground node will also become less, which causes a modulator input-output gain error, and even worse; might cause modulator instability. For this reason, no simulations could be done for $\tau/T_s > 0.1$.

6.8 Excess Loop Delay

A parasitic pole or time delay in the loop can cause loop instability of the modulator. The parasitic pole for instance is caused by a parasitic pole in the loop filter. Time delay is introduced by a deliberate timing difference between quantizer sampling and DAC clocking, to account for the decision time of a (slowly) deciding comparator in the quantizer. The traditional way to compensate for these parasitic effects is increasing current in the circuits. This paragraph will show a less power hungry approach: modifying the loop filter coefficients according to the expected amount of excess phase or time delay. A more extensive analysis on the impact of delay on the modulator DR, and on delay compensation techniques is presented in [38].

Furthermore, it will be shown that a proper choice for the DAC circuit topology also can reduce the total delay in the loop.

6.8.1 Excess Time Delay Compensation

Figure 6.27a shows the block diagram of an Lth order discrete time modulator without excess time delay. The modulator block diagram of a modulator with a full clock cycle of delay $(1/z)$ is shown in Fig. 6.27b. The method described in this paragraph, equates the transfer functions of both block diagrams to calculate the new coefficients b_n based on the old coefficients a_n, and uses an additional feedback path c, in order to compensate for the additional unwanted delay. The feed-forward coefficients a_n of Fig. 6.27b are transformed into coefficients b_n. Furthermore an additional feedback DAC c, of the same resolution as the quantizer is introduced. The loop filter transfer function of Fig. 6.27a is equated to the one of Fig. 6.27b, which yields:

$$\sum_{n=1}^{L} a_n \left(\prod_{m=1}^{n} \frac{1}{z-1} \right) = \frac{1}{z} \left(c + \sum_{n=1}^{L} b_n \left(\prod_{m=1}^{n} \frac{1}{z-1} \right) \right) \tag{6.40}$$

The coefficients b_n and coefficient c then become:

$$\begin{aligned} b_k &= a_k + a_{k+1} \quad (k = 1..L-1) \\ b_L &= a_L \\ c &= a_1 \end{aligned}$$

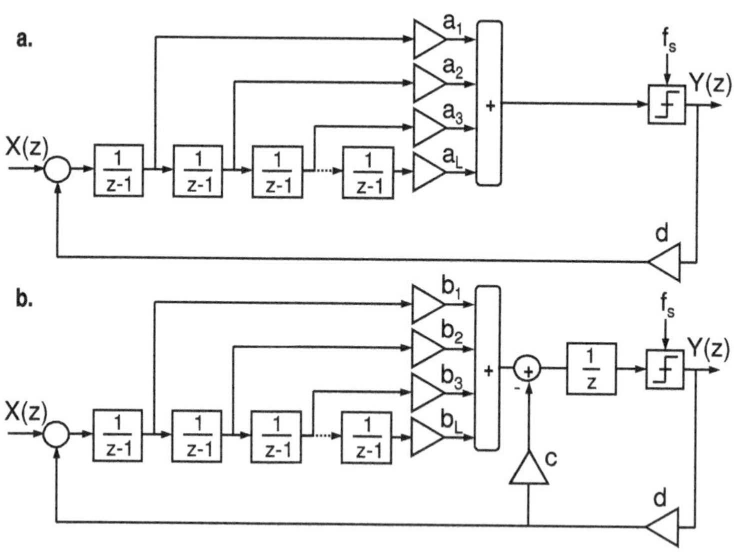

Figure 6.27: ΣΔ *modulator without loop delay (a) and with loop delay and accompanying compensation (b)*

If the loop filter has local feedback coefficients to create notches in the output spectrum of the ΣΔ modulator (like in Fig. 5.3), these coefficients are the same in both situations.

6.8.2 Excess Phase Compensation

A parasitic pole in the loop in the loop, for instance in the summing node of a feed-forward filter can add excess phase shift, causing the ΣΔ modulator instability. In this section a compensation method is introduced, with which the parasitic pole can be compensated, once simulations have uncovered the frequency of such a parasitic pole. If the delay compensation is used to compensate the pole, instead of increasing the current in the amplifier to shift the parasitic pole to higher frequencies, power consumption of the modulator can be reduced [39].

For parasitic excess phase a similar transformation as in Sect. 6.8.1 can be done. The transfer functions of both block diagrams of Fig. 6.28a and b are equated:

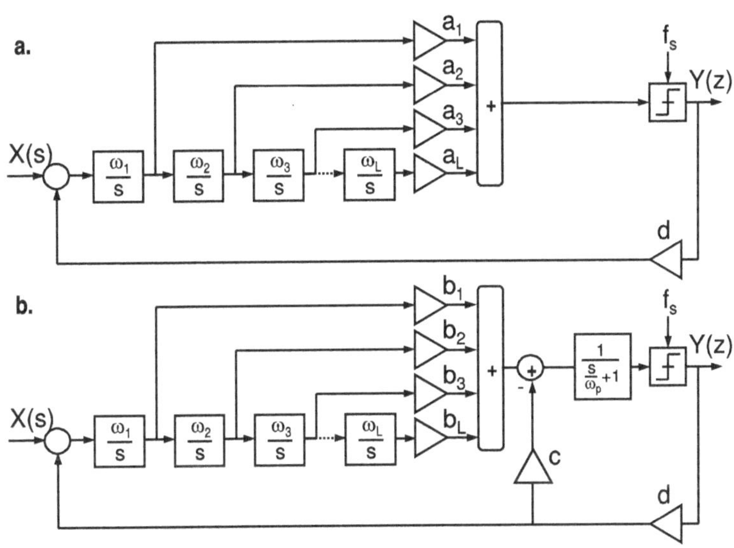

Figure 6.28: *ΣΔ modulator without loop delay (a) and with loop delay and accompanying compensation (b)*

$$\sum_{n=1}^{L} a_n \left(\prod_{m=1}^{n} \frac{\omega_m}{s} \right) = \frac{1}{\frac{s}{\omega_p}+1} \left(c + \sum_{n=1}^{L} b_n \left(\prod_{m=1}^{n} \frac{\omega_m}{s} \right) \right) \tag{6.41}$$

The coefficients b_n and coefficient c then become:

$$b_k = a_k + a_{k+1}\frac{\omega_{k+1}}{\omega_p} \quad (k = 1..L-1)$$
$$b_L = a_L$$
$$c = a_1\frac{\omega_1}{\omega_p}$$

In both cases the integrator unity gains are the same for Fig. 6.28a and b. The feed-forward coefficients a_n change into b_n. This way the integrator output signal swings will stay the same. If the loop filter has local feedback coefficients to create notches in the output spectrum of the $\Sigma\Delta$ modulator (like in Fig. 5.3), these coefficients are the same in both situations. This is because the unity gain frequencies stay the same in both cases (Eq. 5.6). The excess phase compensation has been applied to the $\Sigma\Delta$ modulator implementation of Chap. 9.

Figure 6.29 shows a simulation result of a 5th order 1-bit modulator without the pole (grey), and a simulation with the parasitic pole but using the compensation method described (black). Both the modulator output spectra (Fig. 6.29a), their

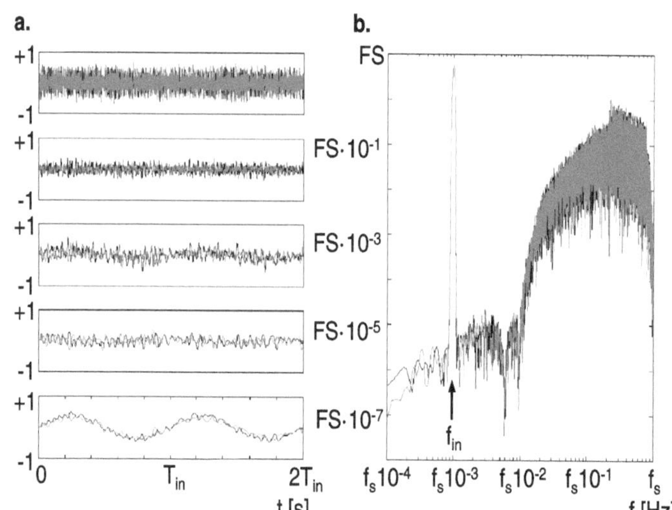

Figure 6.29: *Simulation of a $\Sigma\Delta$ modulator without loop delay (grey) and a simulation of a $\Sigma\Delta$ modulator with loop delay and accompanying compensation (black)*

simulated SQNR, as the signal swings on the output of the integrators (Fig. 6.29b) are the same.

6.8.3 DAC Feedback Pulse Shape and Delay

In Sect. 6.6.3, Fig. 6.12 it is shown that a full-T SI or SR DAC outputs pulses with constant charge $Q_{DAC} = I_{DAC} \cdot T_s$, in which the current is on during the complete feedback period (hence the term full-T). One of the reasons to avoid full-T pulses is to eliminate ISI as described in Sect. 6.6.3. Solutions to ISI are the introduction of RTZ in the DAC, to use an SC DAC instead of an SI/SR DAC or a combination of both. Next to the reduction of ISI, the introduction of RTZ and/or an SC DAC also has the advantage of introducing less delay in the feedback loop. Figure 6.30 shows the effect. In the figure three simulations are done for different SI/SR DACs and different SC DACs. In addition the SC DAC output pulses are given for three

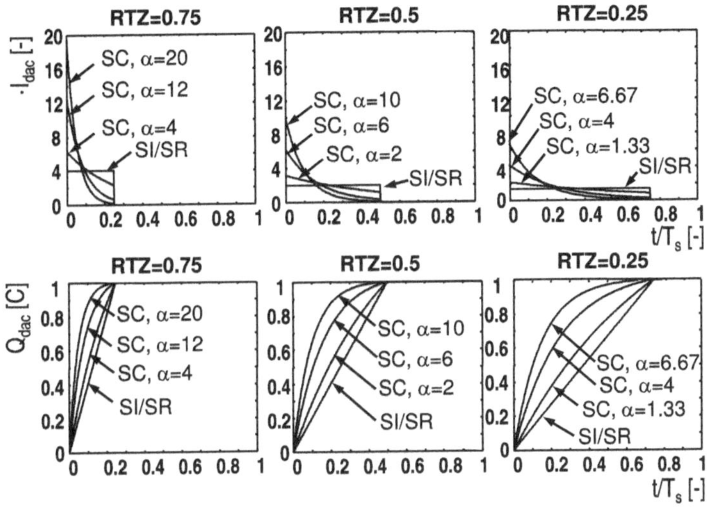

Figure 6.30: *Pulse shape for switched current/resistor and switched capacitor DACs*

different values of α ($\alpha = T_s/\tau$). In the upper plots the DAC output current is given. In the lower plots the integrated charge of the pulses Q_{DAC} for the three different values of RTZ is given, in which Q_{DAC} is normalized to 1. Obviously, the larger the RTZ period, the faster the total charge has arrived at the input stage. For the SC DAC, this effect is increased by enlarging α. The earlier the current arrives at the input stage, the less delay is introduced in the $\Sigma\Delta$ loop by the DAC, and the better its stability, as less delay is introduced in the $\Sigma\Delta$ modulator loop. A disadvantage of increasing RTZ and α is the increasing peak current required for the pulse. This is shown in the upper plots. At an RTZ of 0.25, 0.5 and 0.75 the SI/SR current increases from 1 to 1.33, 2 and 4 respectively. For the SC DAC the peak current increases to 6.3, 12.6 and 20.1. These larger DAC currents have to be digested by the input stage, which means larger bias currents. In general a very short RTZ period has lower peak currents, but larger delay. Very long RTZ periods require large peak currents, but guarantee low delay in the DAC. However, the required clock frequency increases inversely proportional with very short or long RTZ periods. Therefore RTZ=0.5 seems to be a good choice, as it does not need a higher clock, if it is assumed that both the rising and falling clock edge can be used (of course this requires an accurate input clock duty cycle).

6.9 Clock Jitter in CT ΣΔ Modulators

A major concern in ΣΔ modulators is the degradation of the ΣΔ modulator's dynamic range due to clock jitter. Imprudent design of the clock circuitry, which generates the ΣΔ modulator's clock, will induce too much clock jitter on the ΣΔ modulator's sampling operations and will degrade the performance of the modulator. In this chapter the impact of clock jitter on the dynamic range of a CT ΣΔ modulator will be further analyzed.

In a CT ΣΔ modulator, clock jitter is injected into the loop at two nodes: the quantizer and the DAC. In Fig. 6.31 the block diagram (a) and corresponding model (b) of a continuous time modulator are shown. The model consists of an

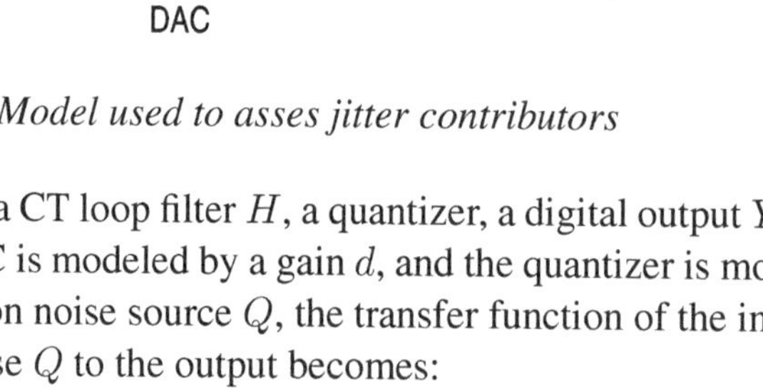

Figure 6.31: *Model used to asses jitter contributors*

analog input X, a CT loop filter H, a quantizer, a digital output Y, and a feedback DAC. If the DAC is modeled by a gain d, and the quantizer is modeled by a gain c and a quantization noise source Q, the transfer function of the input signal X and quantization noise Q to the output becomes:

$$Y = \frac{H}{1+H}X + \frac{1}{1+H}Q \tag{6.42}$$

when it is assumed that c and d are one. The jitter noise on the quantizer clock (J_Q) is shaped by the loop filter gain like the quantization noise. Therefore, it will have a very small contribution to the in-band noise at the output of the ΣΔ modulator and can be neglected. The jitter noise on the DAC clock (J_D) however, is directly at the input of the $\Sigma\Delta$ modulator. Equation 6.42 shows that an in-band error at the input X will manifest itself directly at the output of the modulator as in the signal bandwidth $|H/(1 + H)|$ is close to one. If the jitter noise at the output of the feedback DAC causes frequency components (or noise) in the signal bandwidth, the jitter noise will decrease the dynamic range of the modulator.

In the next two chapters, two clock jitter models are presented: the Time-to-Amplitude-Jitter-Error (TAJE) model and the Time-to-Phase-Jitter-Error (TPJE) model. In the TAJE model the time-shifting of clock edges is transformed into a random amplitude error (white noise) on the unity output charge per clock cycle of the feedback DAC. The more precise TPJE model, on the contrary, describes how the time-shifting of the edges of the sampling clock, due to jitter, influences the clock spectrum. In the elaboration of this model, the jitter at first will be modeled by a single frequency sine wave for sake of simplicity. Next, the model describes how this sine wave induced jittered clock causes amplitude and phase modulation of a sine wave going through a DAC sampled with this clock. It will be shown that the clock jitter spectrum will be present around every output signal of the DAC. Once the impact of a single sine wave jitter component to the DAC output is known, the sine wave jitter will be replaced by white noise jitter.

The derivation of the TAJE and TPJE model will be done for a SC and SI DAC. As the shape of the output pulses of an SI DAC is the same as the shape of the output pulses of an SR DAC, the SI DAC jitter theory presented also can be applied to an SR DAC. Therefore, the SR DAC is not further mentioned in this section.

6.9.1 The TAJE Model

In the TAJE model, white noise jitter on the DAC clock edges is modeled as an amplitude (or charge) error. The white noise clock jitter shifts the DAC output edges up and down with a certain variance causing DAC feedback pulse width variations. This leads to a constantly changing integrated charge of the DAC output pulse from clock cycle to clock cycle, which in the TAJE model, is modeled by an amplitude error. Therefore, the TAJE model only approximates the pulse width variations on the DAC output pulses, and neglects the effect that the pulses are also shifted back and forth in time.

In this section a modulator with a DAC, which outputs either switched current or switched capacitor output pulses will be analyzed using the TAJE model. In [40] modulators, with DACs that output different pulse shapes are analyzed with a TAJE like model.

6.9.1.1 CT 1-bit ΣΔ Modulator with SI DAC

In [30] the TAJE model was introduced (under a different name) to calculate the clock jitter SNR limitation of a ΣΔ modulator with 1-bit non-return-to-zero (NRTZ) SI DAC, which resulted in:

$$\text{SJNR}^{dB}_{SI\ NRTZ\ DAC} = 10 \cdot \log_{10}\left(\frac{1}{16\text{OSR}\sigma_s^2 B^2}\right) \ \text{[dB]} \tag{6.43}$$

in which OSR and B are the over-sampling ratio and bandwidth of the modulator respectively, and σ_s is the variance of the white noise clock jitter. In the next two paragraphs the SNR limitation of the TAJE model will be determined for a ΣΔ modulator with an RTZ SI and RTZ SC DAC.

6.9.1.2 CT 1-bit ΣΔ Modulator with RTZ SI DAC

In Fig. 6.32a the output waveform of a SI DAC is displayed. When there is clock jitter present on the clock this will cause timing errors Δt with variance σ_s^2 causing a variation on the pulse width. The variance of the error charge σ_q^2 transferred per

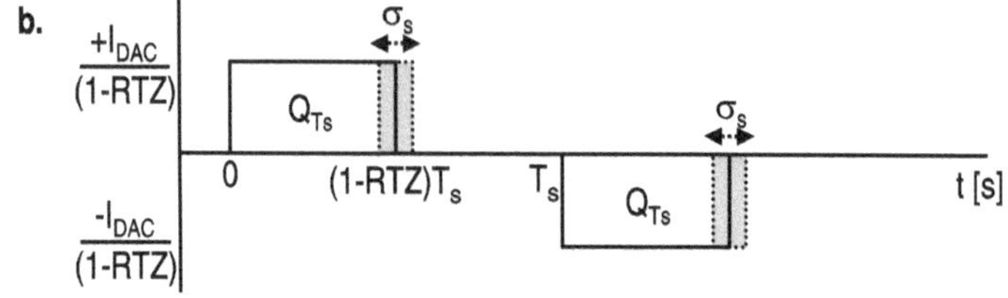

Figure 6.32: *Switched current time signals with and without return-to-zero*

clock cycle T_s can be calculated by:

$$\sigma_q^2 = \sigma_s^2 \cdot I_{DAC}^2 \ [\text{C}^2] \tag{6.44}$$

in which I_{DAC} is the amplitude of the feedback current. In Fig. 6.32b the feedback pulse with RTZ is shown. When RTZ is used, the feedback pulse has to have an amplitude inversely proportional to (1-RTZ) to ensure that the amount of integrated feedback charge Q_{DAC} in one clock cycle is constant, ensuring an RTZ independent gain in the feedback path. This increases the magnitude of σ_q because of the (1/(1-RTZ)) times larger amplitude feedback pulses. The RTZ interval can be added to Eq. 6.44 which leads to:

$$\sigma_q^2 = \frac{\sigma_s^2 \cdot I_{DAC}^2}{(1 - \text{RTZ})^2} \ [\text{C}^2] \tag{6.45}$$

A disadvantage of the RTZ interval is the higher clock frequency needed to create the RTZ interval. When RTZ=0.5, the feedback pulse is two times shorter in time, and to feedback the same amount of charge Q_{Ts}, the pulse has to be twice the amplitude. The error charge due to clock jitter is expected to be two times higher compared to the RTZ=0 pulses, due to the two times larger amplitude pulses. Furthermore, two edges of the clock are used. If these edges are assumed to be uncorrelated (high frequency jitter) the total amount of jitter power has increased by a factor of two. If this is taken into account Eq. 6.45 changes into:

$$\sigma_q^2 = \frac{2\sigma_s^2 \cdot I_{DAC}^2}{(1 - \text{RTZ})^2} \ [\text{C}^2] \tag{6.46}$$

The power of a sine wave with amplitude I_{signal} is $I_{signal}^2/2$. The maximum signal amplitude at the input of the modulator is -3 dB (Sect. 5.2) compared to the DAC current levels, so $I_{signal} = I_{DAC}/\sqrt{2}$. The signal charge per sample period can be calculated to be:

$$Q_{signal}^2 = \frac{I_{DAC}^2 \cdot T_s^2}{4} \ [\text{C}^2] \tag{6.47}$$

The maximum SNR is the maximum signal power divided by the noise power in the signal band. Assuming that the noise power that is introduced by clock jitter is white, the maximum achievable SNR due to pulse width jitter is:

$$\begin{aligned} \text{SJNR}_{SI\ RTZ\ DAC}^{dB} &= 10 \cdot \log_{10}\left(\frac{Q_{signal}^2}{\sigma_q^2} \cdot \frac{f_s}{2 \cdot B}\right) \\ &= 10 \cdot \log_{10}\left(\frac{(1 - \text{RTZ})^2}{32\text{OSR}\sigma_s^2 B^2}\right) \ [\text{dB}] \end{aligned} \tag{6.48}$$

According to Eq. 6.48 and Eq. 14 in [30] (or Eq. 6.43 of the previous paragraph), the SNR of an SI DAC with a return-to-zero period RTZ is a factor $\sqrt{(1 - \text{RTZ})^2/2}$ lower compared to the SNR of a non-RTZ SI DAC.

It should be noted that the choice of a certain RTZ period can influence the clock's σ_s, as a very high or very low RTZ period, will require a higher clock frequency, and therefore the oscillator or PLL needs to generate a higher clock frequency. Therefore, for each RTZ value, the σ_s should be calculated at constant clock generator power consumption, to do an honest comparison.

6.9.1.3 CT 1-bit ΣΔ Modulator with RTZ SC DAC

The TAJE model predicts that if the influence of clock jitter on the pulse width of the DAC output pulse is eliminated, the SNR degradation due to clock jitter is also eliminated. As the output pulse of a switched capacitor feedback DAC has a decaying shape, the influence of clock jitter on the pulse width will be dependent on the settling of the DAC [41]. The feedback current in an SC DAC is shown in Fig. 6.33. The integrated feedback charge per clock cycle Q_{Ts} coming out of

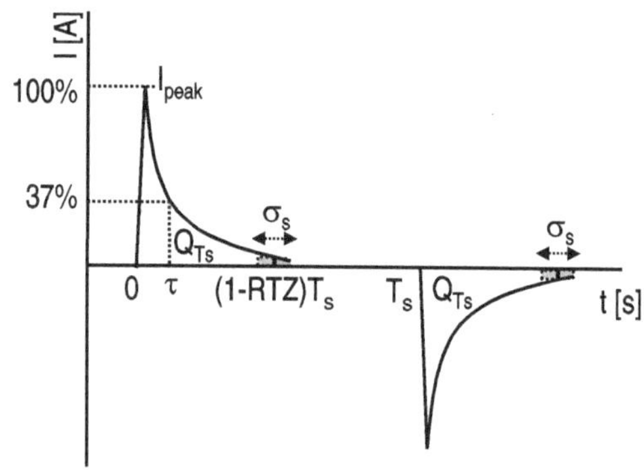

Figure 6.33: *Switched capacitor time signals with return-to-zero*

the DAC, is the same as in the switched current example ($Q_{Ts,\,SC} = Q_{Ts,\,SI}$), because the same amount of charge has to be fed back to the input to assure the same gain from input to output of the modulator. From the figure it can be seen that the error charge due to clock jitter is now dependent on the settling-time constant τ of the DAC. When it is assumed that $\sigma_s \ll T_s$ and $\sigma_s \ll \tau$, the variance of the error charge transferred per clock cycle can be approximated by:

$$\sigma_q^2 = \sigma_s^2 \cdot I_{peak}^2 \left(e^{\frac{-(1-\mathrm{RTZ})T_s}{\tau}} \right)^2 \quad [\mathrm{C}^2] \tag{6.49}$$

where $\tau = R \cdot C_{DAC}$. The resistance R via which the DAC capacitor C_{DAC} is discharged, is generally determined by the DAC switch resistances and the equivalent input impedance of the integrator stage in feedback configuration (Fig. 6.8). The peak current DAC output current I_{peak} is given by Eq. 6.21. From Fig. 6.33

the integrated feedback charge per clock period can be calculated:

$$Q_{T_s} = I_{peak} \int_{0}^{(1-\mathrm{RTZ})\cdot T_s} e^{\frac{-t}{\tau}} \cdot dt = \tau I_{peak} \cdot \left(1 - e^{\frac{-(1-\mathrm{RTZ})T_s}{\tau}}\right) \ [\mathrm{C}] \tag{6.50}$$

The signal charge per clock period Q_{signal} can be calculated in the same way Eq. 6.47 is derived:

$$Q_{signal}^2 = \frac{\tau^2 I_{peak}^2}{4} \cdot \left(1 - e^{-(1-\mathrm{RTZ})\alpha}\right) \ [\mathrm{C}^2] \tag{6.51}$$

with α defined as:

$$\alpha = \frac{T_s}{\tau} = \frac{1}{\tau \cdot f_s} \ [\text{-}] \tag{6.52}$$

which gives the number of settling time constants τ relative to T_s. According to the TAJE model, the SNR limitation due to jitter in a modulator with switched capacitor feedback can be calculated using Eqs. 6.49 and 6.51:

$$\mathrm{SJNR}_{SC\ RTZ\ DAC}^{dB} = 10 \log_{10} \left(\frac{1}{32\mathrm{OSR}\sigma_s^2 B^2} \left(\frac{e^{(1-\mathrm{RTZ})\ [\mathrm{dB}]\alpha} - 1}{\alpha} \right)^2 \right) \ [\mathrm{dB}] \tag{6.53}$$

6.9.1.4 The TAJE Model: SI Versus SC Feedback DAC

Equations 6.53 and 6.48 represent the SNR limitation for a modulator with 1-bit SC and 1-bit SI DAC respectively. Clocked with a white noise jittered clock, the SNR difference using a SC DAC over a SI DAC according to the TAJE model becomes:

$$\Delta\mathrm{SJNR}_{SC/SI}^{dB} = 20 \cdot \log_{10} \left(\left| \frac{\left(e^{(1-\mathrm{RTZ})\alpha} - 1\right)}{\alpha(1-\mathrm{RTZ})} \right| \right) \ [\mathrm{dB}] \tag{6.54}$$

According to the TAJE model, an SC DAC always outperforms an SI DAC. The improvement is only dependent on the multiplication of (1-RTZ) and α which describes the effective settling of the switched capacitor DAC. The SC DAC's performance dominance can also be seen intuitively, because when T_s is far larger than τ, α goes to infinity, and the switched capacitor current would settle completely. This would give an infinite improvement in achievable SNR, because the impact of jitter on the DAC output pulse width is eliminated completely. In Fig. 6.34 the result of Eq. 6.54 is plotted as a function of (1-RTZ) and α. In UMTS, CDMA2000 and GSM mode α is 8.4, 10, 12 respectively, and RTZ is 0.5 in all cases. In these three modes, the SNR improvement using a SC DAC instead of an SI DAC is 23.9, 29.4, and 36.5 dB respectively.

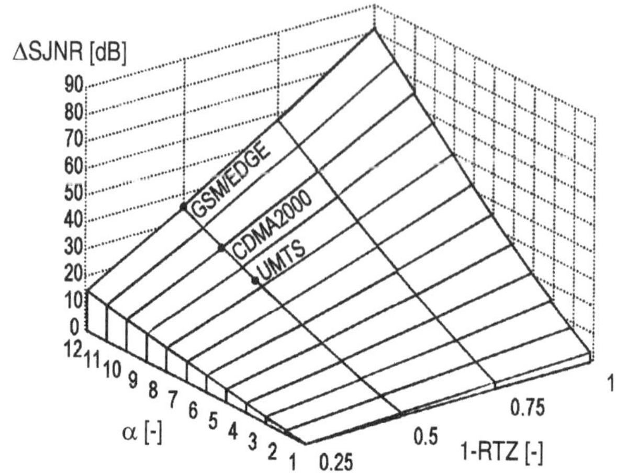

Figure 6.34: *SNR clock jitter limitation difference between an SC and an SI DAC as a function of (1-RTZ) and α according to the TAJE model*

6.9.1.5 TAJE Model Summary

Equations to calculate the impact of time jitter on the SNR of a ΣΔ modulator with SI (SR) or SC DAC (with or without RTZ period) have been derived using the TAJE model. Comparing these equations, the TAJE model predicts that a modulator with SC feedback DAC will be more robust to time jitter compared to modulators with an SI or an SR DAC, as the impact of a time error on the DACs output pulse is reduced by the decaying nature of an SC pulse.

Shortcomings of the TAJE model are:

- In the TAJE model, white noise time jitter is transformed into a random (white noise) amplitude or charge error, which is basically not what happens in reality; the clock jitter shifts the clock edges back and forth in time, which means the clock is phase modulated.
- The TAJE model assumes that the consecutive clock edges on the clock are uncorrelated, while for low jitter frequencies (compared to the clock frequency) the correlation is very strong.
- The TAJE model only considers DAC output pulse width errors. DAC output pulse position errors are neglected.
- In the TAJE model it is assumed that the in-band SNR limitation of time jitter on the clock is independent of the input signal amplitude and frequency

applied to the modulator, which from experience is known to be untrue (as will be shown in Sect. 6.9.2).

- In the TAJE model the spectral shape of the jitter on the clock is not taken into account, and white noise jitter on the clock is assumed always. However, the majority of the jitter on the clock coming from a PLL or oscillator very often is close to the clock carrier. This means that in the calculation of the impact of clock jitter on the SNR of the modulator due to clock jitter, the clock jitter's spectral shape should be taken into account.

Due to these short comings of the TAJE model, the outcome of this model should only be used as an approximation of the impact of time jitter on the modulator's SNR. In the rest of this book the TAJE model is discarded, and is replaced by the more precise TPJE model. In the TPJE model, the mechanism how clock jitter impacts the SNR of a $\Sigma\Delta$ modulator is represented more accurately. Although with limited effort Sects. 6.9.1.2 and 6.9.1.3 could be extended to multi-bit modulators, due to the short comings of the TAJE model, this extension is not further pursued.

6.9.2 The TPJE Model: Sine Wave Induced Jitter

In the TPJE model, first a single jitter noise component on the clock is modeled by a low frequency sine wave phase modulated on the modulator clock[1]. The transfer function of this phase modulated sine wave to the output spectrum of a $\Sigma\Delta$ modulator will be determined for different feedback DACs. After determining these transfer functions, the sine wave jitter will be replaced by noise in Sect. 6.9.3.

In Fig. 6.35 an ideal clock signal is shown. The clock transitions get timing errors when an ideal clock is modulated by a low frequency sine wave representing one frequency component of the jitter. With a slowly varying sine wave, the time deviation within one clock cycle is small, but the accumulated time deviation from the ideal positions measured over a period of time can be large. During the positive part of the modulating sine wave the clock transitions are delayed, during the negative part advanced. The clock is modulated by a sine wave with time error amplitude Δt and modulation frequency ω_m. The time axis of the ideal clock t and the jittered clock $t^{\sim}$ are related by:

$$t^{\sim} = t + \Delta t^{\sim} = t + \Delta t \cdot \sin(\omega_m \cdot t) \ \ [\mathrm{s}] \tag{6.55}$$

[1]Examples of phase modulation based clock jitter models were presented in [42–44]. The clock jitter model presented in Sects. 6.9.2 to 6.9.2.1.3 was presented in [44]. An extension on this model is presented in this book (Sects. 6.9.3.1 to 6.9.5.2). Part of this extension was already published in [45] and [46].

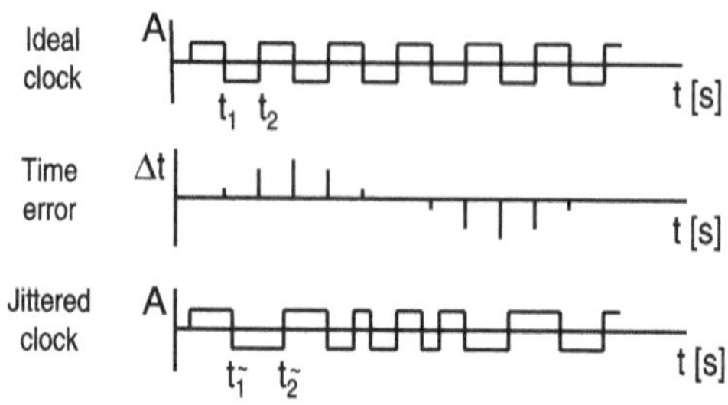

Figure 6.35: *Ideal clock, time dependent time error and jittered clock signal*

The clock is longitudinally modulated in time. This results in phase modulation (PM) of the clock, which mathematically is described by:

$$\begin{aligned} x^{\sim}(t) &= A_s \cdot \cos(\omega_s \cdot t^{\sim}) \\ x^{\sim}(t) &= A_s \cdot \cos(\omega_s \cdot t + \omega_s \cdot \Delta t \cdot \sin(\omega_m \cdot t)) \end{aligned} \tag{6.56}$$

where A_s is the clock carrier amplitude, and only the fundamental frequency of the clock is considered. When $\omega_s \Delta t \ll 1$, the higher order Bessel components can be neglected, which leads to:

$$x(t)^{\sim} \approx A_s \cdot \cos(\omega_s \cdot t) + \frac{A_s}{\Delta_s} \cdot \sin((\omega_s - \omega_m) \cdot t) - \frac{A_s}{\Delta_s} \cdot \sin((\omega_s + \omega_m) \cdot t) \tag{6.57}$$

in which

$$\Delta_s = \frac{2}{\omega_s \Delta t} \text{ [-]} \tag{6.58}$$

Figure 6.36 shows the frequency spectrum of the clock signal.

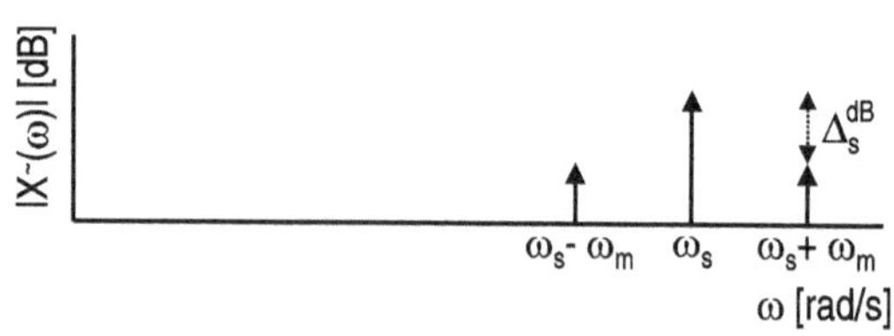

Figure 6.36: *Frequency spectrum of the sine wave induced jittered clock*

6.9.2.1 ΣΔ Modulator with SC DAC

In this section, the influence of clock jitter on the analog output of an SC DAC is investigated, using the clock model of the previous paragraph. As explained earlier, the spectrum coming out of the ΣΔ modulator will contain the effect of the clock jitter because the DAC is in a feedback loop. In a SC DAC, the sine wave induced jitter clock will cause amplitude and phase modulation of the signals passing the DAC.

6.9.2.1.1 Amplitude Modulation Firstly the amplitude modulation is investigated. The period of the jittered clock is determined by measuring the time between two consecutive rising edges. Assume that the time instants of two consecutive rising edges in the ideal case equal t_1 and t_2 where $t_2 = t_1 + T_s$. The corresponding time instants $t_1^{\sim}$ and $t_2^{\sim}$ of the jittered clock equal:

$$\begin{aligned} t_1^{\sim} &= t_1 + \Delta t \cdot \sin(\omega_m \cdot t_1) \ \text{[s]} \\ t_2^{\sim} &= t_1 + T_s + \Delta t \cdot \sin(\omega_m \cdot (t_1 + T_s)) \ \text{[s]} \end{aligned} \tag{6.59}$$

The instantaneous clock period $T_s^{\sim}$ of the jittered clock is found by calculating the difference between both time instants $T_s^{\sim} = t_2^{\sim} - t_1^{\sim}$. Using variable t instead of t_1, the instantaneous clock frequency $f_s^{\sim}$ is found by calculating $1/T_s^{\sim}$. For low modulation frequencies where ω_m is much smaller than ω_s, $\cos(\omega_m \cdot T_s) \approx 1$ and $\sin(\omega_m \cdot T_s) \approx \omega_m \cdot T_s$. Therefore, the instantaneous frequency of the jittered clock can be calculated to be:

$$f_s^{\sim} = f_s \cdot (1 - \omega_m \cdot \Delta t \cos(\omega_m \cdot t)) \ \text{[Hz]} \tag{6.60}$$

The amplitude of the output current of an SC DAC is proportional with the instantaneous frequency so its amplitude modulation for sine wave jitter equals:

$$I_{DAC}^{\sim} = I_{DAC} \cdot (1 - \omega_m \cdot \Delta t \cos(\omega_m \cdot t)) \ \text{[A]} \tag{6.61}$$

The amplitude modulation can also be explained intuitively. If the clock frequency is slowly increasing, more SC pulses will be generated per time and thus the amount of charge transferred per unit time is increasing. If the clock frequency is decreasing, the amount of charge transferred per time is decreasing. This means that the data coming out of the DAC is amplitude modulated by the phase modulated clock.

6.9.2.1.2 Phase Modulation Besides the observed amplitude modulation, the samples of the DAC are also clocked out at wrong time instants, which causes phase modulation of the output y. The ideal modulator output signal y is given by

$$y(t) = I_{DAC} \cdot \sin(\omega_i \cdot t) \tag{6.62}$$

when an input signal of frequency ω_i is applied to its input. The phase modulation become clear when writing the modulator output signal with frequency ω_i as a function of the jittered time axis variable $t^{\sim}$, using Eq. 6.55:

$$\begin{aligned} y(t) &= I_{DAC} \cdot \sin(\omega_i \cdot t) \\ y(t)^{\sim} &= I_{DAC}^{\sim} \cdot \sin(\omega_i \cdot (t^{\sim} - \Delta t \cdot \sin(\omega_m \cdot t))) \end{aligned} \tag{6.63}$$

Rewriting y to a function of $t^{\sim}$ and assuming $\omega_i \cdot \Delta t \ll 1$ yields:

$$y(t)^{\sim} \approx I_{DAC}^{\sim} \cdot \sin(\omega_i \cdot t) - I_{DAC}^{\sim} \cdot \omega_i \Delta t \cos(\omega_i \cdot t) \cdot \sin(\omega_m \cdot t) \tag{6.64}$$

in which the higher order Bessel components are neglected.

6.9.2.1.3 Combination of Amplitude and Phase Modulation The jittered output spectrum of the 1-bit SC DAC (or at the $\Sigma\Delta$ modulator output when the DAC is in the loop) is found by combining the amplitude modulation of Eq. 6.61 with the phase modulation of Eq. 6.64:

$$\begin{aligned} y(t)^{\sim} = I_{DAC} \cdot (\sin(\omega_i \cdot t) - \frac{\Delta t}{2}((\omega_i + \omega_m) \cdot \sin((\omega_i + \omega_m) \cdot t) \\ -(\omega_i - \omega_m) \cdot \sin((\omega_i - \omega_m) \cdot t))) \end{aligned} \tag{6.65}$$

in which the higher order product term is neglected. Combining Eqs. 6.65 and 6.58 results in:

$$\Delta_{i,+} = \frac{\omega_s \cdot \Delta_s}{\omega_i + \omega_m} \text{ and } \Delta_{i,-} = \frac{\omega_s \cdot \Delta_s}{\omega_i - \omega_m} \text{ [-]} \tag{6.66}$$

The sidebands in the output spectrum of the SC DAC have different amplitudes due to the combination of phase and amplitude modulation. Figure 6.37a shows the modulated output spectrum of the clock, and in Fig. 6.37b the output signal of the SC DAC (or $\Sigma\Delta$ modulator) is shown. Every frequency component coming out of the $\Sigma\Delta$ modulator has jitter components at $\omega_i \pm \omega_m$. The amplitude ratio

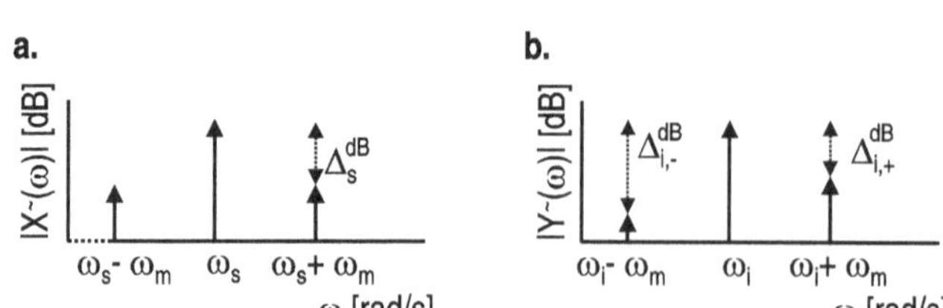

Figure 6.37: *Clock frequency spectrum (a) and SC DAC output frequency spectrum (b)*

of this frequency component to its jitter components is fixed and equal to $\Delta i_{\pm}$. Therefore, the jitter components scale proportionally with the amplitude of the wanted component at ω_i.

6.9.2.2 ΣΔ Modulator with SI DAC

In this paragraph the output spectrum of a switched current (SI) DAC is calculated, when clocked with the phase modulated clock described in Sect. 6.9.2.

6.9.2.2.1 Amplitude Modulation In Fig. 6.38 the ideal clock, the timing error, the jittered clock, and a 1-bit SI DAC output signal are shown. Instead of

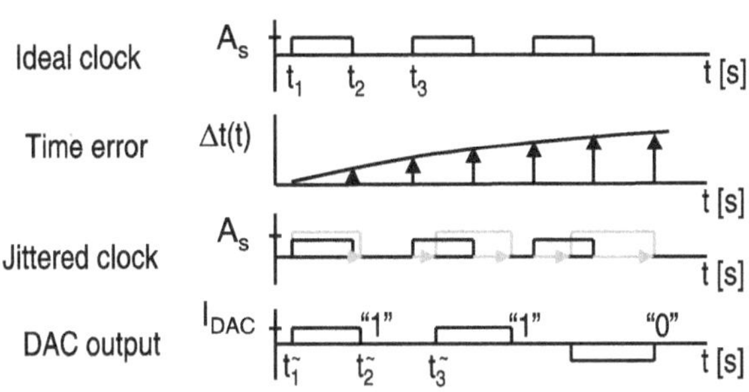

Figure 6.38: *Ideal clock, time dependent time error, jittered clock, and 1-bit SI DAC output signal*

being clocked at the exact time moments t, the DAC output is clocked at a jittered time moment $t^{\sim}$. For the three time instants $t_1^{\sim}$, $t_2^{\sim}$ and $t_3^{\sim}$ it can be written that:

$$\begin{aligned} t_1^{\sim} &= t_1 + \Delta t \cdot \sin(\omega_m \cdot t_1) \ \ [\mathrm{s}] \\ t_2^{\sim} &= t_2 + \Delta t \cdot \sin(\omega_m \cdot t_2) \ \ [\mathrm{s}] \\ t_3^{\sim} &= t_3 + \Delta t \cdot \sin(\omega_m \cdot t_3) \ \ [\mathrm{s}] \end{aligned} \tag{6.67}$$

The pulses at the output of the DAC can have an RTZ period, preventing any ISI which might be present in the signal when full-T pulses (Sect. 6.6.3) are used. To assure that the amount of charge per time period is independent of the DAC output pulse duty cycle, the amplitude of the DAC output pulse has to be multiplied by a factor 1/(1-RTZ). It can be written that:

$$\begin{aligned} t_2 - t_1 &= (1 - \mathrm{RTZ}) \cdot T_s \quad [\mathrm{s}] \\ t_3 - t_1 &= T_s \quad [\mathrm{s}] \end{aligned} \tag{6.68}$$

The jittered pulse width $T_p^{\sim}$ and clock period $T_s^{\sim}$ are:

$$\begin{aligned} T_p^{\sim} &= t_2^{\sim} - t_1^{\sim} = (1 - \text{RTZ}) \cdot T_s - t_{e1} \quad [\text{s}] \\ T_s^{\sim} &= t_3^{\sim} - t_1^{\sim} = T_s - t_{e2} \quad [\text{s}] \end{aligned} \tag{6.69}$$

with t_{e1} and t_{e2} given by:

$$\begin{aligned} t_{e1} &= \Delta t \cdot \sin(\omega_m \cdot (t_1 + (1 - \text{RTZ}) \cdot T_s)) - \Delta t \sin(\omega_m \cdot t_1) \quad [\text{s}] \\ t_{e2} &= \Delta t \cdot \sin(\omega_m \cdot (t_1 + T_s)) - \Delta t \sin(\omega_m \cdot t_1) \ [\text{s}] \end{aligned} \tag{6.70}$$

Assuming that $\omega_m \cdot T_s \ll 1$ and $\omega_m \ll \omega_s$ and neglecting higher order product terms, the output signal amplitude of the DAC (normalized to I_{DAC} in Fig. 6.38) can be written as:

$$\begin{aligned} I_{DAC}^{\sim} &= \frac{1}{(1 - \text{RTZ})} \cdot \frac{T_p^{\sim}}{T_s^{\sim}} I_{DAC} \\ &= \frac{1}{(1 - \text{RTZ})} \cdot \frac{(1 - \text{RTZ}) \cdot T_s \cdot (1 + \Delta t \cdot \omega_m \cdot \cos(\omega_m \cdot t))}{T_s \cdot (1 + \Delta t \cdot \omega_m \cdot \cos(\omega_m \cdot t))} I_{DAC} \\ &= I_{DAC} \ [\text{A}] \end{aligned} \tag{6.71}$$

This means that in an SI DAC using RTZ, there is no first order amplitude modulation. The absence of jitter-induced amplitude modulation in the output signal of an SI DAC can also be explained intuitively. In the upper time axis in Fig. 6.38 the ideal clock is illustrated. The second time axis represents the timing error $\Delta t \cdot \sin(\omega_m \cdot t)$, and the third axis illustrates the ideal clock together with the jittered clock. As can be seen from the third axis of the plot, the error on the instantaneous clock period introduced by the jitter on the clock, is a function of the derivative of the jittered clock. Therefore, the jitter on the DAC output pulse, is also a function of the derivative of the jitter on the clock. For $\omega_m \ll \omega_s$, two consecutive DAC output edges shift almost an equally amount of time, and the charge per time coming out of the DAC is assumed to be constant. Therefore, no amplitude modulation will be present in the output spectrum of an SI DAC.

6.9.2.2.2 Phase Modulation The phase modulation in an SC and an SI DAC is not different. Both pulses are clocked out at the wrong moment. Therefore Eq. 6.64 still holds, and with the help of Eq. 6.58 can be rewritten to:

$$\begin{aligned} y^{\sim}(t) = I_{DAC}^{\sim} \Bigg(&\sin(\omega_i t) - \frac{\omega_i}{\omega_s \Delta_s} \sin((\omega_i + \omega_m)t) \\ &+ \frac{\omega_i}{\omega_s \Delta_s} \sin((\omega_i - \omega_m)t) \Bigg) \end{aligned} \tag{6.72}$$

6.9.2.2.3 Combination of Amplitude and Phase Modulation The combination of the amplitude and phase modulation in a 1-bit SI DAC is equal to Eq. 6.72 because there is no amplitude modulation (Eq. 6.71). The distance from the signal to its jitter components is

$$\Delta_{i,+} = \Delta_{i,-} = \frac{\omega_s \cdot \Delta_s}{\omega_i} \quad [-] \tag{6.73}$$

Figure 6.39a shows the modulated output spectrum of the clock, and in Fig. 6.39b the output signal of the SI DAC (or ΣΔ modulator) is shown. The output spectrum

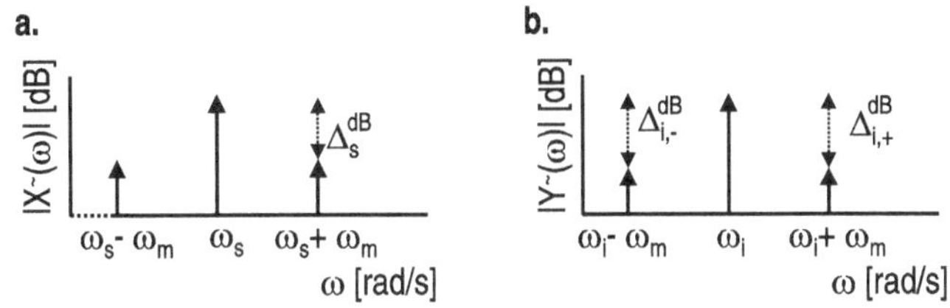

Figure 6.39: *Clock frequency spectrum and SI DAC output frequency spectrum*

of the 1-bit SI DAC is comparable to the spectrum in the SC case, except for the difference that the two spurious tones around the signal are now of the same amplitude.

6.9.2.3 Application of the Sine Wave Induced Jitter Model

If an in-band signal is applied to the modulator, this input signal will contain the side band components around its carrier. The level of the side band components for a modulator with an SC or an SI DAC can be calculated with Eqs. 6.66 and 6.73 respectively. The distance from DAC output signal carrier to its side bands is fixed to $\Delta_{i,\pm}^{dB}$, and is independent of signal amplitude. At decreasing carrier signal amplitudes, the jitter side bands will gradually disappear below the other noise (circuit and quantization noise) present in the ΣΔ modulator. This is shown by Fig. 6.40a and b for a modulator with SI DAC.

The in-band jitter component do not necessarily have to come from an in-band signal. If an interferer is just next to the modulator's bandwidth B_{ADC}, its left jitter side band can fall in the modulator's bandwidth. In Fig. 6.40c an out-of-band interferer is present at the input of the modulator. In case of narrow-band jitter, the left side band does not fall into the signal bandwidth as long as $f_{int} - f_m > B_{ADC}$.

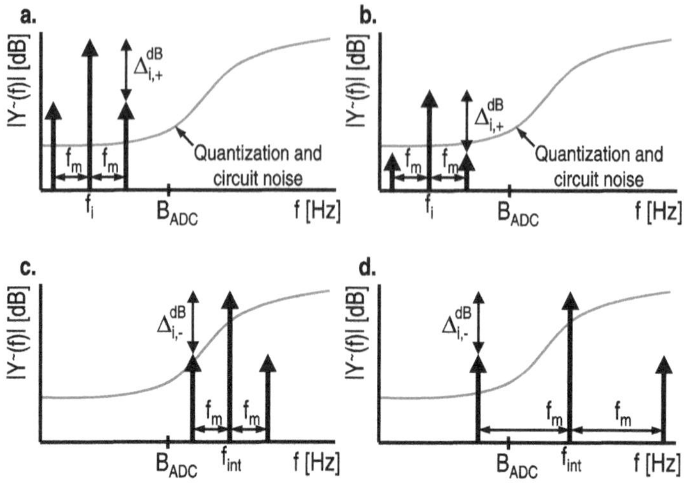

Figure 6.40: *Amplitude and frequency of jitter components as a function of amplitude and frequency of modulator input signal*

At lower interferer frequencies f_{int} and/or larger modulation frequencies f_m, the left side band jitter component might fall into signal bandwidth of the modulator.

It is important to realize that if an interferer at the input of the modulator has a frequency which is attenuated by the modulator's STF, its jitter components visible int the output spectrum of the modulator will go down with the amount of carrier attenuation.

6.9.2.4 Verification of the TPJE Model with Sine Wave Induced Clock Jitter

In this section measurements are presented which are done on three different continuous time $\Sigma\Delta$ modulators, to verify the TPJE model presented in the previous sections. In these measurements the clock is modulated with a sine wave. The combination of frequencies of the input signal and sine wave jitter are chosen such that at least one of the sine wave induced jitter components falls inside the signal bandwidth.

The first modulator is a triple mode 5th order low-pass $\Sigma\Delta$ modulator with 1-bit SC DAC [41], of which the implementation details will be shown in Sect. 9.1.2. The RTZ period of the modulator's SC DAC is half the clock period. The modulator clock frequency f_s is 26 MHz. The input signal frequency f_i is 100 kHz and modulation frequency f_m is 50 kHz. The clock spectrum and $\Sigma\Delta$ modulator output spectrum are shown in Fig. 6.41a and b (note the logarithmic scale in Fig. 6.41b). The spectrum shows the effect of the amplitude modulation predicted by Eq. 6.66. The expected jitter component level is also shown in the figure as a reference. The f/f_s relation is clearly visible. The distance between clock carrier

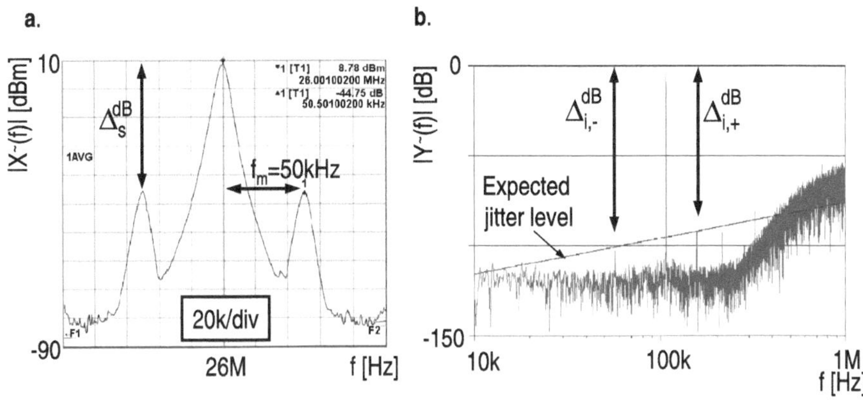

Figure 6.41: *Clock spectrum (a) and ΣΔ modulator output spectrum (b)*

to its jitter components Δ_s is 44.8 dBc. Using Eq. 6.66 it can be calculated that for the jitter components around the wanted signal at frequency f_i, this distance increases by 54.3 dB for the left component at $f_i - f_m$, and by 44.8 dB for the right component at $f_i + f_m$, yielding a $\Delta_{i,-}^{dB}$ and $\Delta_{i,+}^{dB}$ of 99.1 and 89.6 dB respectively. Compared to the measured distances of 100.4 and 89.6 dB, this gives an error of 1.3 and 0 dB respectively. The measurement results together with the outcome of Eq. 6.66 are shown in Table 6.2.

5th order low-pass ΣΔ modulator with 1-bit SC DAC, $f_i = 100$ kHz [41]					
Component freq. [Hz]	Δ_s^{dB} [dBc]	Calculated $\frac{f_s}{f_i \pm f_m}$ [dBc]	Expected $\Delta_{i,\pm}^{dB}$ [dBc]	Measured $\Delta_{i,\pm}^{dB}$ [dBc]	Error [dB]
$f_i - f_m$	44.8	54.3	99.1	100.4	1.3
$f_i + f_m$	44.8	44.8	89.6	89.6	0

Table 6.2: *Calculated and measured sine wave induced jitter performance of a ΣΔ modulator with 1-bit SC DAC*

The same modulator is also tested with an out-of-band input signal f_i of 1.6 MHz. Modulation frequency f_m in this case is 1.5 MHz, which yields an in-band jitter component at 100 kHz. The clock spectrum and ΣΔ modulator output spectrum are shown in Fig. 6.42a and b. Note that the input signal is slightly smaller compared to the first measurement due to the modulator out-of-band filtering. The in-band jitter components due to the phase modulated sampling of the input signal are expected to go down accordingly. The calculation and measurement in

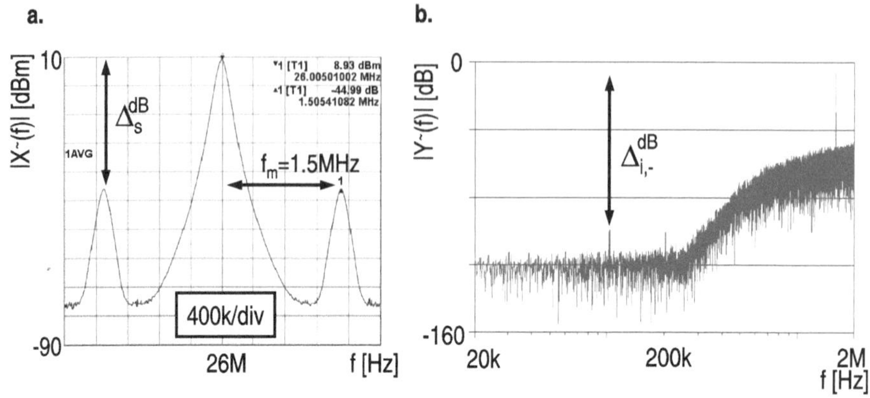

Figure 6.42: *Clock spectrum (a) and ΣΔ modulator output spectrum (b)*

Table 6.3 proves that the ratio between input signal level and its jitter component level indeed is as expected.

5th order low-pass ΣΔ modulator with 1-bit SC DAC, $f_i = 1.6$ MHz [41]					
Component at freq. [Hz]	Δ_s^{dB} [dBc]	Calculated $\frac{f_s}{f_i \pm f_m}$ [dBc]	Expected $\Delta_{i,\pm}^{dB}$ [dBc]	Measured $\Delta_{i,\pm}^{dB}$ [dBc]	Error [dB]
$f_i - f_m$	45	48.3	93.3	92.4	-0.9
$f_i + f_m$	45	18.5	63.3	x	x

Table 6.3: *Calculated and measured sine wave induced jitter performance of a ΣΔ modulator with 1-bit SC DAC*

An I&Q measurement of the complex modulator (Fig. 6.43a and b) shows the behavior of the modulator when used in an I&Q configuration. The input signal component and its jitter components are on the right side of the spectrum, and the jitter components have the same distance to the input signal compared to the first measurement presented in Fig. 6.41b. The image at $-f_i$ is suppressed by the IRR of the I&Q modulator. The jitter components phase modulated on this image are scaled accordingly, and are not visible in the spectrum as these components are below the thermal and quantization noise of the I&Q modulator. This is supported by the calculations presented in Appendix D. The HD2 and HD3 are also indicated in the spectrum. As explained in Appendix A, the HD2 is visible on both sides of the spectrum, where the HD3 is only visible on the left side of the spectrum.

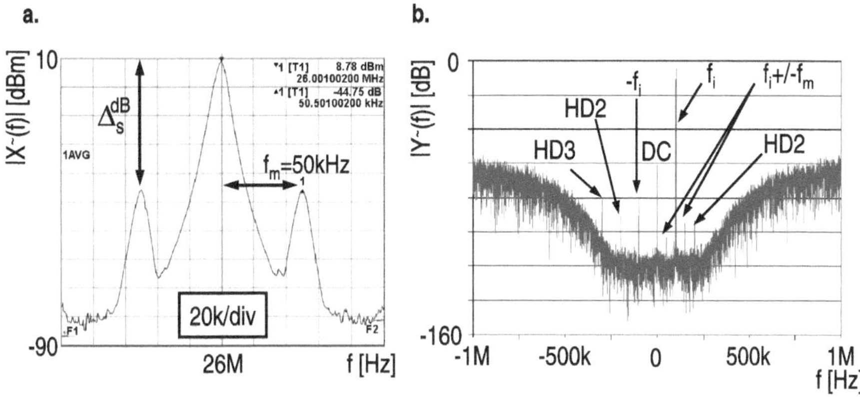

Figure 6.43: *Clock spectrum (a) and complex ΣΔ modulator output spectrum (b)*

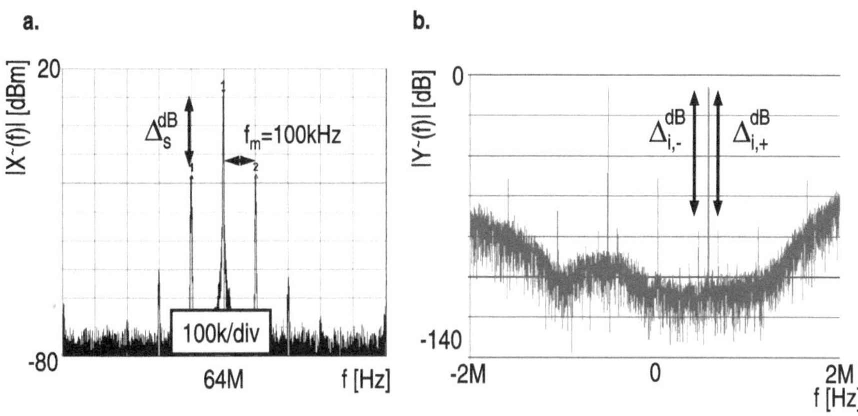

Figure 6.44: *Clock spectrum (a) and ΣΔ modulator output spectrum (b)*

The second DUT is a 4th order complex ΣΔ modulator with 1-bit SI DAC [47]. The RTZ period of the modulator's SI DAC is half the clock period. The modulator clock frequency f_s of 64 MHz is phase modulated with a modulation frequency f_m of 100 kHz. The I&Q modulator input signal is non-complex meaning that both the I&Q modulator have an input signal of the same phase. Input signal frequency f_i is 530 kHz. The clock spectrum and ΣΔ modulator output spectrum are shown in Fig. 6.44a and b respectively. Note that in this case, the higher order Bessel components are visible in both the clock and ΣΔ modulator output spectrum. Due to the fact that a non-complex input signal was used, the input signal and jitter components have equal amplitude on both sides of the complex spectrum. If a perfect I&Q modulator input signal would be used, the input signal and its jitter components in this case would be visible either at the positive or at the negative side of the spectrum (Appendix D), as $f_i - f_m > 0$. Furthermore, it can be seen that the modulator RTZ SI DAC indeed does not have the amplitude modulation ($\Delta i, + = \Delta_{i,-}$), like Eq. 6.73 predicted. Using Eq. 6.73, the expected $\Delta_{i,\pm}$ can be calculated. The measurement results together with the calculations are shown in Table 6.4.

4th order complex ΣΔ modulator with SI 1-bit DAC [47]					
Component at freq. [Hz]	Δ_s^{dB} [dBc]	Calculated $\frac{f_s}{f_i}$ [dBc]	Expected $\Delta_{i,\pm}^{dB}$ [dBc]	Measured $\Delta_{i,\pm}^{dB}$ [dBc]	Error [dB]
$f_i - f_m$	28.4	41.6	70	71.4	1.4
$f_i + f_m$	28.4	41.6	70	70.9	0.9

Table 6.4: *Calculated and measured sine wave induced jitter performance of an I and Q ΣΔ modulator with 1-bit SI DAC*

The last test object is a 2–2 MASH ΣΔ modulator with 4-bit SI DACs with an RTZ period of zero [23]. The clock frequency f_s is 160 MHz, the frequency of modulation f_m is 250 kHz, and the input signal frequency f_i is 1 MHz. The clock spectrum and ΣΔ modulator output spectrum are shown in Fig. 6.45a and b respectively. Note that there is no (first order) amplitude modulation of the jitter components like expected ($\Delta i, + = \Delta_{i,-}$). The measurement results together with the outcome of Eq. 6.73 are shown in Table 6.5.

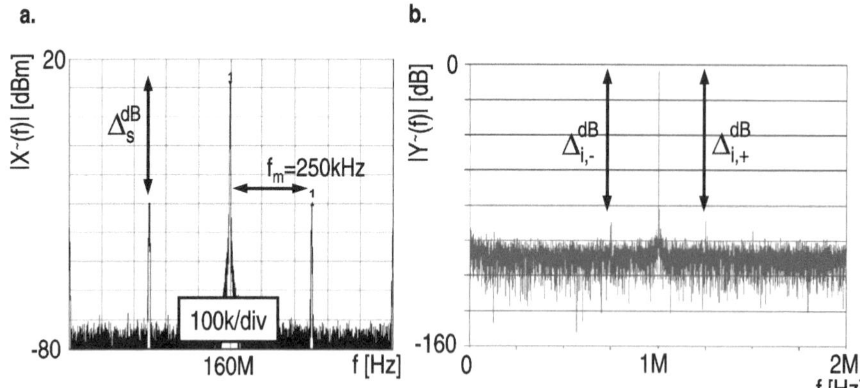

Figure 6.45: *Clock spectrum (a) and ΣΔ modulator output spectrum (b)*

2–2 MASH ΣΔ modulator with 4-bit SI DAC [23]					
Component at freq. [Hz]	Δs^{dB} [dBc]	Calculated $\frac{f_s}{f_i}$ [dBc]	Expected $\Delta i_{\pm}^{dB}$ [dBc]	Measured $\Delta i_{\pm}^{dB}$ [dBc]	Error [dB]
$f_i - f_m$	40.3	44.1	84.4	86	1.6
$f_i + f_m$	40.3	44.1	84.4	85.3	0.9

Table 6.5: *Calculated and measured sine wave induced jitter performance of a ΣΔ modulator with multi-bit SI DAC*

From Tables 6.2 to 6.5, it can be concluded that the calculations and measurements for all modulators match within 1.6 dB. Furthermore, it is proven that the same equations hold for both single and multi-bit modulators.

6.9.3 The TPJE Model: Substitution of White Noise Jitter in the Sine Wave Induced Jitter Model

From the transfer function of a single frequency jitter component on the clock to the output spectrum of the feedback DAC, the SNR limitation of a modulator clocked with a clock, which is white noise phase modulated, can be calculated. To achieve this, the single frequency jitter component is replaced by white noise jitter.

6.9.3.1 ΣΔ Modulator with SC DAC

The DAC clock spectrum is drawn in Fig. 6.46a, and the expected DAC output spectrum for two different input signal frequencies are drawn in Fig. 6.46b and c. The noise level at frequency f_1 is determined by the amount of white noise on

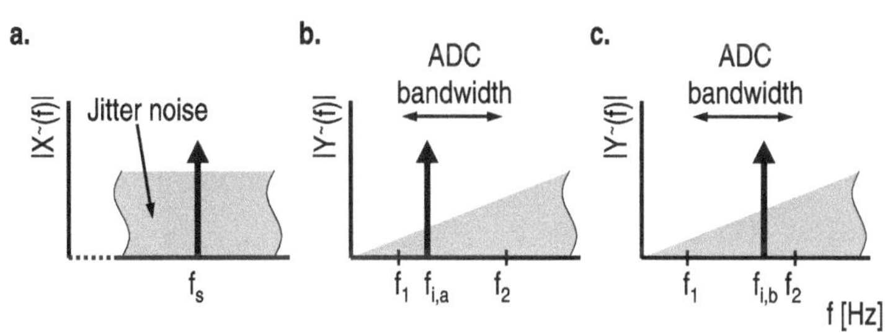

Figure 6.46: *SC DAC output spectrum when the jitter on the clock is white noise*

the clock times the ratio $(f_i - f_m)/f_s$. This ratio can be rewritten to f_1/f_s, and is the same for Fig. 6.46b and c. Therefore, the shape and level of the jitter is

independent of the input signal frequency f_i (Eq. 6.66). Furthermore, this means that the noise at the output of the modulator has the shape f/f_s.
The carrier to noise ratio of the clock is described by:

$$\mathrm{CNR}^2_{clk}(f) = \frac{A^2_{clk}}{\int S_{clk}(f)df} \quad [\mathrm{Hz}] \tag{6.74}$$

The SNR of a ΣΔ modulator with SC DAC will be limited by the convolution of the jitter (J) noise and the input signal (S). For a modulator with a SC DAC, with a single frequency sine wave at its input and clocked with a white noise phase modulated jitter clock, the signal-to-jitter-noise-ratio ($\mathrm{SJNR}_{JS,SC}$) can be calculated using Eq. 6.66:

$$\mathrm{SJNR}^{dB}_{JS,SC} = 10\log_{10}\left(\frac{A^2_{clk}}{\int_{f_1}^{f_2} S_{clk}(f)(\frac{f}{f_s})^2 df}\right) \quad [\mathrm{dB}] \tag{6.75}$$

Note that the $\mathrm{SJNR}_{JS,SC}$ will be the same for a single and multi-bit modulator, if the same input signal power is assumed.

Normally, the spectral noise density $S_{clk}(f)$ of the output clocks of oscillators or phase locked loops, is a function of f_m. For sake of simplicity, the clock spectrum for now is assumed to be white. Therefore, $S_{clk}(f)$ becomes S_{clk}, and Eq. 6.75 can be rewritten to:

$$\begin{aligned}\mathrm{SJNR}^{dB}_{JS,SC} &= 10\log_{10}\left(\mathrm{CNR}^2_{clk}\right) - 10\log_{10}\left(\int_{f_1}^{f_2}\left(\frac{f}{f_s}\right)^2 df\right) \\ &= 10\log_{10}\left(\mathrm{CNR}^2_{clk}\right) - 10\log_{10}\left(\frac{f_2^3 - f_1^3}{3f_s^2}\right) \quad [\mathrm{dB}]\end{aligned} \tag{6.76}$$

Note that the $\mathrm{SJNR}_{JS,SC}$ is relative to the input signal amplitude, and will not change at smaller input amplitudes, as the jitter noise goes down with its signal carrier amplitude.

Equation 6.76 describes the SNR limitation for a modulator with SC DAC with a single sine wave at its input. As said earlier, the jitter level around a single-frequency input signal applied to a ΣΔ modulator with either a 1-bit or a multi-bit DAC is the same. This is illustrated in Fig. 6.47. An input signal of frequency f_i, is equally large for the 1-bit and multi-bit modulator. Therefore, the expected jitter level around this input signal is equal. However, in a ΣΔ modulator, due to its noise shaping function, significant quantization noise power is present out-of-band. An individual quantization noise frequency component f_Q can cause

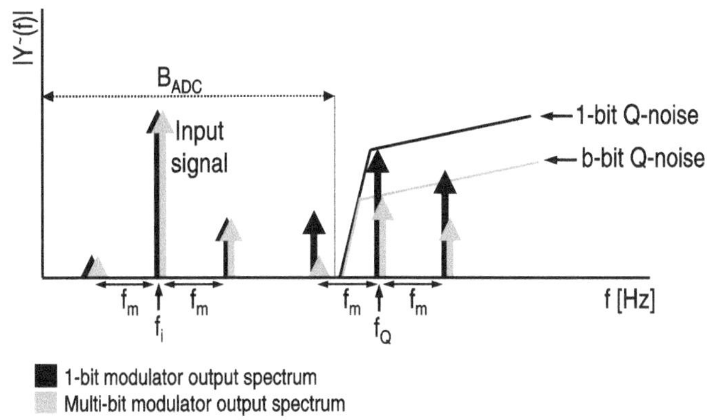

Figure 6.47: *Jitter in a 1-bit and multi-bit modulator*

in-band jitter noise at $f_Q - f_m$ (Fig. 6.47) due to cross modulation of jitter and quantization noise. In a multi-bit modulator the quantization noise power at the output of the modulator is lower compared to a 1-bit modulator. Therefore, the jitter components related to the convolution of jitter and quantization noise, will be lower. As explained earlier, the in-band jitter is independent of signal frequency (or quantization noise shape) and only is dependent on its power. If a 1-bit modulator is assumed, which toggles between $+1$ and -1, its total output power is 1. If there is no input signal applied to the modulator, this power will be quantization noise only. In a b-bit modulator with N quantization levels, the quantization noise power reduces by $(N-1)^2$ or $(2^b-1)^2$. Therefore, the maximum in-band SJNR due to the convolution of jitter (J) and quantization (Q) noise for a b-bit modulator with SC DAC can be calculated to be:

$$\begin{aligned}\mathrm{SJNR}^{dB}_{JQ,SC} &= 10\log_{10}\left(\mathrm{CNR}^2_{clk}\right) - 10\log_{10}\left(\frac{f_2^3 - f_1^3}{3f_s^2}\right) \\ &\quad - 10\log_{10}((N-1)^2) - 10\log_{10}\left(\left(\frac{(N-1)}{0.7+N-2}\right)^2\right) \quad [\mathrm{dB}]\end{aligned} \tag{6.77}$$

Note that the SJNR is related to full scale input signal power (and not to the quantization noise power) and hence the maximum input signal scaling factor is introduced (Eq. 5.8). As the maximum input power of a 2 level modulator is 0.5, this scaling factor becomes -3 dB in Eq. 6.77.

In reality, the in-band jitter will be a combination of $\mathrm{SJNR}_{JS,SC}$ and $\mathrm{SJNR}_{JQ,SC}$. For a single bit modulator the maximum input signal power is 0.5. Therefore, at maximum input signal the power of the quantization noise reduces to 0.5, as the total power in the output spectrum is of a ± 1 toggling modulator is 1. For a multi-bit modulator the quantization noise power is $(N-1)^2$ times lower. Therefore, its maximum input signal power is slightly higher as Eq. 5.8 already predicted, which means a slightly better SJNR_{JQ} for multi-bit modulators.

Because the jitter on the clock is assumed to be white, Eq. 6.76 can be rewritten to an equation in which the noise of the clock is described with a certain variance σ_s only. The CNR of the clock can be written as:

$$\text{CNR}^2_{clk} = \frac{2\sqrt{2}^2 \sigma_s^2 \omega_s^2}{2^2 \Delta B} \text{ [Hz]} \tag{6.78}$$

realizing that a modulating sine wave with a frequency f_m and a time amplitude Δt, causes two side bands on the clock with an amplitude Δ_s (Eq. 6.58) times lower than the carrier. If $\Delta B \gg f_s$, and using Eq. 6.76 and 6.78, for a single frequency input signal one can write that:

$$\begin{aligned} \text{SJNR}^{dB}_{JS,SC} &= -10 \cdot \log_{10}\left(\int_{f_1}^{f_2} \left(\frac{4\pi^2 \sigma_s^2 f_s^2}{\frac{1}{2} f_s} \right) \left(\frac{f}{f_s} \right)^2 df \right) \\ &= 10 \cdot \log_{10}\left(\frac{3 f_s}{8\pi^2 \sigma_s^2 \left(f_2^3 - f_1^3 \right)} \right) \text{ [dB]} \end{aligned} \tag{6.79}$$

The author discourages the reader to use the σ_s based equation for calculation of modulator jitter sensitivity, as the maximum achievable SJNR is heavily dependent on the spectral shape of the clock. The σ_s based equation is only mentioned for completeness. A similar equation can be derived for modulators with SC DAC that have significant out-of-band quantization noise at their output, by rewriting Eq. 6.77 using Eq. 6.78.

6.9.3.2 ΣΔ Modulator with SI DAC

From the transfer function of a single frequency jitter component on the clock to the output spectrum of the SI feedback DAC, the SNR limitation for a sine wave DAC input signal can be calculated. The single frequency jitter component is replaced by white noise jitter. The DAC clock spectrum is drawn in Fig. 6.48a, and the expected DAC output spectrum for two different input signal frequencies are drawn in Fig. 6.48b and c. The level of jitter around the signal is now dependent on the input signal frequency (Eq. 6.73). This is shown in Fig. 6.48. The jitter on the

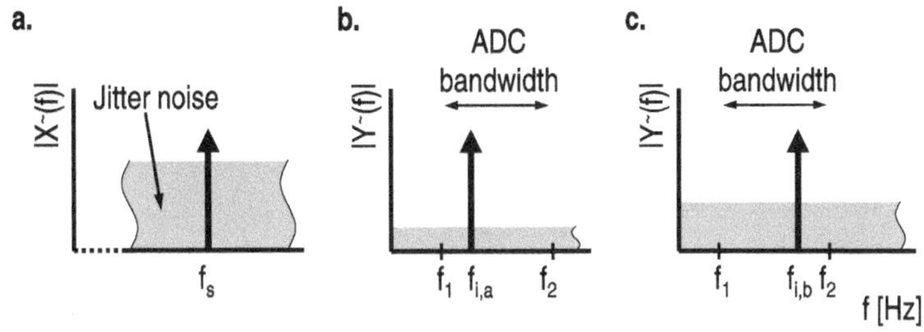

Figure 6.48: *SI DAC output spectrum when the jitter on the clock is white noise*

clock is present around the input signal which is applied to the DAC. The input signal in Fig. 6.48a is of lower frequency compared to the signal in Fig. 6.48b. Therefore, according to Eq. 6.73, the jitter level around the signal of frequency $f_{i,a}$ will be lower than the jitter level around the signal with frequency $f_{i,b}$. For a given input signal frequency f_i at the input of the $\Sigma\Delta$ modulator with SI DAC, the amplitude of the sideband components is independent of the jitter frequency f_m, thus white noise jitter on the clock will cause white noise jitter in the output spectrum of the DAC. This means that for a certain input signal frequency f_i, the jitter (J) noise and input signal (S) convolution products that appear in the signal bandwidth for a $\Sigma\Delta$ modulator with SI DAC, can be calculated by integrating the white noise in the $\Sigma\Delta$ modulator's bandwidth ($f_2 - f_1$). To calculate the $\text{SJNR}_{JS,SI}$, Eq. 6.75 is changed into:

$$\begin{aligned} \text{SJNR}_{JS,SI}^{dB} &= 10 \cdot \log_{10}\left(\text{CNR}_{clk}^2\right) - 10\log_{10}\left(\left(\frac{f_i}{f_s}\right)^2 \int_{f_1}^{f_2} df\right) \\ &= 10 \cdot \log_{10}\left(\text{CNR}_{clk}^2\right) - 10\log_{10}\left(\left(\frac{f_i}{f_s}\right)^2 (f_2 - f_1)\right) \quad \text{[dB]} \end{aligned} \tag{6.80}$$

Unlike in a modulator with SC feedback DAC where the in-band jitter components are only dependent on the total input signal power, the in-band jitter of a modulator with SI feedback DAC is dependent on input signal power *and* its spectral shape. Therefore, the limitation on the maximum achievable SNR due to in-band jitter (J) and quantization (Q) noise convolution products, the SJNR_{JQ}, is much harder to calculate for a b-bit modulator with SI DAC. To simplify the calculation it is assumed that the majority of the power in the output spectrum of the modulator is at high frequencies, which is justified by the fact that the modulator tries to shape the quantization noise to higher frequencies, and is idling at $\frac{1}{2}f_s$. Therefore, using

Eq. 6.80 it can be calculated that the maximum achievable SNR, limited by the convolution of quantization and jitter noise for an N-level (or b-bit) modulator becomes:

$$\begin{aligned}\mathrm{SJNR}^{dB}_{JQ,SI} &= 10\cdot\log_{10}\left(\mathrm{CNR}^2_{clk}\right) - 10\log_{10}\left(\left(\frac{1}{2}\right)^2(f_2-f_1)\right)\\ &\quad + 10\log_{10}\left((N-1)^2\right) - 10\log_{10}\left(\left(\frac{(N-1)}{0.7+N-2}\right)^2\right)\\ &= 10\cdot\log_{10}\left(\mathrm{CNR}^2_{clk}\right) - 10\log_{10}(f_2-f_1)\\ &\quad + 10\log_{10}\left((N-1)^2\right) - 10\log_{10}\left(\left(\frac{(N-1)}{0.7+N-2}\right)^2\right) + 6 \ [\mathrm{dB}]\end{aligned} \tag{6.81}$$

As white noise jitter on a clock can be described by a variance σ_s, the SNR limitation of a modulator with SI DAC for a single frequency input signal can be described by:

$$\mathrm{SJNR}^{dB}_{JS,SI} = 10\cdot\log_{10}\left(\frac{f_s}{8\pi^2\sigma_s^2 f_i^2(f_2-f_1)}\right) \ [\mathrm{dB}] \tag{6.82}$$

using Eqs. 6.76 and 6.78. If $f_1 = 0$ and f_2 is $\frac{1}{2}f_s$, this transforms in the well known SNR jitter limitation frequently used for Nyquist ADCs:

$$\mathrm{SJNR}^{dB} = 10\cdot\log_{10}\left(\frac{1}{4\pi^2\sigma_s^2 f_i^2}\right) \ [\mathrm{dB}] \tag{6.83}$$

Again the author discourages the reader to use these σ_s based equations. Therefore, the equations which describe the convolution of jitter and quantization noise SNR limit are not derived for a modulator with significant out-of-band quantization noise.

6.9.3.3 Verification of the TPJE Model with White Noise Induced Clock Jitter

In this section measurements are presented which are done on two different continuous time ΣΔ modulators, to verify the TPJE model when the clock is modulated with white noise.

The first modulator is a triple mode 5th order low-pass ΣΔ modulator with 1-bit SC DAC of [41] of which the implementation details are presented in Sect. 9.1.2. The modulator clock frequency f_s is 76.8 MHz in CDMA mode. The input signal frequency f_i is 400 kHz. The clock is modulated with white noise which is bandwidth limited to about ±100 kHz. The clock spectrum is shown in Fig. 6.49a. The clock carrier amplitude is 8.5 dBm; within the ±100 kHz bandwidth the noise is at −36 dBm (RBW=1 kHz), which leads to a CNR_{clk}^{dB} of 74.5 dBc$\sqrt{\mathrm{Hz}}$ in the ±100 kHz bandwidth. Outside the bandwidth the CNR_{clk} is much higher. In Fig. 6.49b the modulator output spectrum is plotted. As expected with a signal frequency of 400 kHz, the jitter in the modulator output spectrum is visible in a bandwidth of 300 kHz to 500 kHz. The jitter-quantization noise convolution products that appear in the signal bandwidth are insignificant, as the jitter noise bandwidth is limited to ±100 kHz. Using Eq. 6.76 with $f_1 = 300$ kHz and

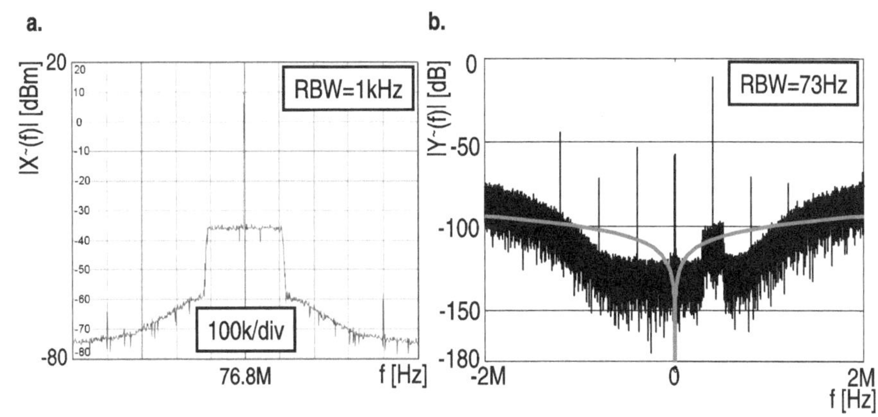

Figure 6.49: *Clock spectrum (a) and ΣΔ modulator output spectrum (black) and expected jitter shaping (grey) (b)*

$f_2 = 500$ kHz, the expected SNR at the output of the modulator can be calculated, which gives $74.5 - 7.4 = 67.1$ dB. The measured SNR is 67.9 dB in the 300 k to 500 kHz bandwidth, which gives only an error of 0.8 dB compared to the calculated value. In this measurement the jitter noise is dominant above the other noise sources (quantization and thermal noise), and therefore these are neglected. The jitter noise in the 200 kHz bandwidth around the input signal in the output spectrum of the modulator follows the f/f_s relation as expected. This relation is represented by the grey line in Fig. 6.49b (the absolute level of this line is chosen arbitrary).

Figure 6.50 shows a measurement of the triple mode modulator in UMTS mode. Input signal frequency is 500 kHz. The jitter bandwidth is limited to ± 1.2 MHz and is modulated on a 153.6 MHz clock. The clock carrier amplitude is 8.5 dBm, and the noise in the ± 1.2 MHz bandwidth is at -39 dBm (RBW=2 kHz). This leads to a CNR_{clk} of 80.5 dBc$\sqrt{\mathrm{Hz}}$. By applying Eq. 6.76 two times, once for

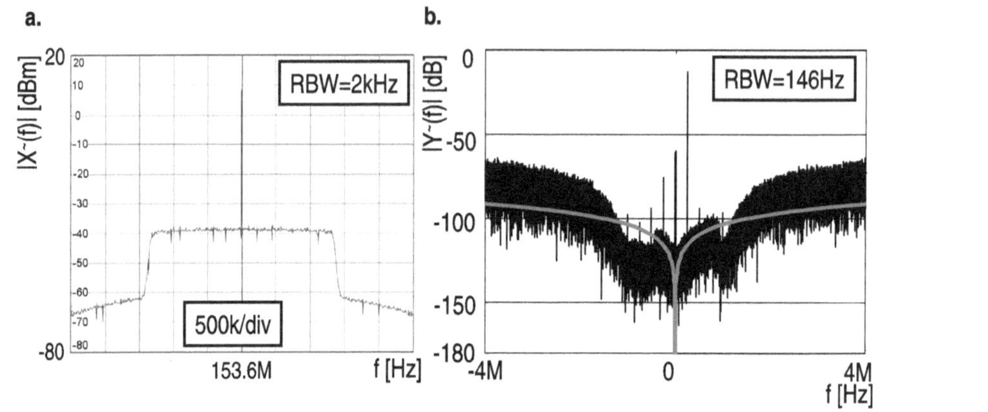

Figure 6.50: *Clock spectrum (a) and ΣΔ modulator output spectrum (black) and expected jitter shaping (grey) (b)*

the left side (-700 kHz to 0 Hz) and once for the right side (0 Hz to 1.7 MHz) of the spectrum, the expected SNR can be calculated, which gives $80.5 - 18.7 = 61.8$ dB. The measured SNR in the -700 kHz to 1.7 MHz bandwidth is 62.6 dB, which is only 0.8 dB different from calculation. In the plot the f/f_s relation (grey line) between jitter on the clock and jitter in the output spectrum is clearly visible. The in-band power expected from the jitter-quantization noise convolution products is limited due to the limited bandwidth of the jitter on the clock, and therefore is neglected, as is the quantization and circuit noise.

In Fig. 6.51 the in-band jitter-quantization noise convolution product power is measured for the triple mode modulator in CDMA mode. The black spectrum in Fig. 6.51b shows the modulator output spectrum when clocked with a very low jitter clock (clock spectrum is not shown), and no significant in-band jitter is expected. The grey spectrum shows the modulator output spectrum when it is clocked with a clock of which the spectrum is shown in Fig. 6.51a. The jitter bandwidth is limited to ± 50 MHz and is modulated on a 76.8 MHz clock. The clock carrier amplitude is 8.7 dBm, and within the ± 50 MHz the noise is at -42 dBm (RBW=50 kHz), which leads to a CNR_{clk} of 97.7 dBc$\sqrt{\mathrm{Hz}}$. Using two times Eq. 6.77 for the positive and negative frequency band, it can be calcu-

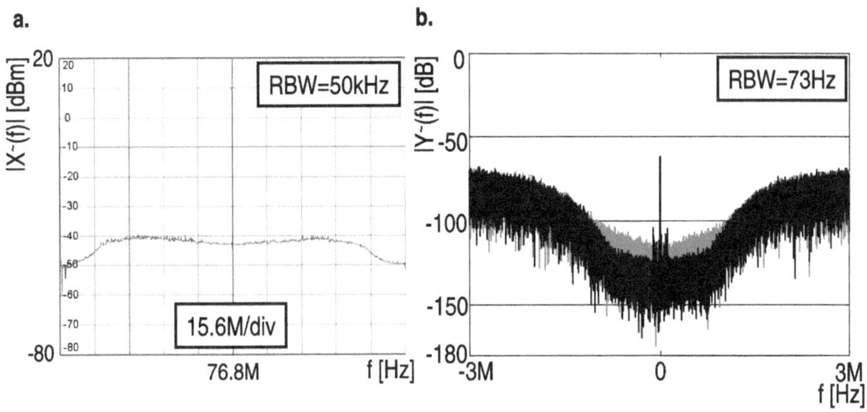

Figure 6.51: *Clock spectrum (a) and ΣΔ modulator output spectrum (b)*

lated that the theoretical maximum achievable SNR due to white noise jitter in ±600 kHz bandwidth will be $97.7 - 13.8 - 3 - 3 = 77.9$ dB compared to a full scale I and Q input signal. The extra 3 dB again stems from the fact that we have an I and Q modulator which are both clocked with a wide-band jittered clock. Therefore, the jitter-quantization noise convolution products of the left side of the spectrum coincide with the jitter-quantization noise convolution products of the right side, and vice versa. These products are uncorrelated which means a 3 dB increase of the total jitter noise (Appendix D). The measured noise in ±600 kHz is at a level of -80 dB. Note that in this case the measured in-band jitter noise is referred to the total quantization noise power, which in this case is 2, as the I and Q ADC both generate a ±1-bitstream. Theoretically, this results in a maximum achievable SNR of 77 dB in ±600 kHz when a full scale quadrature signal would have been applied to the modulator, which is 0.9 dB different from calculation. For reference, the black spectrum predicts a theoretical SNR of 87.1 dB for a full scale quadrature input signal. In this case the total noise is a combination of jitter, quantization and circuit noise.

Similar measurements are done on a the complex ΣΔ modulator with a 1-bit SI DAC of [47]. The modulator clock frequency f_s is 64 MHz. The input signal frequency f_i is 1.1 MHz. The clock is modulated with white noise which is bandwidth limited to about ±100 kHz. The clock spectrum is shown in Fig. 6.52a. The clock carrier amplitude is 7.6 dBm, within the ±100 kHz the noise is at -36 dBm (RBW=2 kHz), which leads to a CNR_{clk} of 76.6 dBc$\sqrt{\text{Hz}}$ in the ±100 kHz bandwidth. In Fig. 6.52b the modulator output spectrum is plotted. Using Eq. 6.80 the expected SNR at the output of the modulator can be calculated, which gives

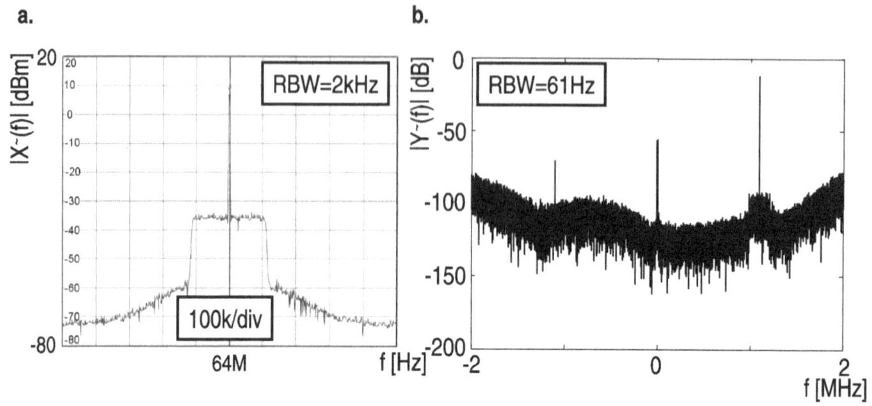

Figure 6.52: *Clock spectrum (a) and ΣΔ modulator output spectrum (b)*

$76.6 - 17.7 = 58.9$ dB. In this calculation, the jitter-quantization noise convolution products which fall in the signal bandwidth can be neglected as the clock jitter is bandwidth limited. The measured SNR in the 1 MHz to 1.2 MHz bandwidth is 59.7 dB, which is only 0.8 dB different from calculation.

In Fig. 6.53b a jitter-quantization noise convolution measurement is shown for the complex modulator with SI DAC. The black spectrum in Fig. 6.53b shows the modulator output spectrum when clocked with a very low jitter clock. The grey spectrum shows the modulator output spectrum when it is clocked with the clock of which the spectrum is shown in Fig. 6.53a. The jitter bandwidth is limited to ± 50 MHz and is modulated on a 64 MHz clock. The clock carrier amplitude is 7.8 dBm, the noise within the $\pm$50MHz bandwidth is at -45 dBm (RBW=50 kHz), which leads to a CNR_{clk} of 99.8 dBc$\sqrt{\text{Hz}}$. Using Eq. 6.81, the expected maximum achievable SNR limited by the jitter-quantization noise convolution products in 0 to 1 MHz bandwidth can be estimated, which yields $99.8 - 10\log_{10}(1e6) + 3 - 3 = 39.8$ dB. The extra -3 dB again stems from the fact that we have an I and Q modulator which are both clocked with a wide-band jittered clock. The measured in-band jitter noise is referred to the total quantization noise power, which in this case is 2, as both I and Q modulator output a ± 1-bitstream. This results in a theoretical measured SNR of 40.3 dB in 0–1 MHz if a full scale quadrature signal would have been applied to the modulator input. As comparison of calculation and measurement shows only 0.5 dB difference, the choice to replace f_i/f_s by $\frac{1}{2}$ in Eq. 6.81 is justified.

Unfortunately, at the time of the white noise clock jitter measurements there was no multi-bit modulator available to check Eqs. 6.77 and 6.81 for modulators with more than 2 levels.

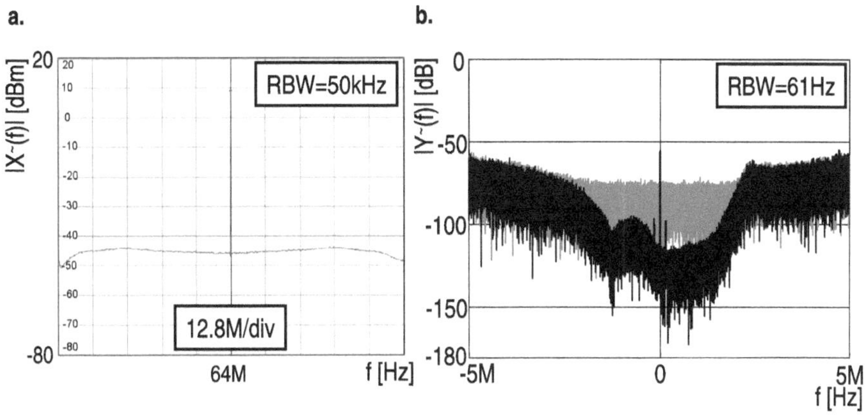

Figure 6.53: *Clock spectrum (a) and ΣΔ modulator output spectrum (b)*

6.9.4 The TPJE Model: SI Versus SC Feedback DAC

Equations 6.76 and 6.80 represent the SNR limitation according to the TPJE model for a single frequency input signal applied to a modulator with SC and SI DAC (no significant quantization noise). Clocked with a white noise jittered clock, the SNR difference using a SC DAC over a SI DAC becomes:

$$\Delta\text{SJNR}^{dB}_{JS,SC/SI\ DAC} = 10 \cdot \log_{10}\left(\frac{3f_i^2(f_2 - f_1)}{f_2^3 - f_1^3}\right) \quad [\text{dB}] \tag{6.84}$$

For a low-pass modulator in which $f_1 = 0$, Eq. 6.84 transforms into:

$$\Delta\text{SJNR}^{dB}_{JS,SC/SI} = 10 \cdot \log_{10}\left(\frac{3f_i^2}{f_2^2}\right) \quad [\text{dB}] \tag{6.85}$$

For input signals greater than $0.577f_2$, a modulator with SC DAC will outperform a modulator with SI DAC. Note that Eqs. 6.84 and 6.85 are the same for both 1-bit and multi-bit modulators.

For the in-band jitter (J) quantization (Q) noise convolution products of a modulator, it can be calculated that the difference in maximum achievable SNR for a b-bit modulator with SC or SI DAC is given by

$$\Delta\text{SJNR}^{dB}_{JQ,SC/SI} = 10 \cdot \log_{10}\left(\frac{3f_s^2(f_2 - f_1)}{4\left(f_2^3 - f_1^3\right)}\right) \quad [\text{dB}] \tag{6.86}$$

which can be calculated from the combination of Eqs. 6.77 and 6.81. For a low pass $\Sigma\Delta$ modulator in which $f_1 = 0$, Eq. 6.86 transforms into:

$$\Delta \mathrm{SJNR}^{dB}_{JQ,SC/SI} = 10 \cdot \log_{10}\left(\frac{3f_s^2}{4f_2^2}\right) = 10\log_{10}\left(3 \cdot \mathrm{OSR}^2\right) \ \ [\mathrm{dB}] \qquad (6.87)$$

6.9.5 The TPJE Model: An Application Driven Choice Between SI Versus SC Feedback DAC

The application of the $\Sigma\Delta$ modulator can play an important role in the optimization of the modulator's feedback DAC circuit architecture when it comes to clock jitter. In this paragraph the application area of $\Sigma\Delta$ modulators is split into two areas; applications in which the top-end DR of the modulator is determined by in-band signals, and applications in which the top-end DR of the modulator is determined by out-of-band signals. For both areas it will be analyzed how jitter can influence the performance of the system the modulator is in.

6.9.5.1 Modulators with a Top-End DR Determined by In-band Signals

An example application in which the top-end DR of the modulator normally is determined by in-band signals is audio. For audio the considerations concerning time jitter are transparent. The maximum input signal to the ADC is determined by the audio signal itself and the dynamic range required for the ADC is determined by the dynamic range of the human ear. The jitter requirement therefore is determined by the jitter around the wanted signal and the in-band jitter-quantization noise products.

For modulator input signals which lie in the bandwidth, the choice between SC and SI DAC is not directly obvious, as the performance degradation difference due to jitter is dependent on f_i, f_1 and f_2 (Eq. 6.84). For audio f_1 equals 0 Hz and f_2 is equal to the audio bandwidth. To calculate the worst case jitter degradation for the SI DAC, f_i is chosen at the edge of the signal bandwidth f_2. This reduces Eq. 6.84 to:

$$\Delta \mathrm{SJNR}^{dB}_{JS,SC/SI} = 10 \cdot \log_{10}\left(\frac{3f_i^2}{f_2^2}\right) = 10 \cdot \log_{10}(3) = 5 \ \mathrm{dB} \qquad (6.88)$$

and predicts that a $\Sigma\Delta$ modulator with an SC DAC used in an audio application always outperforms a $\Sigma\Delta$ modulator with an SI DAC by 5 dB for the worst case scenario where the input signal frequency equals the signal bandwidth. It must be noted though that if the input signal frequency is well below the signal bandwidth, a $\Sigma\Delta$ modulator with SI DAC will outperform a modulator with SC DAC. The in-band noise break-even point between a SC or SI DAC is at $f_i = \sqrt{f_2^2/3}$ which equals $0.577 f_2$ or $0.577 B_{ADC}$. For input signals above $0.577 B_{ADC}$, the SI DAC causes more in-band jitter noise than the SC DAC. If it is assumed that in audio applications the modulator has to have equal performance independent from input signal frequency, an SC DAC is preferred. In reality though, the high audio frequencies have lower probability, and the question which DAC topology to take is not so easy to answer.

The spectral shape of the clock can have a major impact on the way jitter noise on the clock influences the dynamic range of the modulator. In a narrow-band application area such as audio, it is normally required to get the full dynamic range out of the modulator when a full scale in-band signal is applied to its input. This gives a high requirement on the clock jitter close to the carrier. If the clock of the modulator is coming from an oscillator or PLL, the jitter on the clock carrier most likely will not be white, and close to the carrier will be dominated by 1/f-noise. This is stylistically illustrated in Fig. 6.54 (for a modulator with SI DAC). In Fig. 6.54b,

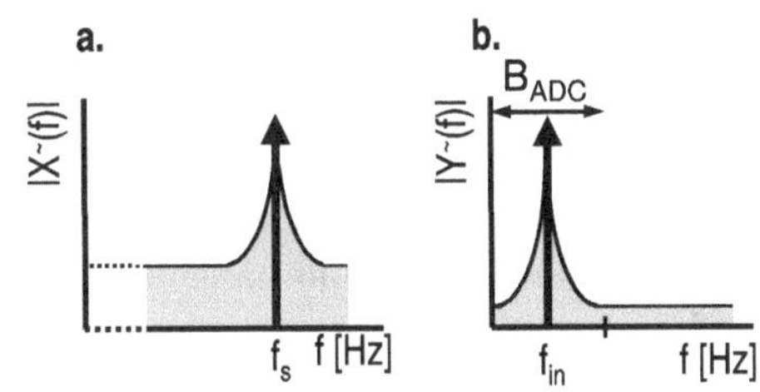

Figure 6.54: *Low frequency jitter in an audio application: clock spectrum (a) and modulator output spectrum (b)*

a clock spectrum is shown, with a significant amount of narrow-band jitter close to the carrier. Figure 6.54b shows the expected modulator output spectrum. The narrow-band jitter will appear in the signal bandwidth around the wanted signal and will be dominant above the wide-band noise on the clock, which is assumed to be at a much lower level and is neglected. In this case, the jitter shaping of an SC DAC does not help that much, as the expected in-band noise is significant, and either SC or SI DAC will perform almost equally. Therefore, a low noise clock generator and/or a sufficiently high over-sampling ratio of the modulator is required to reduce the in-band jitter noise around this signal.

In case of significant wide-band, white noise jitter on the clock, the question which DAC topology to choose becomes easier to answer when the jitter-quantization noise convolution products are observed. Equation 6.87 predicted that the expected in-band power of these products is 3OSR^2 times less for a modulator with an SC DAC, compared to the same modulator with an SI DAC. This is because the modulator with an SC DAC now exploits its jitter shaping, as in $\Sigma\Delta$ modulators the OSR is always much larger than one. These jitter-quantization noise convolution products can be further reduced by using a multi-bit feedback DAC.

6.9.5.2 Modulators with a Top-End DR Determined by Out-of-Band Signals

As seen in Chap. 4, modern highly digitized radio receivers demand a high dynamic range and high linearity from the $\Sigma\Delta$ modulator due to the large dynamic range of signals at the antenna, and the limited filtering in front of the modulator. In this section it will be analyzed if these receivers also put higher requirements on the jitter of the modulator clock.

Figure 4.18b of Chap. 4 showed that the in-band SNR requirements for radio systems which use digital modulation schemes are very low. This means that according to the TPJE model, the jitter requirements for the clock derived from the jitter components around the wanted signal itself are low. An interferer or neighboring channel which is close to the wanted signal however, can have a major impact on the jitter requirements on the clock, as the clock jitter modulated on this interferer partly falls into the signal bandwidth and reduces receiver sensitivity.

In a receiver, the quality of the clock needed for the $\Sigma\Delta$ modulator generally depends on:

- Signal dynamics (in-band as well as out-of-band)
- Filtering in front of the modulator
- Modulation scheme used in the system (QPSK, GMSK, etc.), and the required BER/SN(D)R
- The modulator's feedback DAC topology (SI or SC, 1-bit or multi-bit)
- Over-sampling ratio of the $\Sigma\Delta$ modulator
- Spectral shape and level of the jitter on the clock

In the analysis of how much constraints clock jitter around an interferer puts on the clock of a $\Sigma\Delta$ modulator in a receiver, the in-band jitter-quantization noise convolution products will be neglected. After the derivation of the requirement for CNR_{clk}, in a later stage it should be checked if the in-band jitter-quantization noise convolution products are low enough using a clock with this derived requirement by the use of Eqs. 6.77 and 6.81.

Figure 6.55 shows an input signal model [46] of a $\Sigma\Delta$ modulator in a receiver application which is used to analyze the effect of jitter on the in-band dynamic range of the $\Sigma\Delta$ modulator. The input signal of the modulator consists of a wanted signal at f_{IF} and an interfering signal (sine wave) at frequency f_{int}. The modulator is clocked with a frequency f_s. For simplicity a real ADC (no I&Q) is assumed. Jitter on the clock signal is modeled like in Sect. 6.9.2. Figure 6.56a shows the

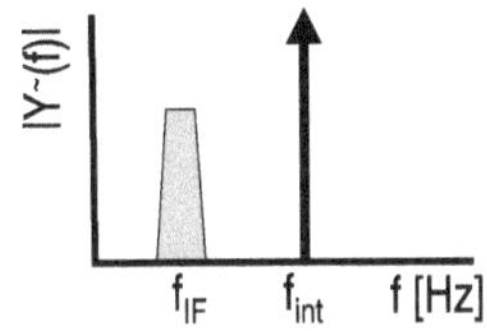

Figure 6.55: *Input signal model used to calculate the in-band jitter noise, caused by an interferer*

modulator output signal in the case of a $\Sigma\Delta$ modulator with SI feedback, while Fig. 6.56b shows the output signal in the case of SC feedback. As only a very limited in-band SNR is required as explained earlier, the jitter around the wanted signal is neglected, because these are low compared to the wanted signal. For both the SI and SC DAC the left sideband of the interferer due to the jitter falls onto the IF and can reduce the receiver sensitivity. For the SI DAC, f_i now becomes

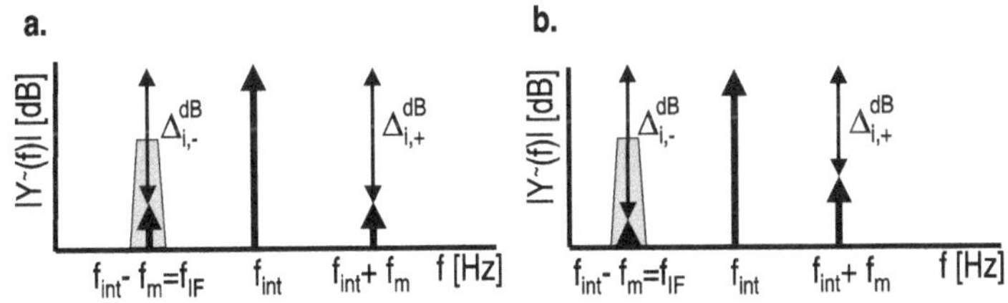

Figure 6.56: *Input signal model and in-band jitter noise for a modulator with SI (a) or SC (b) DAC*

f_{int}. For a spurious tone on the clock Eq. 6.73 changes into:

$$\Delta_s^{dB} = \Delta i_-^{dB} - 20 \cdot \log_{10}\left(\frac{f_s}{f_{int}}\right) \quad [\text{dB}] \tag{6.89}$$

In case of white noise jitter, Eq. 6.80 changes into:

$$\mathrm{CNR}_{clk}^{dB} = \mathrm{SJNR}_{JS,SI}^{dB} + 10 \cdot \log_{10}\left(\left(\frac{f_{int}}{f_s}\right)^2 (B)\right) \quad [\mathrm{dB}] \tag{6.90}$$

The CNR_{clk} is independent of IF frequency, as the in-band noise is white (no shaping like in the SC DAC case), and the level of this noise is only determined by the ratio of interferer frequency and sample frequency.

For the SC DAC, the left sideband component in Fig. 6.56b is much smaller, and clock requirements are relaxed. The input signal f_i becomes f_{int}. Furthermore, the frequency component which falls onto the IF, is at a distance f_m from the interferer at f_i. Therefore, it can be said that $f_i - f_m = f_{int} - f_m = f_{IF}$. Equation 6.66 changes into Eq. 6.91, and shows that the in-band spurious component level is independent of interferer frequency as expected.

$$\Delta_s^{dB} = \Delta_{i,-}^{dB} - 20 \cdot \log_{10}\left(\frac{f_s}{f_{IF}}\right) \tag{6.91}$$

In case of white noise and using $f_1 = f_{IF} - \frac{1}{2}B$ and $f_2 = f_{IF} + \frac{1}{2}B$, Eq. 6.76 changes into:

$$\mathrm{CNR}_{clk}^{dB} = \mathrm{SJNR}_{JS,SC}^{dB} + 10 \cdot \log_{10}\left(\frac{3Bf_{IF}^2 + \frac{1}{4}B^3}{3f_s^2}\right) \quad [\mathrm{dB}] \tag{6.92}$$

Equation 6.92 shows that the larger the IF bandwidth, the more noise will be in-band. Furthermore, the higher the IF frequency, the less one benefits from the jitter noise shaping in an SC DAC (compared to an SI DAC). In both cases the requirements on clock jitter will be higher.

In Table 6.6 the spurious tone and jitter noise that can be allowed on the clock is calculated for a ΣΔ modulator used in a GSM receiver according to the TPJE model for a modulator either an SI or SC DAC. In the example of Table 6.6 it is assumed that the wanted signal level is -109 dBm and the in-band SNR=6.5 dB (BER of 2%), like in Sect. 4.7. The clock frequency of the modulator f_s is 26 MHz, and an IF frequency of 100 kHz is assumed. The IF bandwidth B is 200 kHz. The relevant interferers are copied from Table 4.6. It is assumed that the close-by interferers with a frequency less than the clock frequency reach the modulator without filtering, which results in a worse case specification for the modulator clock. The far-off interferers ($> f_s = 26$ MHz) are assumed to be attenuated to a negligible level by the filtering (in front) of the modulator and by the aliasing factor described in Sect. 6.7, and therefore it is assumed their in-band jitter contribution can be neglected. In the calculation of the spurious tone clock requirement, it is assumed that the spurious tone falls exactly in the middle of the IF bandwidth, thus at f_{IF}. If in reality the spurious tone falls on the low side of the bandwidth ($f_{IF} - \frac{1}{2}B$), the clock specification will be a little bit more relaxed; if the spurious tone falls onto the high side of the bandwidth ($f_{IF} + \frac{1}{2}B$), the spurious tone clock requirement will be a little more stringent.

ΣΔ modulator DAC topology	GSM interferer definitions	Frequency offset to wanted [MHz]	Interferer level [dBm]	Wanted signal level [dBm]	Maximum allowed noise or spurious level in IF band [dBm]	ADC correction factor in case of single spurious tone [dB]	Max. allowed spurious level on clock [dBc]	ADC correction factor in case of white noise [dB]	CNR_{clk} [dBc$\sqrt{Hz}$]
SI DAC	Interferer 1	0.2	−73	−109	−115.5	38.8	3.7	14.3	56.8
	Interferer 2	0.4	−41	−109	−115.5	34.3	40.2	18.7	93.2
	Interferer 3	1.6	−33	−109	−115.5	23.7	58.8	29.3	111.8
SC DAC	Interferer 1	0.2	−73	−109	−115.5	48.3	−5.8	6	48.5
	Interferer 2	0.4	−41	−109	−115.5	48.3	26.2	6	80.5
	Interferer 3	1.6	−33	−109	−115.5	48.3	34.2	6	88.5

Table 6.6: *Clock jitter specification derivation for a ΣΔ modulator in a GSM receiver with an IF of 100 kHz based on the relevant out-of-band interferers defined in the GSM specification*

Figure 6.57a shows the carrier-to-spurious-tone-ratio clock requirement (Δ_s), as a function of clock carrier offset frequency. Figure 6.57b shows the carrier-to-noise-ratio CNR_{clk} of the clock as a function of clock carrier offset frequency. The calculations are done for three different intermediate frequencies of 100 kHz, 300 kHz, and 1 MHz. If the results of the individual intermediate frequencies are compared, the TPJE model predicts that, compared to a modulator with SI DAC, a modulator with SC DAC always has a much lower clock jitter requirement for both spurious tones and white noise jitter. This is because the modulator with SC feedback DAC benefits from the jitter noise shaping due to the amplitude modulation described by the TPJE model. Note that if the jitter on the clock is white,

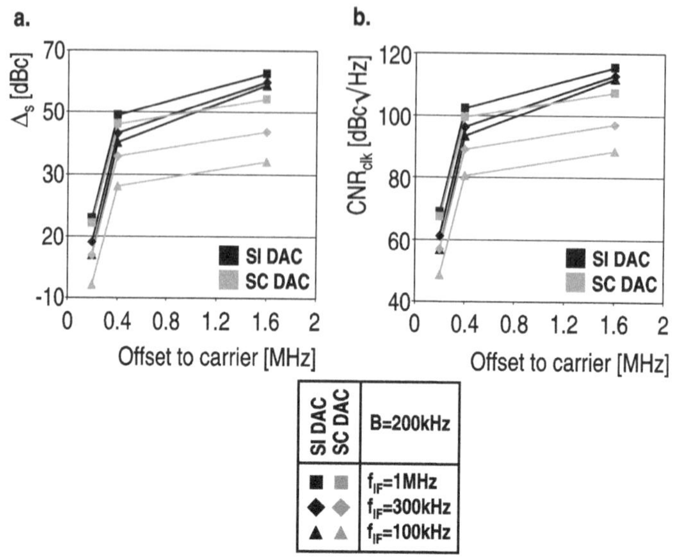

Figure 6.57: *Phase noise specification for the clock of the modulator used in a GSM receiver*

the most far-off interferer at an offset frequency of 1.6 MHz determines the clock requirement. The requirement this interferer puts on the clock is about 40 dB higher than the close-by interferer at 200 kHz offset. In reality, the clock of the modulator is coming from an oscillator or PLL, and the jitter on the clock carrier most likely will not be white, and will for instance contain 1/f-noise.

According to the TPJE model, the way jitter influences the performance of a ΣΔ modulator is heavily dependent on the spectral shape of the jitter. In case of low frequency jitter (close to the clock carrier), the clock jitter on the clock of a modulator in a receiver, can become irrelevant. The clock jitter of an in-band signal are already very low because of the limited in-band SNR required in receiver systems. The narrow-band jitter around an interferer will not fall in the signal bandwidth. This is illustrated in Fig. 6.58 for a modulator with SI DAC (the discussion is similar for a modulator with SC DAC). As long as the jitter noise around the out-of-band signal is not interfering too much with the wanted signal in the modulator's bandwidth (B_{ADC}), the requirements on the jitter of the clock according to the TPJE model due to this mechanism are very limited. The narrow-band jitter requirements for oscillators and PLLs used in receivers to clock the ADC, fortunately are relaxed by the dropping interferer power level closer to the IF bandwidth (Table 6.6), as can be seen in Fig. 6.57b.

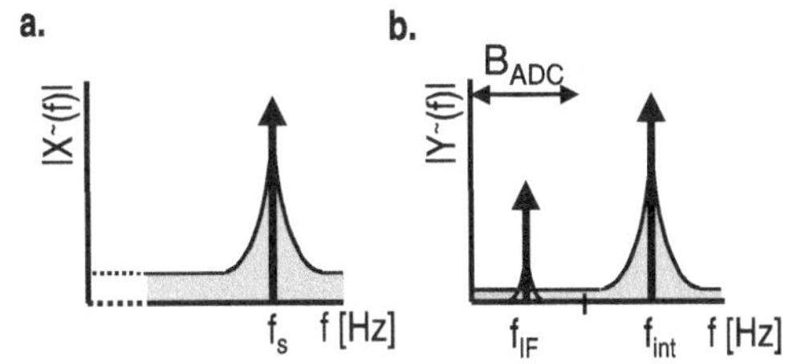

Figure 6.58: *Low frequency jitter in a receiver application. Clock spectrum (a) and modulator output spectrum (b)*

In [46] the ΣΔ modulator clock requirements are calculated more extensively, for a modulator used in a GSM and a Bluetooth receiver. In this paper it is proven that the required phase noise requirement on the clock of the ΣΔ modulator is much lower compared to the phase noise requirements for the local oscillator due to the over-sampling ratio of the modulator.

In the discussion above, the jitter-quantization noise convolution products were neglected. It depends on the modulator's out-of-band quantization noise power (1-bit or multi-bit quantizer) and the spectral shape and level of the jitter on the clock whether this simplification is justified. For example, a 1-bit modulator toggling between ± 1, has an out-of-band quantization power of 0.5, when a full scale input signal is applied to the modulator's input. Therefore, the convolution products can have a large impact on the modulator's in-band SNR. When the jitter is narrow-band, the simplification is justified, but in this case as described, the in-band jitter from an interferer next to the signal bandwidth can also be neglected. Therefore, the discussion above is only relevant for a modulator, with a sufficient number of quantization levels.

6.10 Conclusions

In this chapter the robustness of several ΣΔ modulator architectures has been tested on circuit imperfections. The ΣΔ modulator's sensitivity to gain variations, noise, linearity, loop delay, aliasing and clock jitter have been discussed. From this analysis relations between several important specifications and their cost have been derived. In Chap. 8 some of these relations will be used for benchmarking the ΣΔ modulators presented in this book with ΣΔ modulators published. Chapter 9 will present ΣΔ modulator implementations that are optimized using these relations. Below the conclusions are summarized per section.

Technology:

- Technology scaling can have a major impact on the performance of analog circuit blocks. The expected performance of analog blocks in a successive technology node at best stays the same.

- A way to reduce the sensitivity to the changes in analog design parameters is to replace the highly sensitive analog circuits blocks with digital circuits where possible.

- To decrease time-to-market, analog circuits should be designed like digital circuits. A library of analog functions with their p-cell layout should be created, with which every analog block in the system can be built. When going to the next technology node, only this library has to be ported. From there the analog IP blocks can be generated using the digital layout tools.

Gain:

- Single loop $\Sigma\Delta$ modulators have the lowest requirements on absolute loop filter gain, as the quantizer only needs to decide if its input signal is larger of smaller than its reference level.

- In-band loop filter gain should at least be large enough to sufficiently push the quantization noise out-of-band.

Noise:

- An SC and SR DAC with RC input stage are preferred over an SI DAC as the amount of 1/f noise introduced is independent of the maximum switching speed of the DAC.

- If it is assumed that the $\Sigma\Delta$ modulator input signal and the DAC reference voltage are chosen as close to the available supply as possible, a reduction in supply voltage will mean a quadratically lower required circuit impedance for the same $\mathrm{SNR}_{circuit\ noise}$, and thus a quadratically higher power consumption. If the input signal and the DAC reference voltage are chosen smaller than possible within the supply, even more current will be required to achieve the same $\mathrm{SNR}_{circuit\ noise}$.

Linearity:

- From the proposed input stages, the RC integrator input stage is preferred as it can provide very high linearity at moderate bias currents.

- As long as technology does not limit the over-sampling of the modulator to get sufficiently low quantization noise, a 1-bit feedback DAC is preferred over a multi-bit DAC as it can be implemented without suffering from unit cell mismatch, which drastically reduces the complexity of the modulator. One positive exception is a $\Sigma\Delta$ modulator which employs a 1.5-bit quantizer and DAC, which can still be made very linear, without adding too much complexity.

Delay:

- To compensate for excessive delay (phase shift or time delay) in the modulator, a delay compensation scheme is introduced. The loop filter coefficients are recalculated to compensate for the delay, rather than increasing currents in the circuitry. This opens up the road for $\Sigma\Delta$ modulators with the advantages of feed-forward loop filters in high speed applications, as the additional delay in the feed-forward summation node can be compensated.

- SC DACs inherently have less delay compared to SI DACs for the same RTZ period due to the shape of the feedback pulse, but at the cost of larger peak currents.

Aliasing:

- The quantizer aliasing is not an issue in higher order modulators as the aliases caused by the sampling in the quantizer are attenuated by the loop filter gain.

- The aliasing due to the sampling in the DAC is highly dependent on the implementation details of the DAC and the input stage. In this book the aliasing for an RC integrator input stage with different feedback DACs is analyzed. For all feedback DAC topologies the AAD improves with the Gm of the input stage. For the SI DAC, the AAD further improves with higher DAC current source output impedances. Furthermore, a very small or very large RTZ period improves the AAD, but is not practical as it implies a higher clock frequency. For the SR DAC, the AAD improves with switch impedance matching, and smaller RTZ periods. For the SC DAC, the only degree of freedom is the Gm of the input stage, as f_s and C_{DAC} are dictated by modulator architecture and SNR.

Clock Jitter:

- Two ways of modeling clock jitter in a $\Sigma\Delta$ modulator are presented. Both the TAJE model and the TPJE model predict that a $\Sigma\Delta$ modulator with SC feedback DAC always out-performs an SI DAC when it comes to sensitivity to clock jitter. As the TAJE model is seen as inaccurate, its use is dissuaded, and it is advised to only use the more precise TPJE model.

- In the sine wave induced TPJE model it is described how a sine wave, deliberately modulated onto the clock, influences the output spectrum a DAC. The sine wave induced jitter on the clock repeats itself around every frequency component passing through the DAC, creating two side band components, one on each side of its carrier. The ratio between carrier power and jitter side band power is fixed for each carrier, which means that the jitter spectrum around each signal passing the DAC goes up and down with the amplitude of that carrier.
- Sine wave induced jitter on the clock causes amplitude and phase modulation of the incoming signals in a modulator with SC DAC. This causes the left jitter component around a signal passing the DAC to be smaller than the right jitter component. In a modulator with SI DAC, only phase modulation of the incoming signals is observed, and both jitter components around the DAC's output signal have equal amplitude.
- If an interferer at the input of the modulator has a frequency which is attenuated by the modulator's STF, its jitter components will go down with the amount of carrier attenuation.
- Verification of the sine wave induced model with measurements has shown that the model very accurately describes how the jitter on the clock repeats itself around a signal component that passes the feedback DAC clocked with this sine wave modulated clock. Calculations and measurements match within about 1 dB.
- If white noise jitter is substituted in the sine wave induced jitter model, the in-band SNR limitation of a $\Sigma\Delta$ modulator clocked by this white noise phase modulated clock can be calculated. The calculation of the modulator's SJNR can be done for a modulator with either a 1-bit SC DAC, a multi-bit SC DAC, a 1-bit SI DAC or a multi-bit SI DAC.
- The equations derived can also be used the other way around; for a chosen modulator SNR, signal bandwidth, and DAC type a phase noise clock specification can be derived.
- A low-pass modulator with SC DAC benefits from the jitter noise shaping. The in-band jitter noise of a modulator with SI DAC is much more dependent on out-of-band signal power (either interferer or noise).

- The TPJE model applies to multi-bit as well as 1-bit modulators. The mechanism how jitter on a clock transform into jitter around an input signal, is the same for 1-bit and multi-bit modulators. However, the TPJE model predicts that the expected (in-band) jitter-quantization noise convolution products in a multi-bit modulator's output spectrum, are lower compared to the jitter-quantization noise convolution products in the output spectrum of a 1-bit modulator, when both are clocked with a wide-band, white noise jittered clock. This is explained by the fact that the quantization noise power of a multi-bit $\Sigma\Delta$ converter is b*6 dB (b = number of bits) lower compared to a single bit modulator.

- It is proven by measurements that the white noise substitution in the sine wave induced jitter model is valid. Calculations and measurements match within about 1 dB. Therefore, it can be concluded that the model very accurately describes how white noise jitter on the clock repeats itself around a frequency component (either, wanted signal, interferer or quantization noise component) that passes the feedback DAC.

- If a large out-of-band signal is applied to the modulator, the jitter noise around this signal appears in the signal bandwidth. The level of in-band jitter around this input signal will be equal in the 1-bit and multi-bit modulator.

- In a receiver, a $\Sigma\Delta$ modulator with SC DAC is preferred over a modulator with SI DAC, as the expected in-band jitter spurious or noise due to an out-of-band interferer is lower because of the jitter shaping in an SC DAC.

- If the dominant clock jitter is very close to the clock carrier (narrow-band jitter), it is very likely that the in-band jitter-quantization noise convolution products can be neglected. In this case, a single or multi-bit modulator will perform equally. The narrow-band jitter will only be visible in the signal bandwidth of the modulator when a strong in-band signal is applied to the modulator, like in Figs. 6.49 and 6.52.

From this chapter it has become clear that a 1-bit, CT feed-forward loop filter is preferred because of its low power consumption, large signal stability, large quantizer alias suppression and low loop filter gain accuracy requirement. Furthermore, a 1-bit feedback DAC and a RC integrator input stage are preferred because of linearity. Due to lower sensitivity to clock jitter and its inherent lower delay due to the feedback pulse shape, an SC feedback DAC is preferred over an SI (or SR) DAC. In return, the degrees of freedom to reduce aliasing in the DAC are limited, but this problem can be tackled by careful design of the DAC and input stage. In summary a CT feed-forward ΣΔ modulator with RC integrator input stage and 1-bit SC feedback DAC is the most robust ΣΔ modulator implementation.

Furthermore, technology scaling has more advantages for digital circuits than for analog circuits. Therefore, it can be advantageous to introduce or increase digitization in the modulator loop.

Chapter 7

$\Sigma\Delta$ Modulator Flexibility

Chapters 5 and 6 showed that a high order 1-bit highly over-sampled $\Sigma\Delta$ modulator is a very suitable choice for the ADC in a direct conversion receiver. The modulator can be made with sufficiently high S(Q)NR and linearity at low power consumption, which are the most important specifications for the receiver ADC. Furthermore, Chap. 6 showed that a 1-bit CT feed-forward $\Sigma\Delta$ modulator with SC feedback DAC is preferred because of its built in robustness. In this chapter it will be investigated if such a modulator can be made scalable to fit into a multi-standard radio.

Goal is to come to a scaling method with which a $\Sigma\Delta$ modulator IP block can be designed, which can cope with any standard currently known. Once such an IP block is available, it can serve multiple receiver applications. The same $\Sigma\Delta$ modulator can be used for different receivers with different combinations of modes, e.g. a GSM, Bluetooth and GPS receiver, or a UMTS and WLAN receiver. Such a multi-mode $\Sigma\Delta$ modulator not only enables multi-mode receiver architectures but also leads to faster market introduction of receiver systems, as exactly the same ADC can be re-used in different products.

7.1 Receiver Dictated Flexibility Requirements

In a multi-standard radio the signal dynamics at the antenna are different in each standard. The required SNR at sensitivity changes because other digital modulation techniques are used. Furthermore, changing bit-rates will require different RF front-end and ADC bandwidths. Figure 7.1 shows the channel bandwidth requirements for a number of standards. From the plot it can be seen that the required channel bandwidth changes over two decades.

R.H.M. van Veldhoven, A.H.M. van Roermund, *Robust Sigma Delta Converters*,
Analog Circuits and Signal Processing, DOI 10.1007/978-94-007-0644-6_7,

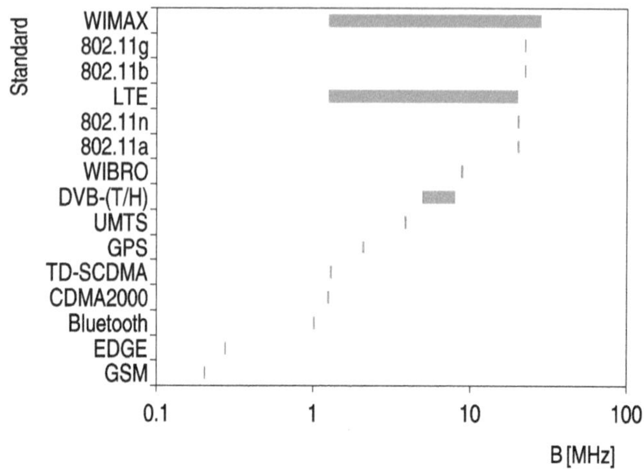

Figure 7.1: *ADC bandwidth requirements in the telecommunication standards*

The changing signal dynamics, channel bandwidth and noise requirements can require scaling of filtering, gain and/or noise impedances of in the receiver. Chapter 4 showed how the ADC and the RF front-end specifications relate.

Changing the noise impedance of the ΣΔ modulator in a multi-standard radio, would mean as many different noise impedances as radio standards, which makes the circuit design very complicated, as it requires switching in the input stage of the ΣΔ modulator, and could lead to non-linearity. In a highly digitized receiver the ADC requirements are already pushed to the limit. The coverage of two decades of ADC bandwidth already gives a large increase in circuit complexity. A scalable noise impedance will lead to further circuit overhead and increase of circuit complexity. Furthermore, the accompanying circuit parasitics will go at the cost of the maximum achievable ADC performance.

In the multi-standard RF front-end, noise impedances have to be scaled anyway to achieve the required noise figure. As the receiver is highly digitized only a limited amount of filter cut-off frequency and gain programmability has to be added to the receiver to accommodate all standards. As this only involves the switching of capacitors and resistors this is relatively easy. This way signal levels at the input of the ADC can be equalized. The signal levels at the input of the ADC should be kept as high as possible (Sect. 6.5.4), which means that the equalization level of the modulator input signals is limited by the supply voltage. At the same time the ADC noise impedance is kept constant, reducing circuit complexity thus implementation risks.

The choice of optimizing the receiver this way, divides the flexibility requirements over the RF front-end and ADC. The RF front-end AGC, filtering and noise impedance scaling in combination with the scaling of the ADC bandwidth, is the only way to come to a competitive receiver power consumption.

In Sect. 9.1 it will be shown that the signal level equalization and fixed ADC noise impedance approach leads to an ADC with a competitive power consumption. The ADC has a fixed architecture on both ΣΔ modulator loop and circuit topology level. This way the ADC design complexity is reduced.

The next few sections will discuss the impact of the signal equalization, fixed noise impedance and bandwidth scaling approach on the design of the ΣΔ modulator, being the choice of ADC architecture.

7.2 ΣΔ Modulator Clock Flexibility

The huge variation of required ADC bandwidth comes with a large variation of the ADC clock. In this section receiver architectures with different ADC clock strategies are explored which can cope with such a large variation in clock frequency. The section will mainly focus on the implications of the chosen clock generator architecture on the ADC and will not address the implications at system level like frequency locking, etcetera.

A generalized block diagram of a highly-digitized receiver is shown in Fig. 7.2. It contains an antenna, an LNA, a mixer, a ΣΔ ADC, decimation filtering, a base

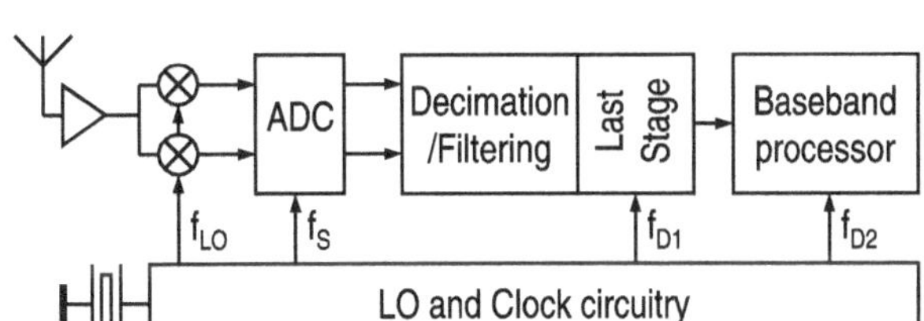

Figure 7.2: *Overview of the relevant clocks in the receiver chain*

band processor, an LO and clock circuitry. The differences between the architectures are confined to the generation of the clocks for the ADC and decimation stages. Different possibilities are dealt with in the subsequent sections.

7.2.1 Receiver Architecture with LO-Dependent ADC Clock

In this architecture the clock of the ADC is integer-divided from the local oscillator (LO). The main advantage of this system is that the high-quality frequency synthesizer is reused for the clock of the ADC. This ensures a sufficiently good clock in terms of jitter, as the requirement on the quality of clock for down-mixing in an (N)ZIF receiver already is higher than for the ADC clock, because a $\Sigma\Delta$ ADC is over-sampled (Sect. 6.9). A disadvantage is that the output frequency of the synthesizer is dependent on the channel selected for down conversion. This means that the clock frequency of the ADC is dependent on the channel selected. In continuous-time $\Sigma\Delta$ modulators this has implications for the values of the coefficients used in the loop filter. Although the relative change of frequency is reasonably small, the modulator needs to be stable for all applied clock frequencies. This can increase the required number of coefficients to be programmed in the loop filter. Furthermore, sample-rate conversion is required after the ADC, if the LO frequency is not a multiple of the bit-rate which is generally the case. In future mobile phones, multiple transceivers will be on the same chip. This needs multiple and potentially non-integer clock frequencies for the different transceivers which is unattractive for reasons of electromagnetic (in)compatibility.

Pros:

- No extra PLL needed for the A/D conversion
- Clock generator has low jitter because of the already present LO requirement
- Power efficient for when clock frequency is optimized to the ADC bandwidth
- As LO and ADC sample frequency are related substrate interference is also related, which gives the advantage that sensitive signal processes can be done when substrate is expected to be quiet

Cons:

- Integer LO division gives limited clock flexibility
- If the relative variation of the LO frequency is large, programmability of the coefficients is required in the loop filter of the $\Sigma\Delta$ modulator
- Sample-rate conversion is required after the ADC
- The sample-rate conversion factor is dependent on the LO frequency

- Harmonic relations between sample frequency and LO can cause in-band aliases due to ADC sampling
- EMC in a multi-pipe transceiver

7.2.2 Receiver Architecture with a Flexible and Independent Clock for the ADC

Another possibility is to generate the ADC clock with a dedicated PLL locked to the main reference oscillator for the receiver. Unless a fractional-N PLL is used, the risk is that multiple crystals are needed to generate a clock frequency which is an integer multiple of the bit-, symbol- or chip-rates for each of the many different radio standards to be covered by the radio, to avoid sample-rate conversion. A large number of crystals is highly undesirable on grounds of size and cost. If a fractional-N PLL can be used, then the need for multiple crystals is removed and this architecture becomes an attractive option. The flexibility in output frequency allows for the most desirable ADC clock frequency to be generated for any given standard. However, the addition of the PLL, either integer-N or fractional-N, increases power consumption. It also demands great care in the design of the PLL to keep phase noise below acceptable limits. In addition, a fractional-N PLL adds spurious tones in its output spectrum, due to its non-integer divider ratio. Next to the phase noise, these tones can degrade the ADC performance (Sect. 6.9). The variable ADC clock frequency in this architecture complicates the ADC design as it necessitates the adaptation of the integrator coefficients in the loop filter. For a wide range of clock frequencies, this can lead to a difficult and bulky ADC design.

Pros:

- Power efficient ADC and decimation stages as clock frequency is optimized

Cons:

- Extra PLL needed with sufficiently low phase noise (extra power consumption)
- Huge number of crystals or fractional-N PLL/sample-rate conversion needed

    Huge number of crystals is not feasible due to cost and size

    Fractional-N PLL next to phase noise introduces spurious tones

- Extra coefficient set needed in the loop filter of the $\Sigma\Delta$ modulator for each optimized clock frequency

- In a multi-pipe receiver multiple PLLs are required if the ADCs in these pipes are not clocked at the same frequency (or an integer multiple)
- EMC in a multi-pipe transceiver

7.2.3 Receiver Architecture with Fixed, Independent ADC Clock

A third possibility involves running the ADC at a fixed clock frequency for all receiver modes. The value of the clock frequency is determined by the mode with the highest signal bandwidth. In this mode the clock frequency is chosen such that the quantization noise is low enough to allow the receiver to achieve its intended noise figure, and the ADC has sufficient dynamic range to handle the largest interferers that find their way through the front-end mixers and IF pre-filters. When dealing with modes with a small signal bandwidth, the modulators loop filter can be programmed to concentrate more gain at lower frequencies if necessary (Sect. 7.4), thereby lowering the level of the quantization noise in the vicinity of the narrow-band signal. Leaving the clock frequency at a fixed high frequency ensures a high over-sampling ratio and, in return, a very high dynamic range (at least in terms of quantization noise). There are considerable advantages in using a fixed-frequency clock generator, as only a single frequency clock generator in a multi-pipe receiver terminal is required. Because of the high frequency clock, the ADC area reduces as it only needs integrator capacitor values related to the high frequency clock which are small. If required, the noise shaping can be optimized as a function of the required signal bandwidth by programming the local feedback coefficients (Sects. 5.1 and 7.4) which determine the notch frequencies in the noise shaped output spectrum of the modulator. An example is give in Fig. 7.3 for signal bandwidths of 1, 3.3 and 10 MHz. Furthermore, the

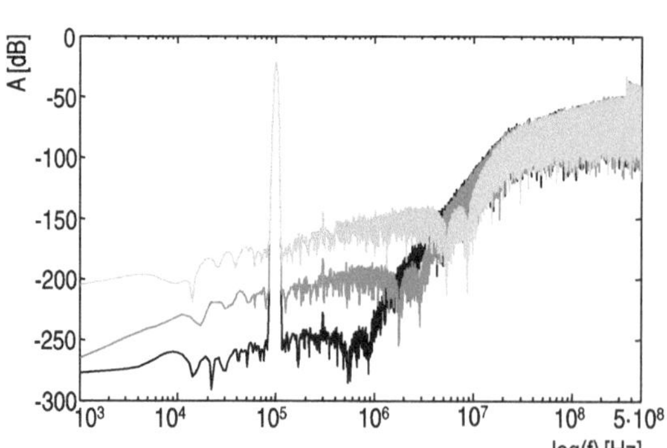

Figure 7.3: *Notch frequency programming to optimize the modulators noise shaping to the required signal bandwidth*

simplicity of the fixed-frequency clock generator and the requirement for only a single-crystal reference oscillator or the use of the frequency divided LO are also attractive aspects. The disadvantage of the sample-rate conversion required to obtain the relevant bit-rate, symbol-rate or chip-rate for any given mode, is canceled by the fact that the receiver pipe takes on an extremely "open" design concept that can deal with virtually any radio standard. It is a substantially future-proof concept whose only main threat would come from the increased digital processing power. Low-band modes need a higher decimation factor, as in these modes the sampling rate at the output of the ADC is very high in relation to the signal bandwidth. As the requirements on the first decimation filter stages are limited as their required roll-off is limited, the additional power and area spending will be limited. As this architecture requires sample-rate conversion this will cause additional power spending. The increase of power in the digital processing, is (partly) compensated by the smaller feature size in nm-technologies as the efficiency of digital circuits increases with $1/s_T^3$ (Sect. 6.1). This makes this clocking strategy extremely suitable for small feature size technologies.

Pros:

- Only one set of integrator coefficients needed in the loop filter of the modulator
- Small ADC silicon area because only the highest bandwidth mode coefficients have to be implemented. In this mode the integrator loop filter capacitor values needed are the smallest
- Very flexible system. An "open-pipe" concept which is substantially future proof
- LO divided clock or low-power clock generator requiring only a single reference crystal is needed
- In a multi-pipe receiver, the ADCs and decimation filters in each pipe are slave to the same clock frequency which gives a better controlled EMC performance
- As the area and power of digital circuits shrink with technology this is a future proof architecture

Cons:

- If LO can not be used, an extra PLL (but no fractional-N) will be needed (extra power consumption)

- Sample-rate conversion needed
- Risk of higher power consumption in digital filters for low-bandwidth modes
- Higher ADC power consumption in the low bandwidth modes
- Additional digital hardware required

In Sect. 9.1.3 a 121-mode ADC is presented, which could be used in a receiver with fixed ADC clock. The ADC power penalty in this case is that the ADC at the highest sample frequency consumes 6.6 mW, whereas the ADC consumes only 1.44 mW in the lowest bandwidth mode, which is significant. Therefore, the power consumption at high sample frequencies has to be lowered. This can be done by using the delay compensation technique presented in Sect. 6.8 to compensate the pole in the summing node. This way the current in the summing's node OTA can be reduced, as it has a significant contribution to the total ADC power. Another solution is to use a limited set of sample frequencies instead of only one sample frequency.

7.2.4 Choice of Clock Strategy

For future multi-mode, multi-pipe transceivers, a fixed ADC clock strategy results in the most flexible and transparent system. It comes at the cost of higher power consumption for the ADC and digital processing especially in the low-bandwidth ADC modes as the overhead is created in the ADC. This clock strategy is especially suited for nm-technologies as the area and power of digital circuits shrink with technology scaling, and for analog IP blocks the transistors become faster. Applying the fixed or LO divided ADC clock strategy furthermore requires extensive knowledge about sample-rate conversion in relation to power consumption of the digital part and in relation to EMC aspects of a multi-pipe receiver. Unfortunately, this is out of the scope of this book.

The ADCs presented in this book are designed for the independent and additional PLL generated ADC clock based systems. The main reason for this is that the multi-mode ADCs presented in this book are designed in an IC technology, in which digital processing still has its cost (comparably large feature size), and high clock rate decimation and fractional decimation therefore are avoided. Furthermore, the ADCs were designed for a multi-mode rather than a multi-pipe receiver. This means that only one additional PLL is required to generate the ADC clock, and EMC is less of an issue.

7.3 Input Stage and DAC Flexibility

The feedback current in a $\Sigma\Delta$ modulator with RC integrator input stage and SC feedback DAC is proportional with $V_{ref} f_s C_{dac}$. The input signal is converted into a current by the input resistor R_{in}. To deliver the same amount of current to compensate the input current at each clock frequency, the DAC capacitor has to be scaled inversely proportional to the clock frequency. This means that at each clock frequency an exact DAC capacitor is required, to avoid an input-output gain change of the modulator.

The input referred noise density of a $\Sigma\Delta$ modulator with RC integrator input stage and SC feedback DAC is given by Eq. 6.7. If the reference voltage noise and OTA noise can be neglected (which is normally the case), the equivalent input referred noise density reduces to:

$$S_{\Sigma\Delta M,in,SR\ or\ SC} \approx 8kTR_{in}\left(1 + R_{in} f_s C_{dac}\right)\ [\mathrm{V^2/Hz}] \qquad (7.1)$$

It can be seen that the equivalent input referred noise density is also proportional to $f_s C_{dac}$. This means that if f_s and C_{dac} are scaled for constant current feedback, the equivalent input noise density of the modulator stays constant, which is preferred (Sect. 7.1).

For a switched current (or switched resistor) DAC, no scaling of the DAC elements is required if the input resistor is also not changed.

7.4 Loop-Filter Flexibility

If the ADC noise impedance, mainly determined by the input resistor and feedback DAC, is kept constant, the scaling of the $\Sigma\Delta$ modulator bandwidth is fairly easy. In Sect. 5.1, the $\Sigma\Delta$ modulator parameters were introduced which determine its SQNR. From the same section it can be concluded, for to add flexibility to the $\Sigma\Delta$ modulator, two things need to be done:

1. The integrator unity gain frequencies of the modulator scale proportional to the sample frequency, if the input signal swing, the integrator signal swings and the loop stability criterion are kept constant (Eqs. 5.3 and 5.4). This way, more DR can be achieved in the same bandwidth (OSR increases), or the same DR can be achieved in a larger bandwidth (OSR fixed).
2. By changing the local feedback coefficient b_n (coefficients k_n are constant), the effective SQNR in a certain modulator bandwidth can be optimized

(Sect. 5.1). If Eq. 5.6 is rewritten, the b_n coefficients should be changed according to:

$$b_n = \frac{\pi^2 k_n^2 l_n}{i_{n-1} i_n l_{n-2} \mathrm{OSR}^2} \tag{7.2}$$

As said in the introduction of this chapter, the ΣΔ modulator has to be re-usable for different receiver systems, of which it is not known up-front (at the time of the ΣΔ modulator design) which bandwidths it has to convert. This means that the modulator has to cover a large sample frequency range. A trade-off has to be made between the allowed integrator swing variation due to the varying sample frequency and the required resolution of the unity gain programmability. This is shown by Eqs. 5.3 and 5.4. The clip level l_n sets the maximum allowed signal at the output of the integrator n. When f_s is changed, either the clip levels l_n have to change, or the unity gain frequencies ω_n have to change (i_n is fixed by the stability criterion and $A_{X,in,max}$ is chosen to be constant). If f_s is lowered, signal swings on the integrator output will increase, but are limited by the supply and the required biasing headroom of the integrator stage transistors. At high sample frequencies the integrator output signals become unrealistically small, which might lead to noise problems and makes the design of the quantizer more difficult. Therefore, the total ADC sample frequency range $[f_{s,min}, f_{s,max}]$ required to convert all bandwidths determined by the variety of modes, should be split up in x_{fs} sub-ranges, which results in x_{fs} unity gain frequencies. Within each frequency range the integrator unity gain frequencies are constant, and with the changing f_s the output swings on the integrators vary with a factor y_{fs} at maximum. The most efficient implementation is achieved when an exponential relation is used between y_{fs} and x_{fs}. The number of sample frequency ranges required is given by:

$$x_{fs} = \frac{\ln(\frac{f_{s,max}}{f_{s,min}})}{\ln(y_{fs})} \quad [\text{-}] \tag{7.3}$$

The integrator unity gain frequency of each range should always be calculated at the lowest sample frequency within the range. This way the maximum integrator output signal will never exceed the intended signal levels, to preserve circuit linearity. At higher sample frequencies within the sub-range, the integrator output signals become smaller. If it is assumed that the resistors of the RC integrator stages are kept constant (no noise impedance scaling), the integrator capacitor values have to change for different sample frequencies according to:

$$C_{ug,n} = C_{ug,min} \cdot y_{fs}^n \quad [\text{F}] \tag{7.4}$$

where $C_{ug,min}$ is determined by the highest sample frequency range, $n = [1..x_{fs}]$, and $C_{ug,max} = C_{ug,min} \cdot y_{fs}^{x_{fs}}$. It should be noted that for each sample frequency sub-range an additional DAC capacitor is required. This gives a maximum modulator input-output gain variation of y_{fs} times over the particular sample frequency sub-range.

As said, with the local feedback coefficients b_n, the bandwidth of the modulator can be optimized (Eq. 7.2). The value of the feedback coefficients are related to i_n, l_n, k_n and the OSR. The choice of OSR completely determines the value of b_n, as the other coefficients are fixed. This means in the scaling of the b_n coefficients, only the OSR range has to be considered. The number of coefficients is determined by the resolution with which the bandwidth of the modulator has to be trimmed. The number of coefficients required is given by:

$$x_{\mathrm{OSR}} = \frac{\ln(\frac{\mathrm{OSR}_{max}}{\mathrm{OSR}_{min}})}{\ln(y_{\mathrm{OSR}})} \quad [\text{-}] \tag{7.5}$$

in which y_{OSR} is determined by the required programming resolution, which leads to an ADC with $(x_{fs} \cdot x_{\mathrm{OSR}})$ modes.

7.5 Quantizer Flexibility

The only flexibility required in the quantizer is that it needs to work at all sample frequencies. Figure 7.4 shows a latch frequently used in the quantizer of the modulator. For this latch, the probability of occurrence of a meta-stable state whose

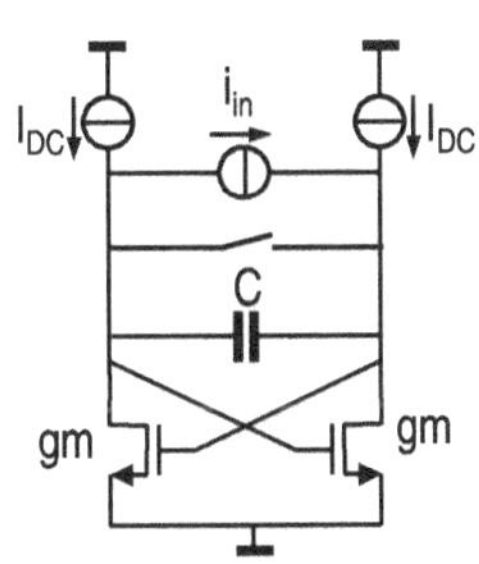

Figure 7.4: *General model of a latch frequently used in the quantizer of a ΣΔ modulator*

duration is longer than a time T_e is [48]:

$$\mathrm{BER} = P(t > T_e) \approx e^{\frac{-gm}{Cf_s}} \quad [\text{-}] \tag{7.6}$$

In which gm is the gm of the latch transistors, C is the total capacitance between node 1 and 2, and f_s is the frequency with which the quantizer is sampled. Close to the meta-stable state, which only occurs at small quantizer input signals, the quantizer is prone to make an error in its decision whether to output a one or a minus one as in this state the output signal of the quantizer latch is still very small at time T_e. On the other hand, due to the small quantizer latch input signal, the error made is small, as the input signal is very close to the quantizer reference. On top of that, the coincidental error is shaped by the modulators loop filter like the quantization noise, and generally this will lead to a very limited performance reduction of the $\Sigma\Delta$ modulator. What is of utmost importance though, is that the quantizer is forced to decide either to output a digital one or minus one. To decrease the probability of a bit error, re-clocking of the latch output can be done by an additional non-transparent edge triggered flip-flop as it introduces additional gain. Alternatively the quantizer can be designed such, that if a meta-stable state occurs, the quantizer outputs the same value as was done in the previous clock cycle. In both cases it is very important that the quantizer output bit is taken over by the digital output *and* feedback DAC correctly. If the output bit of the modulator is different from the bit fed back to the feedback DAC, the performance of the modulator will decrease very rapidly, as this error translates back to a bit error in the DAC.

In a multi-mode ADC with flexible clock frequency the gm of the transistors in the latch can be reduced proportional to the sample frequency to have the same BER, to save current in the low speed modes. The actual derivation of the requirement of the quantizer BER is out of the scope of this book.

7.6 Conclusions

For future multi-mode, multi-pipe transceivers a fixed ADC clock strategy results in the most flexible system. It comes at the cost of higher power consumption for the ADC and digital processing because of the always high clock frequency and required sample-rate conversion, and therefore is especially suited for nm-technologies.

The multi-mode modulators presented in this book in principle are designed for a system with a fully flexible ADC clock. For such systems a CT 1-bit $\Sigma\Delta$ modulator with SC feedback DAC is proven to be scalable, putting minimum additional requirements on the circuits. If the noise density of the modulator can be kept constant, changing the clock frequency requires only the programming of capacitors in the loop filter and DAC. Furthermore, by changing the local feedback coefficients in the loop filter, the bandwidth of the modulator can be optimized, increasing the modulators noise shaping efficiency. This way, a multi-mode $\Sigma\Delta$ modulator IP block can be designed, which can be programmed to different conversion bandwidths to implement the multi-mode capability. At the same time the $\Sigma\Delta$ modulator IP block can be used in different receiver applications, decreasing time-to-market of products containing receivers, which require a $\Sigma\Delta$ modulator..

It should be noted that the extremely scalable ADC of Sect. 9.1.3 can be used for the fixed clock system and the LO divided systems as well, for prove of concept of these systems.

Chapter 8

$\Sigma\Delta$ Modulator Efficiency

A way of determining the quality of an A/D converter design is to evaluate its performance parameter-cost ratios. A Figure-of-Merit (FOM) relates $\Sigma\Delta$ modulator performance and cost parameters. With a FOM it can be determined whether a design efficiently uses its secondary inputs compared to other designs presented in literature. A benchmark over existing $\Sigma\Delta$ modulator implementations yields the state-of-art FOM with their individual performance and cost parameters as inputs.

The most important performance parameters for $\Sigma\Delta$ modulator are DR or peak SNR and HD3D. The most important $\Sigma\Delta$ modulator cost parameters are power consumption and silicon area. As will be shown in this chapter, a conventional FOM (FOM_{DR}), that relates power with bandwidth and ENOB (or DR) is available. This FOM is frequently misused to compare noise and distortion of different ADC implementations, but the mechanisms between power consumption and noise and the relation between power spending and distortion are completely different. Unfortunately, separate FOMs to benchmark $\Sigma\Delta$ modulator linearity or area is not available. This chapter will evaluate the conventional FOM, and will introduce new FOMs to benchmark the ADC dynamic range and linearity with the consumed power and area in separate FOMs. The FOMs discussed in this chapter are:

- Conventional power efficiency FOM: FOM_{DR}
- New power efficiency FOM: $\text{FOM}_{eq,th}$
- New distortion FOM: FOM_{HD3D}
- New area FOM: FOM_{area}

These FOMs will be used to benchmark the $\Sigma\Delta$ modulators presented in this book with the $\Sigma\Delta$ modulators published in literature.

R.H.M. van Veldhoven, A.H.M. van Roermund, *Robust Sigma Delta Converters*, Analog Circuits and Signal Processing, DOI 10.1007/978-94-007-0644-6_8,

The 117 modulators used in the various benchmarks performed in this chapter, are modulators published at the ISSCC, the symposium on VLSI Circuits or in the JSSC in the time frame 1997–2009. Single and multi-bit modulators are distinguished from each other, and number 54 and 73 respectively. Furthermore, continuous time modulators (CT) are distinguished from discrete time (DT) modulators, and number 60 and 67 respectively. Low-pass as well as bandpass converters are included, but are not distinguished (the number of bandpass modulators included in the benchmark is only 6). The legend for the modulator benchmark figures presented in this chapter is shown in Fig. 8.1 together with the number of included modulators.

□ 1-bit CT ΣΔ modulator published at ISSCC/JSSC/VLSI	34
◇ 1-bit DT ΣΔ modulator published at ISSCC/JSSC/VLSI	20
+ Multi-bit CT ΣΔ modulator published at ISSCC/JSSC/VLSI	26
× Multi-bit DT ΣΔ modulator published at ISSCC/JSSC/VLSI	47

Figure 8.1: *Legend of the ΣΔ modulator benchmark figures presented in this chapter*

Some of the benchmarks presented in this chapter are a function of the minimum transistor length L_{min} available in the technology the particular modulator was designed in. As the introduction rate of new IC technologies is about 2 years, the L_{min} x-axis also can be interpreted as an alternative time-axis.

To indicate the performance matrix of the modulators benchmarked, Fig. 8.2 shows the peak SNR vs. bandwidth of all modulators. Bandwidth ranges from 1 kHz up to 40 MHz, peak SNR ranges from 118 dB down to 50 dB.

Multi-bit modulators tend to represent more extreme performance data points (high bandwidth, high peak SNR) as they can achieve a higher algorithmic accuracy at a lower sample frequency in a given technology compared to single bit converters. The extreme high bandwidth area (at about 20 MHz) in the figure is dominated by multi-bit CT modulators. These modulators have less stringent circuit bandwidth requirements compared to their SC counterparts, which makes it easier to achieve high modulator bandwidth at low power consumption.

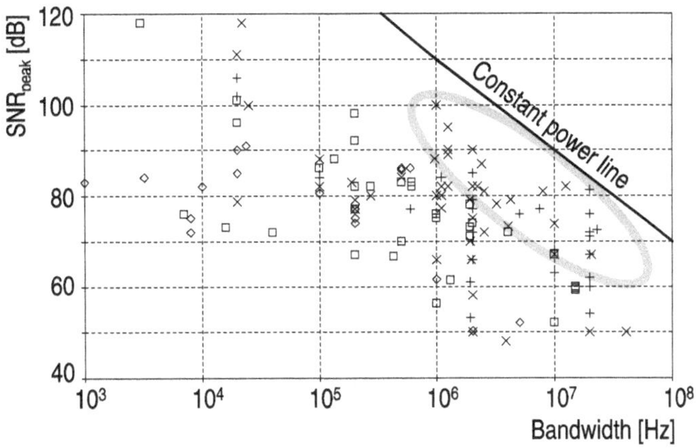

Figure 8.2: *Modulator peak SNR versus bandwidth for all benchmarked modulators*

8.1 Power Efficiency FOM: FOM_{DR}

Equation 8.1 gives an example of a well established FOM for ADC power consumption, normally used for Nyquist ADCs:

$$\text{FOM}_{ENOB} = \frac{P}{2^{\text{ENOB}}2B} \quad \text{or} \quad \text{FOM}_b = \frac{P}{2^b 2B} \;\; [\text{J/conversion}] \qquad (8.1)$$

in which P is the power consumption of the ADC, ENOB is the effective number of bits, and B is the ADC bandwidth. For $\Sigma\Delta$ modulators, the effective number of bits is interchanged by DR (DR is defined in the voltage domain, not in the power domain!), as it is not common to use ENOB as a performance indicator for $\Sigma\Delta$ modulators. The DR represents the noise performance of the $\Sigma\Delta$ modulator. This changes Eq. 8.1 into:

$$\text{FOM}_{DR} = \frac{P}{\text{DR} \cdot 2B} \;\; [\text{J/conversion}] \qquad (8.2)$$

The state-of-art FOM can be used to do a first order parameter estimation for an ADC to be developed once the other parameters in the FOM are known. For example, from Eq. 8.2 a first order estimate for the power consumption for a $\Sigma\Delta$ modulator to be designed can be calculated once its required bandwidth and DR and the state-of-art FOM are known. Extensive analysis on FOM_{DR} has been done in [49].

FOM_{DR} relates the ADC power consumption, with its bandwidth and DR. The FOM_{DR} lacks the referencing of the ADC dynamic range to absolute signal levels (Sect. 6.5.4). This is illustrated in Fig. 8.3. The FOM_{DR} has its DR as an input

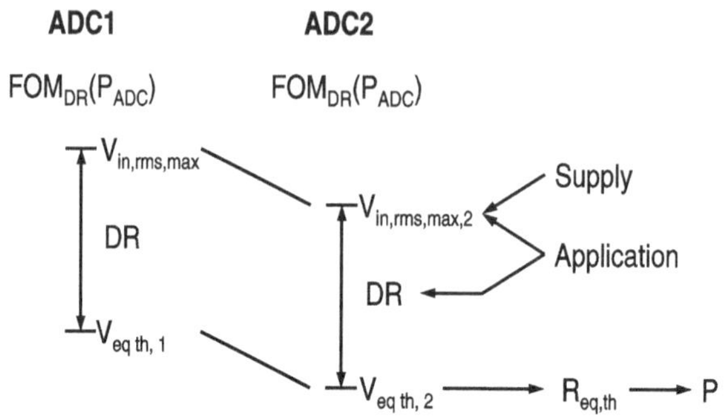

Figure 8.3: *FOM_{DR} and absolute signal referencing*

parameter, but the dynamic range is not related to the absolute signal levels at the input of the ADC, but the input signal level to the ADC can be limited by the supply voltage and/or by ADC application. This will lead to lower noise requirements for the ADC. In Fig. 8.3, ADC1 has a much higher input signal ($V_{in,rms,max,1}$) compared to the input signal of ADC2 ($V_{in,rms,max,2}$). Both ADCs have the same dynamic range determined by the application. Therefore, the absolute noise level $V_{eq\ th,1}$ of ADC1 that can be tolerated is much higher than the noise level $V_{eq\ th,2}$ of ADC2. When the ADC SNR is thermal noise limited, which is mostly the case in ΣΔ modulators, the relation between input signal and ADC power consumption is:

$$\text{SNR} = \frac{V_{in,rms,max}}{V_{eq,th}}, \quad V_{eq,th} \propto \sqrt{R_{eq,th}}, \quad P \propto \frac{1}{R_{eq,th}}$$

$$\Rightarrow P_{ADC} \propto \frac{\text{SNR}^2}{V_{in,rms,max}^2} \tag{8.3}$$

If the maximum ADC input signal is halved, the power of the ADC is expected to increase by a factor of 4, as the noise density and the according equivalent input impedance of the ADC $R_{eq,th}$ has to reduce by a factor of four. This will make the design of ADC2 much more challenging than the design of ADC1, if the power of both ADCs is limited to the same amount, which is not taken into account in the FOM_{DR} (Eq. 8.2).

Another disadvantage of FOM_{DR}, is that the ADC resolution ENOB does not unambiguously take into account noise and/or distortion. Often resolution 2^b in

Eq. 8.1 is interchanged with SNR$_{peak}$, DR, SNDR or even SFDR in literature. This makes the comparison of ADC power efficiencies inconsistent and dubious. The lowering of both noise and distortion of an ADC will cost power, but their relation to and mechanism of power spending is completely different. Therefore, this chapter splits the power efficiency FOMs for noise and distortion. First, a FOM based on noise impedance rather than SNR$_{peak}$, DR or 2^b is presented in Sect. 8.2. Secondly, a power efficiency FOM to solely benchmark harmonic distortion will be presented in Sect. 8.3. In this benchmark the third order harmonic distortion will be related to power consumption, instead of the replacing 2^b by SNDR or SFDR in Eq. 8.1.

8.1.1 Benchmarking with FOM$_{DR}$

In Fig. 8.4 the FOM$_{DR}$ is plotted for all modulators. The state-of-art FOM$_{DR}$ has

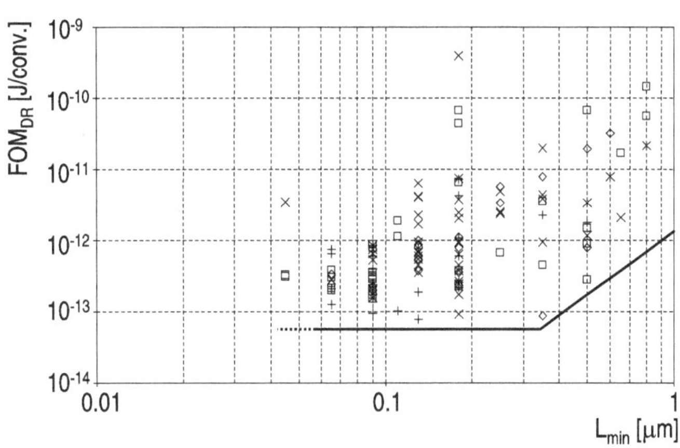

Figure 8.4: *FOM$_{DR}$ versus minimum transistor length L_{min}*

saturated to about 0.1 pJ/conversion at 0.35 µm technologies. In deep sub-micron technologies CT modulators are preferred above SC modulators. This is probably due to the decreasing technology native supply voltage, making it more difficult to make low impedance switches required for SC loop filter implementations.

Table 8.1 presents the top thirty state-of-art $\Sigma\Delta$ modulators according to the FOM$_{DR}$. The implementations presented in Sects. 9.1.1 and 9.1.2 are among the top thirty most efficient $\Sigma\Delta$ modulators.

Ranking [number]	FOM_{DR} [J/conv.]	Technology [μm]	Multi-bit? [yes/no]	Author	Year	Reference
1	7.92E-14	0.13	yes	Mitteregger	2006	[50]
2	8.79E-14	0.35	no	Chae	2007	[51]
3	9.21E-14	0.18	yes	Kwon	2006	[52]
4	9.60E-14	0.09	yes	Bos	2009	[53]
5	1.03E-13	0.11	yes	Matsukawa	2009	[54]
6	1.25E-13	0.065	yes	Huang	2009	[55]
7	1.52E-13	0.09	no	Crombez	2009	[56]
8	1.55E-13	0.09	yes	Koh	2005	[57]
9	1.73E-13	0.09	no	Crombez	2009	[56]
10	1.74E-13	0.18	yes	Park	2008	[58]
11	1.75E-13	0.09	yes	Malla	2008	[59]
12	1.88E-13	0.13	yes	Doerrer	2005	[60]
13	1.89E-13	0.13	yes	Park	2009	[61]
14	1.97E-13	0.09	no	Burger	2001	[62]
15	2.01E-13	0.065	no	Putter	2007	[63]
16	2.09E-13	0.065	yes	Dhanasekaran	2009	[64]
17	2.11E-13	0.09	no	Crombez	2009	[56]
18	2.16E-13	0.065	no	Putter	2007	[63]
19	2.22E-13	0.18	yes	Yaghini	2005	[65]
20	2.24E-13	0.18	no	***Veldhoven***	2003	[66]
21	2.26E-13	0.09	no	Yao	2004	[67]
22	2.27E-13	0.09	yes	Malla	2008	[59]
23	2.36E-13	0.18	no	***Veldhoven***	2003	[66]
24	2.39E-13	0.18	no	***Veldhoven***	2003	[66]
25	2.41E-13	0.09	yes	Bos	2009	[53]
26	2.45E-13	0.18	yes	Lee	2006	[68]
27	2.61E-13	0.18	yes	***Veldhoven***	2002	[69]
28	2.67E-13	0.18	yes	Lee	2008	[70]
29	2.79E-13	0.065	yes	Shu	2008	[71]
30	2.80E-13	0.5	no	Das	2005	[72]

Table 8.1: *Top 30 of* FOM_{DR} *benchmark*

8.2 Power Efficiency FOM: $\mathrm{FOM}_{eq,th}$

The power efficiency FOM of the previous section benchmarks the ΣΔ modulator's DR and bandwidth against power consumption. In the FOM of this section, the noise performance is judged by relating the ADC's load impedance to its equivalent noise impedance:

$$\mathrm{FOM}_{eq,th} = \frac{R_{load}}{R_{eq,th}} \text{ [-]} \tag{8.4}$$

in which R_{load} is the resistive load the ADC shows on the supplies and $R_{eq,th}$ is the equivalent ADC input noise impedance.

The load impedance R_{load} is calculated from the ADC power supply voltage and power consumption.

$$R_{load} = \frac{V_{supply}^2}{P_{ADC}} \quad [\Omega] \tag{8.5}$$

The equivalent noise impedance $R_{eq,th}$ is the noise impedance the ADC shows at its input. It can be calculated from the ADC peak SNR and its maximum input signal.

$$R_{eq,th} = \frac{V_{in,rms,max}^2}{4kTB \cdot \text{DR}^2} \quad [\Omega] \tag{8.6}$$

Combining Eq. 8.5 and Eq. 8.6 leads to:

$$\text{FOM}_{eq,th} = \frac{V_{supply}^2 4kTB \cdot \text{DR}^2}{V_{in,rms,max}^2 \cdot P_{ADC}} \quad [\text{-}] \tag{8.7}$$

Note that the FOM$_{eq,th}$ should be as large as possible, unlike the FOM$_{DR}$ which should be as small as possible.

Theoretically, FOM$_{eq,th}$ can become greater than one, if the noise impedance is dominated by the differential pair of the input stage. In this case the applied degeneration is very limited or absent (Fig. 6.9a). If the input transistors are biased in weak inversion (Eq. 6.19), the maximum gm of $\frac{qI_D}{kT} \approx 40 \cdot I_D$ [A/V] is achieved. As there are two transistors, the equivalent noise impedance becomes $R_{eq,th} = \frac{2}{gm} = \frac{1}{20I_D}$ and the load impedance $R_{load} = V_{supply}/(2I_D)$ when it is assumed that all current is spent in the input transistors. This leads to a FOM$_{eq,th}$ of $10V_{supply}$. As supply voltages are in the range of 1 to 5 V, this means an upper limit of 10 to 50. In reality this FOM level will be impossible to achieve, as not all current is spent in the input transistors, and the total noise of the modulator is not determined by the input transistors only. Furthermore, if the transistors are biased in weak inversion achieving the maximum gm, their V_{gt} will be extremely low, leading to a highly non-linear input stage for reasonable input signal swings.

Also for a degenerated transistor input pair (Fig. 6.9b), the upper limit for the FOM$_{eq,th}$ can be calculated. The equivalent noise impedance of a degenerated differential pair is $R_{eq,th} = \frac{2}{gm} + 2R_{in}$. If it is assumed that the degeneration resistors R_{in} are larger than $1/gm$, this can be simplified to $R_{eq,th} = 2R_{in}$. Furthermore, like for a non-degenerated differential pair the input transistors are assumed to be biased in weak inversion (Eq. 6.19), which gives the maximum $gm \approx 40I_D$. As there are two transistors, the load impedance $R_{load} = V_{supply}/(2I_D)$ when it is assumed that all current is spent in the input transistors. This leads to a $\text{FOM}_{eq,th}$ limit of $\frac{10V_{supply}}{gmR_{in}}$. The upper boundary of the $\text{FOM}_{eq,th}$ for a non-degenerated pair is reduced by the degeneration factor. As degeneration factors of 10 or larger are very common, the upper boundary of the $\text{FOM}_{eq,th}$ will be close to or lower than one.

Properties of $\text{FOM}_{eq,th}$ are:

1. The $\text{FOM}_{eq,th}$ is dimensionless.
2. The $\text{FOM}_{eq,th}$ should be as high as possible, unlike FOM_{DR}, which should be as low as possible.
3. The $\text{FOM}_{eq,th}$ is equal to one when load and noise impedance are equal.
4. Theoretically, the $\text{FOM}_{eq,th}$ for a modulator with a non-degenerated transistor input pair, has an upper boundary of $10V_{supply}$; in reality however, this will be difficult (if not impossible) to achieve, and will lead to a highly non-linear input stage for reasonable input signal swings. For a degenerated input pair the upper boundary is reduced by the degeneration factor gmR_{in}.
5. The ADC input signal is now included in $\text{FOM}_{eq,th}$, which results in power scaling. If the input signal of the ADC is halved, the equivalent input impedance should go down with a factor of 4 to maintain the same equivalent input noise. This will lead to a higher power consumption as ADCs with a small input signal and a large DR, are expected to be more power hungry than ADCs with same DR but a large input signal. In the system supply domain selection, signal equalization and system power budgeting, a more trustworthy figure for the ADC power consumption can be estimated using the $\text{FOM}_{eq,th}$.
6. $\text{FOM}_{eq,th}$ is linear with bandwidth. When the ADC bandwidth is doubled and the total integrated noise is kept the same, the noise impedance should be halved to keep the same SNR which means power is expected to increase by a factor of two.
7. The $\text{FOM}_{eq,th}$ is defined at a comfortable room temperature of 300 K.

Furthermore it should be noted that:

$$\text{FOM}_{eq,th} = \frac{1}{\text{FOM}_{DR}} \frac{V_{supply}^2 2kT \cdot \text{DR}}{V_{in,rms,max}^2} \quad [\text{-}] \tag{8.8}$$

8.2.1 Benchmarking with FOM$_{eq,th}$

Figure 8.5 shows the FOM$_{eq,th}$ (Eq. 8.4) for all ΣΔ modulators in the benchmark, as a function of minimum transistor L_{min}. As mentioned, the FOM should be as high as possible; the upper boundary is $10V_{suppy}$. From Fig. 8.5 it can be seen that the FOM has not improved much over newer technologies, like the old power efficiency FOM of Eq. 8.2. The FOM$_{eq,th}$ predicts that for most state-of-

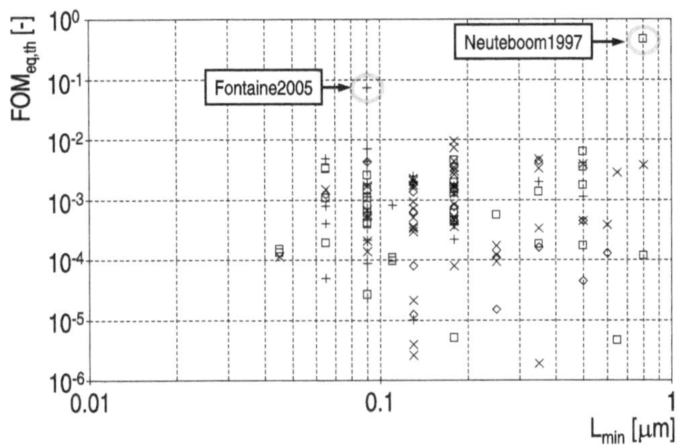

Figure 8.5: *FOM$_{eq,th}$ versus minimum transistor length L_{min}*

art modulators, the noise impedance is always about 100 times larger than the load impedance between the supply. There are two positive exceptions: Fontaine 2005 [73] and Neuteboom 1997 [74], which achieve a FOM$_{eq,th}$ of 0.075 and 0.46 respectively. The first design uses very simple non-degenerated gm-C integrator stages [75], leading to a minimum of circuitry overhead. The majority of the current is spent in the input stage allowing for a very low equivalent noise impedance, and a high linearity (HD3D $=$ 83 dB) at a small input signal level of $50mV_{rms}$. The second design uses a limited amount of resistive degeneration in its input stage and achieves an HD3D of 50 dB at an input signal swing of $14mV_{rms}$. Both designs use limited or no degeneration. Therefore, their noise level is dominated by input stage transistors, which can lead to a higher FOM$_{eq,th}$ as explained. It should be noted though, that choosing a low input signal swing also leads to a low equivalent noise impedance, and accompanying high power consumption. The low signal swing allows for limited or no degeneration in the input stage. Therefore, one can benefit from the lower noise impedance per unit current, however the requirement on the noise impedance increases quadratically with $1/V_{in,rms,max}$. The power inefficiency of both modulators, becomes clear when their FOM$_{DR}$ is calculated which are 0.71 and 2.68 pJ/conversion, which is at least 7 times worse than state-of-art. Conclusion is, that a small maximum input signal level for the ADC should only be used, when it is forced by application or technology.

Table 8.2 presents the top thirty state-of-art ΣΔ modulators according to the $FOM_{eq,th}$. The implementations presented in Sects. 9.1.2 and 9.1.3 are among the top thirty most efficient ΣΔ modulators. If [73] and [74] are excluded from the benchmark for reasons described earlier, Park 2008 [58] sets the state-of-art im-

Ranking [number]	$FOM_{eq,th}$ [-]	Technology [µm]	Multi-bit? [yes/no]	Author	Year	Reference
1	4.60E-01	0.8	no	Neuteboom	1997	[74]
2	7.47E-02	0.09	yes	Fontaine	2005	[73]
3	9.33E-03	0.18	yes	Park	2008	[58]
4	7.28E-03	0.18	yes	Kwon	2006	[52]
5	7.04E-03	0.09	yes	Bos	2009	[53]
6	6.24E-03	0.5	no	Zwan	1997	[76]
7	4.86E-03	0.065	yes	Huang	2009	[55]
8	4.74E-03	0.35	yes	Yang	2003	[77]
9	4.48E-03	0.18	no	***Veldhoven***	2003	[66]
10	4.41E-03	0.35	no	Chae	2007	[51]
11	4.33E-03	0.09	yes	Malla	2008	[59]
12	4.31E-03	0.18	yes	Putter	2004	[78]
13	4.24E-03	0.09	no	Burger	2001	[62]
14	3.80E-03	0.5	yes	Fujimori	2000	[79]
15	3.78E-03	0.8	yes	Leung	1997	[80]
16	3.65E-03	0.18	no	Sauerbrey	2002	[81]
17	3.50E-03	0.5	no	Das	2005	[72]
18	3.39E-03	0.065	no	Putter	2007	[63]
19	3.36E-03	0.18	yes	Lee	2006	[68]
20	3.30E-03	0.35	yes	Oliaei	2002	[82]
21	3.25E-03	0.065	no	Putter	2007	[63]
22	3.01E-03	0.18	yes	Lee	2008	[70]
23	2.83E-03	0.65	yes	Geerts	2000	[83]
24	2.67E-03	0.18	yes	Jiang	2002	[84]
25	2.59E-03	0.09	no	***Ouzounov***	2007	[85]
26	2.46E-03	0.13	yes	Doerrer	2005	[60]
27	2.26E-03	0.13	yes	Kim	2006	[86]
28	2.01E-03	0.13	yes	Gomez	2002	[87]
29	2.01E-03	0.13	yes	Park	2009	[61]
30	1.99E-03	0.35	yes	Nguyen	2005	[88]

Table 8.2: *Top 30 of* $FOM_{eq,th}$ *benchmark*

plementation loss of 99%. This means that the load impedance measured between the supply lines is 100 times smaller compared to the modulator's equivalent noise impedance.

Figure 8.2 showed that extreme performance modulators are mostly implemented with multi-bit modulators. Multi-bit modulators can achieve a high SQNR at a low OSR. To achieve a high DR, the input referred noise normally dominated by the circuit noise also should be low, which means a low $R_{eq,th}$. Figure 8.6 proves that multi-bit modulators are most commonly used for low OSR, and low equivalent noise impedance modulators. For narrow band systems the technology speed limitations are normally no issue, and the OSR can be chosen high to achieve a sufficiently high SQNR with a 1-bit modulator. As signal bandwidth is limited, $R_{eq,th}$ does not have to be extremely low to achieve a high DR, and 1-bit modulators can be used.

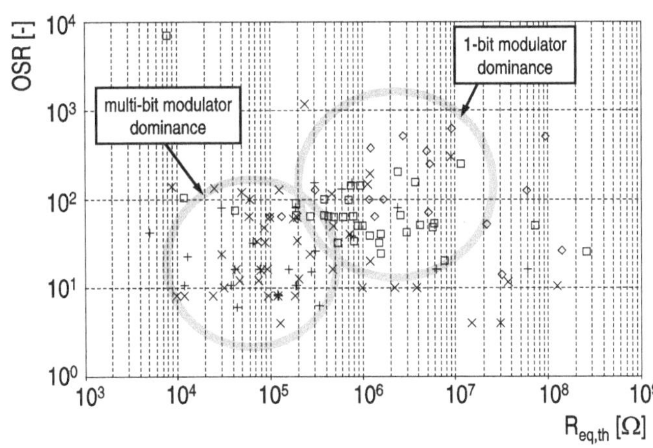

Figure 8.6: *Modulator OSR versus* $R_{eq,th}$

For the benchmarked modulators, $R_{eq,th}$ is hardly scaling over technology. This is shown in Fig. 8.7. The modulator's equivalent noise impedance limit in the benchmark is about 5 kΩ. This lower boundary is set by the maximum current to be spent. If an implementation loss of 99% (Park 2008 [58]) is assumed, the equivalent load impedance seen by the supply is 50 Ω. This means 20 mA load current per volt supply. As most of the modulators in the benchmark are for portable applications this is already quite high. The only way to cross this lower boundary of $R_{eq,th}$ is the discovery of more efficient modulator (circuit) architectures with lower implementation losses like the one in [58].

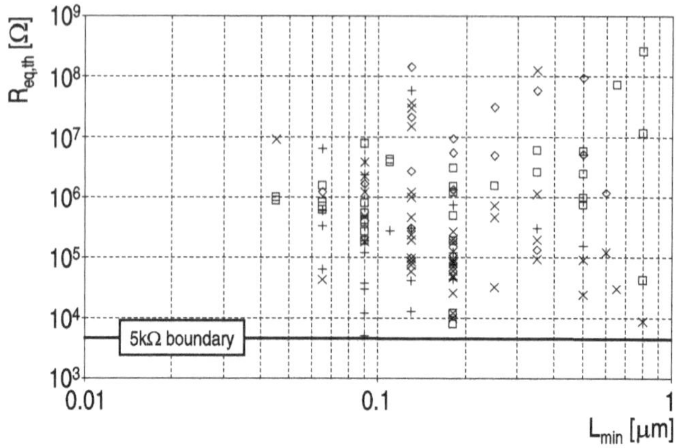

Figure 8.7: *Equivalent modulator noise impedance $R_{eq,th}$ versus minimum transistor length L_{min}*

8.3 Distortion FOM: FOM$_{HD3D}$

In this section a FOM will be introduced for the distortion of the ΣΔ modulator. For distortion only the HD3 is considered, as the HD2 is absent in a perfectly balanced circuit and is determined by the matching of e.g. the input differential pair transistors rather than power consumption.

The FOM$_{HD3D}$ is defined as the ratio between the ΣΔ modulators measured HD3D and a upper boundary calculated from theory, or:

$$\text{FOM}_{HD3D} = \frac{\text{HD3D}_{measured}}{\text{HD3D}_{theory}} \; [\text{-}] \tag{8.9}$$

For the calculation of the FOM$_{HD3D}$, the following assumptions are made:

1. An RC integrator input stage is taken as a reference as this topology is expected to have the best linearity of all possible implementations (Sect. 6.6.1.2) because it uses feedback. Its HD3D$_{theory}$ (Eq. 6.19) is:

$$\text{HD3D}_{theory} = \frac{24 \cdot gm \cdot I_D^2 R_{in}^3}{\left(1 + \frac{R_{in}}{R_{DAC}}\right) V_{in,rms,max}^2} \; [\text{-}] \tag{8.10}$$

2. The input pair is assumed to be in weak inversion (Eq. 6.19), as in weak inversion the gm per unit current is largest. To set the upper boundary, the gm is taken to be the absolute maximum in weak inversion which is $\frac{qI_D}{kT} \approx 40 \cdot I_D$ [A/V].

3. If it is assumed that all the current is spent in the differential input stage, I_D in Eq. 8.9 can be replaced by $\frac{1}{2}V_{supply}/R_{load}$. This gives an upper limit boundary. This assumption is false for single ended circuits, but as single ended modulators are very rare this is neglected.

4. Normally $R_{in} \approx R_{DAC}$; here it is assumed that $R_{in} = R_{DAC}$.

5. It is assumed that the input and DAC resistor R_{in} and R_{DAC} dominate noise and therefore $R_{in} = \frac{1}{4}R_{eq,th}$. This gives an upper limit boundary.

The combination of the assumptions above together with Eq. 8.10 and Eq. 8.9 yields the FOM_{HD3D}:

$$\text{FOM}_{HD3D} = \text{HD3D}_{measured}\frac{128}{120} \cdot \text{FOM}_{eq,th}^3 \cdot \frac{V_{in,rms,max}^2}{V_{supply}^3} \quad [\text{-}] \tag{8.11}$$

For a non-degenerated and degenerated differential input pair (Figs. 6.9a and 6.9b) the further elaboration is different. The HD3D of a non-degenerated input pair is given by Eq. 6.14 which is:

$$\text{HD3D}_{non\text{-}degenerated} = \frac{48 \cdot I_D^2}{V_{in,rms,max}^2 \cdot gm^2} \quad [\text{-}] \tag{8.12}$$

The HD3D for a degenerated input pair according to Eq. 6.15 is:

$$\text{HD3D}_{degenerated} = \frac{48 \cdot I_D^2 \cdot R_{in}^2}{V_{in,rms,max}^2} \quad [\text{-}] \tag{8.13}$$

Following the same reasoning as in the derivation of the upper boundaries for the $\text{FOM}_{eq,th}$, for both type of input pairs $R_{load} = V_{supply}/(2I_D)$. $R_{eq,th}$ in case of a non-degenerated input pair can be approximated by $2/gm$ (with $gm = 40I_D$), in case of a degenerated input pair it can be approximated by $2R_{in}$. Using Eq. 8.9, this in both cases leads to a FOM_{HD3D} of:

$$\text{FOM}_{HD3D} = \frac{\text{HD3D}_{measured}}{3} \cdot \text{FOM}_{eq,th}^2 \cdot \frac{V_{in,rms,max}^2}{V_{supply}^2} \quad [\text{-}] \tag{8.14}$$

Although the outcome of the FOM_{HD3D} for both type of input pairs seems the same, the maximum achievable $\text{FOM}_{eq,th}$ for a modulator with a degenerated input pair is expected to be lower, as it has a lower upper boundary compared to a modulator with a non-degenerated input pair. Therefore, the expected FOM_{HD3D} for a modulator with a degenerated input pair is expected to be worse than the FOM_{HD3D} for a modulator with a non-degenerated input pair. On the other hand, the input signal for a non-degenerated input pair will be smaller, than for a degenerated input pair, which will give a modulator with a degenerated input pair a better the FOM_{HD3D}. Properties of the FOM_{HD3D} are:

- FOM_{HD3D} is dimensionless and should be as high as possible.
- FOM_{HD3D} can be used in different ways. The modulator's HD3D can be benchmarked to a state-of-art (in terms of linearity) input stage circuit topology, which in this case is the RC integrator input stage. Another way of using the FOM_{HD3D}, is that each modulator is benchmarked to the circuit topology which is really used as the modulator's input stage (either a non-degenerated input pair, a degenerated input pair or an RC integrator input stage).
- A modulator with a high $FOM_{eq,th}$, likely also has a high FOM_{HD3D}, as both FOMs are optimized when most of the current is spent in the input stage.

8.3.1 Benchmarking with FOM_{HD3D}

The FOM_{HD3D} (Eq. 8.11) for the modulators in the benchmark is plotted in Fig. 8.8. All the modulators in the benchmark, are compared to an RC integrator input stage as it has the best linearity per power consumption, even if a different input stage topology is used. Most of the state-of-art modulators that achieve a FOM_{HD3D} of close to 1E-3, have an RC integrator input stage, in which the input and DAC resistors are implemented with a resistor or a switched capacitor

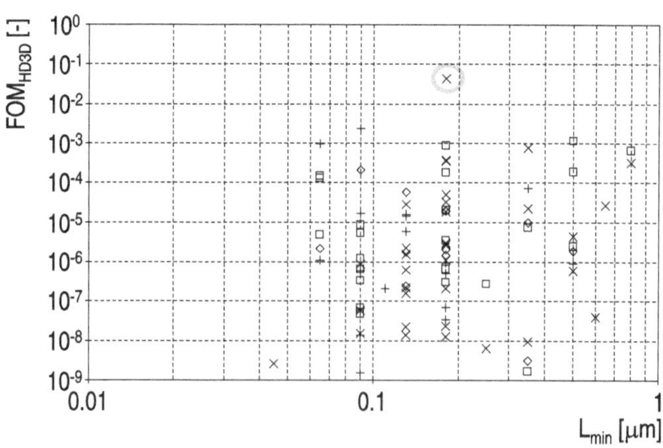

Figure 8.8: *FOM_{HD3D} versus minimum transistor length L_{min}*

equivalent. This confirms the assumption that an RC integrator input stage has the best linearity of all possible implementations published in literature. There is one modulator with an extremely high FOM$_{HD3D}$ of 4.29E-2 (Park 2008 [58]). This modulator uses incomplete settling in its SC integrator stages, which reduces power without decreasing linearity. This way very competitive FOM$_{eq,th}$ and FOM$_{HD3D}$ are achieved.

Table 8.3 presents the top thirty state-of-art $\Sigma\Delta$ modulators according to the FOM$_{HD3D}$. From the table, it can be seen that the modulators with limited or no

Ranking [number]	FOM$_{HD3D}$ [-]	Technology [μm]	Multi-bit? [yes/no]	Author	Year	Reference
1	4.29E-02	0.18	yes	Park	2008	[58]
2	2.34E-03	0.09	yes	Fontaine	2005	[73]
3	1.14E-03	0.5	no	Zwan	1997	[76]
4	9.60E-04	0.065	yes	Huang	2009	[55]
5	8.70E-04	0.18	no	***Veldhoven***	2003	[66]
6	7.46E-04	0.35	yes	Yang	2003	[77]
7	6.48E-04	0.8	no	Neuteboom	1997	[74]
8	3.64E-04	0.18	yes	Kwon	2006	[52]
9	3.54E-04	0.18	yes	Lee	2008	[70]
10	3.18E-04	0.8	yes	Leung	1997	[80]
11	2.08E-04	0.09	no	Burger	2001	[62]
12	1.91E-04	0.5	no	Zwan	1997	[76]
13	1.79E-04	0.18	no	Silva	2006	[89]
14	1.50E-04	0.065	no	Putter	2007	[63]
15	1.32E-04	0.065	no	Putter	2007	[63]
16	7.13E-05	0.35	yes	Nguyen	2005	[88]
17	5.79E-05	0.13	no	Tsang	2006	[90]
18	5.01E-05	0.18	yes	Putter	2004	[78]
19	3.25E-05	0.18	yes	Lee	2008	[70]
20	2.77E-05	0.13	yes	Kim	2006	[86]
21	2.68E-05	0.65	yes	Geerts	2000	[83]
22	2.24E-05	0.35	yes	Oliaei	2002	[82]
23	2.01E-05	0.18	no	***Veldhoven***	2003	[66]
24	1.96E-05	0.18	no	Chae	2008	[91]
25	1.84E-05	0.18	yes	Lee	2006	[68]
26	1.64E-05	0.09	yes	Malla	2008	[59]
27	1.62E-05	0.13	yes	Park	2009	[61]
28	1.45E-05	0.13	yes	Mitteregger	2006	[50]
29	9.78E-06	0.35	no	Chae	2007	[51]
30	8.67E-06	0.09	no	***Ouzounov***	2007	[85]

Table 8.3: *Top 30 of FOM$_{HD3D}$ benchmark*

degeneration (Fontaine 2005 [73], Neuteboom 1997 [74]) have a high FOM_{HD3D}. This is because their relatively high $FOM_{eq,th}$. It should not be forgotten though, that this $FOM_{eq,th}$ was achieved at high power consumption, as input signal swing for both modulators was very small leading to a low equivalent noise impedance, and thus high power consumption. If their HD3D is benchmarked with Eq. 8.14 as they should be, their FOM_{HD3D} improves even further to 1.47E-2 and 9.47E-4 respectively. This is due to the fact that the expected linearity of an RC integrator stage is higher than for a (non-)degenerated input pair (Sect. 6.6.1). This confirms the assumption that for a high performance modulator, the majority of the current should be spent in the input stage to achieve low noise and high linearity.

As expected there is no modulator implementation published with a FOM_{HD3D} greater than one. This validates the choice of an RC integrator input stage as a reference input stage in the FOM_{HD3D}. Next to that, it is assumed that all current is spent in the input stage, which of course is not true in reality. The implementations presented in Sects. 9.1.2 and 9.1.3 are along the top thirty of modulators with the highest FOM_{HD3D}.

Figure 8.9 shows the $FOM_{eq,th}$ versus FOM_{HD3D}. A fourth order relation between $FOM_{eq,th}$ (Eq. 8.7) and FOM_{HD3D} (Eq. 8.11) is found. The figure shows, that if a good $FOM_{eq,th}$ is found for a certain modulator implementation, the expected FOM_{HD3D} will also be good and vice versa.

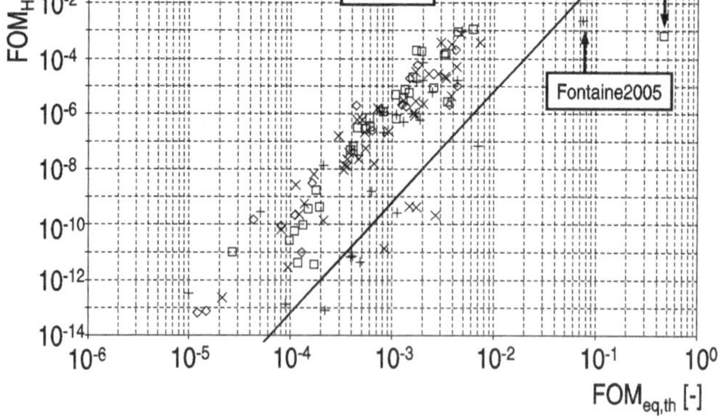

Figure 8.9: *$FOM_{eq,th}$ power versus FOM_{HD3D}*

8.4 Area FOM: FOM$_{area}$

Before determining a FOM for modulator area, first some relations between circuit area and power consumption are considered.

Higher power means larger transistor bias currents, which means larger transistor widths and thus larger area, or:

$$P \propto I_D \propto W \propto A \tag{8.15}$$

Equation 8.3 concluded that $R_{eq,th} \propto 1/P$. Therefore, at higher power, $R_{eq,th}$ gets lower. To achieve the same integrator unity gain, larger capacitors are required, which leads to a larger area, or:

$$P \propto 1/R_{eq,th} \propto C_{int} \propto A \tag{8.16}$$

Equation 8.3 predicts that power has a quadratical relation to modulator accuracy. If the accuracy of the modulator is increased by a factor of two, its power is expected to get four times higher, in the case the modulator performance is thermal noise dominated. In an optimized multi-bit modulator design there is no overhead in the offset of the comparators in the quantizer and therefore, if a two times higher accuracy is required, also a two times lower offset is required in the comparators. Therefore it can be said that:

$$\frac{1}{P} \propto V_{eq,th}^2 \propto \sigma_{offset}^2 \propto \frac{1}{A} \tag{8.17}$$

A similar relation can be found for 1/f noise. As a conclusion of the foregoing, from circuit theory perspective, a linear relation between power and area is expected.

In Fig. 8.10 the performance of all modulators is plotted as a function of modulator area A. Performance in this case is defined as FOM$_{eq,th} \cdot P$. In the figure state-of-art modulators achieve a very high performance in an as small as possible area. From the figure it is estimated that, that state-of-art modulator area scales with A.

In Fig. 8.11, performance FOM$_{eq,th} \cdot P$ is plotted as a function of P. A state-of-art modulator has an as large as possible performance for a minimum amount of power. From the plot it can be read that the state-of-art modulators have a linear relation between performance and power. Therefore, it can be said that:

$$\begin{aligned} &\text{FOM}_{eq,th} \cdot P \propto A \quad \text{and} \quad \text{FOM}_{eq,th} \cdot P \propto P \\ &\Rightarrow A \propto P \end{aligned} \tag{8.18}$$

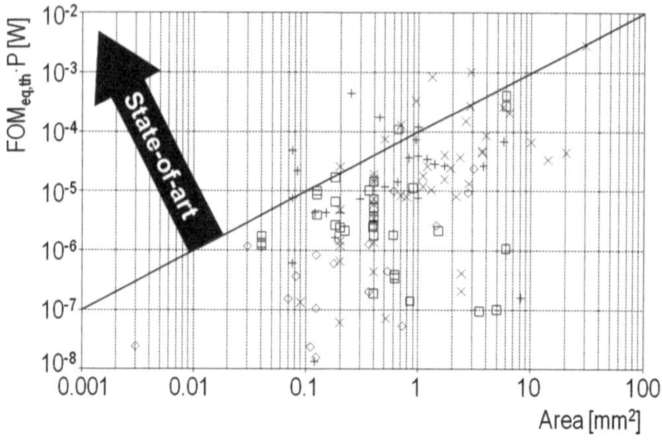

Figure 8.10: *Modulator performance versus area*

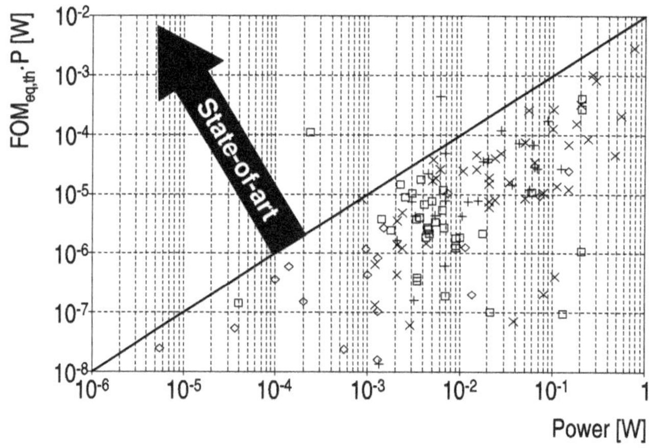

Figure 8.11: *Modulator performance versus power consumption*

which means that from the above it can be concluded that state-of-art modulators have a linear relation between area and power. As $P \propto A$, FOM_{area} can be defined as a relation between performance and area using the power efficiency FOM of Eq. 8.7, which leads to:

$$\text{FOM}_{area} = \text{FOM}_{eq,th} \cdot \frac{P}{A} = \frac{V^2_{supply} 4kTB \cdot \text{DR}^2}{V^2_{in,rms,max} \cdot A} \quad [\text{W/m}^2] \tag{8.19}$$

As an example an operational amplifier with area A and power consumption P is taken. If two of these amplifiers are switched in parallel, the area increases to $2A$ and the power consumption changes to $2P$. At the same time its SNR increases with $\sqrt{2}$ (Eqs. 8.3 and 8.7). When both amplifier designs sequentially are substituted in Eq. 8.19 the FOM_{area} will be the same as expected.

As said in Sect. 8.2, for a power efficient state-of-art modulator, the modulator's input signal should be made as large as possible. Therefore, if the maximum modulator performance per area is analyzed, $V_{in,rms,max}$ and supply voltage should be kept out of Eq. 8.19, as they have no direct relation to absolute performance. Therefore, Eq. 8.19 becomes:

$$\mathrm{FOM}_{area,\ SoA} = \mathrm{FOM}_{eq,th} \cdot \frac{P}{A} = \frac{4kTB \cdot \mathrm{DR}^2}{A} \quad [\mathrm{W/m^2}] \tag{8.20}$$

The modulator's absolute performance ($FOM_{eq,th} \cdot P \cdot V^2_{in,rms,max}/V^2_{supply}$) as a function of modulator area is presented in Fig. 8.12. The modulator's absolute

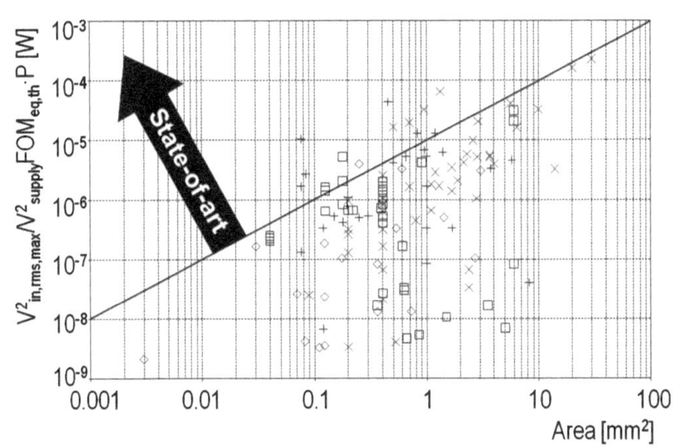

Figure 8.12: *Modulator performance versus area*

performance as a function of modulator power is presented in Fig. 8.13. Combination of the two figures reconfirm the linear relation between A and P. Note that in the remainder of the book only FOM_{area} (and not $FOM_{area,\ SoA}$) is used, to identify the most area optimized modulator design taking into account its design constraints ($V_{in,rms,max}$ and supply voltage).

In Fig. 8.14 modulator area A is plotted as function of modulator power consumption P. The figure indeed confirms that a linear relation between modulator power and area is a reasonable assumption (note that state-of-art can not be indicated in this figure). From the figure it can be seen that multi-bit modulators dominate the high power and area part of Fig. 8.14, where single bit modulators dominate the

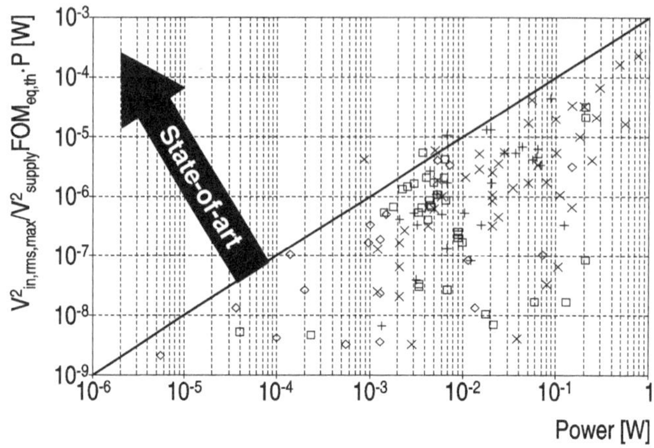

Figure 8.13: *Modulator performance versus power consumption*

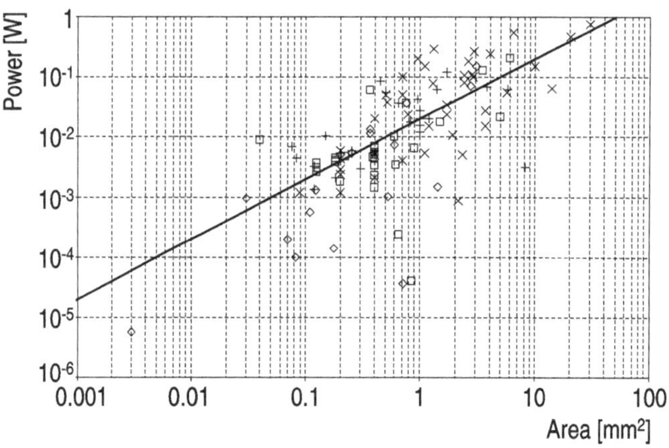

Figure 8.14: *Modulator power versus area*

low power, low area part of Fig. 8.14. This can be explained when the relation to Fig. 8.2 is observed. Multi-bit modulators more often represent more extreme data points, which will make them larger because of an increase in modulator complexity, and more power hungry due to higher circuit complexity and lower noise requirements.

Note that for a particular implementation, the linear relation assumed between P and A is subject to modulator architecture, technology, implementation efficiency (state-of-art design), implementation constraints, etcetera, which explains the cloud of implementations that are not exactly on the linear power-area relation line in Fig. 8.14.

In summary properties of the FOM$_{area}$ are:

- FOM$_{area}$ can be used at the start of the modulator design trajectory to estimate the state-of-the-art area that can be achieved, once the state-of-art FOM$_{area}$ is determined.
- FOM$_{area}$ is based on the assumption that there is a linear relation between modulator power and area, which seems plausible from the analysis made.
- FOM$_{area}$ should be as large as possible, as it strives for the highest modulator performance per unit area.
- FOM$_{area}$ has dimension [W/m^2].

If the highest absolute performance per modulator area is needed $FOM_{area,\ SoA}$ can be used. In this FOM, design constraints $V_{in,rms,max}$ and supply voltage are left out of the equation. Again, note that in the remainder of the book only FOM_{area} (and not $FOM_{area,\ SoA}$) is used.

8.4.1 Benchmarking with FOM$_{area}$

Figure 8.15 shows the area benchmark using the FOM of Eq. 8.20. Surprisingly, Fig. 8.15 shows that the FOM$_{area}$ for state-of-art multi-bit modulators is higher

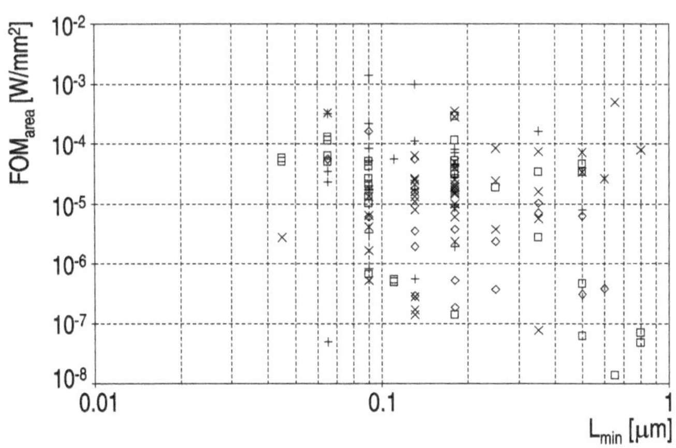

Figure 8.15: *FOM_{area} versus minimum transistor length L_{min}*

than the FOM_{area} for state-of-art 1-bit modulators. For the multi-bit modulators, Bos 2009 [53] has the best FOM_{area} of 1.38E-04 W/mm^2. For 1-bit modulators, Veldhoven 2003 [66] has the best FOM_{area} of 2.92e-5 W/mm^2 in GSM mode.

Table 8.4 presents the top thirty most area efficient ΣΔ modulators according Eq. 8.20. The implementations presented in Sects. 9.1.2, 9.2 and 9.3, are along the top thirty ΣΔ modulators which have the highest performance per area.

Ranking [number]	FOM_{area} [W/mm^2]	Technology [μm]	Multi-bit? [yes/no]	Author	Year	Reference
1	1.38E-04	0.09	yes	Bos	2009	[53]
2	9.70E-05	0.13	yes	Park	2009	[61]
3	4.92E-05	0.65	yes	Geerts	2000	[83]
4	3.45E-05	0.18	yes	Balmelli	2004	[92]
5	3.34E-05	0.065	yes	Shu	2008	[71]
6	3.20E-05	0.065	yes	Huang	2009	[55]
7	2.92E-05	0.18	no	***Veldhoven***	2003	[66]
8	2.85E-05	0.18	yes	Yang	2008	[93]
9	2.18E-05	0.09	yes	Bos	2009	[53]
10	1.61E-05	0.35	yes	Nguyen	2005	[88]
11	1.58E-05	0.09	no	Yao	2004	[67]
12	1.29E-05	0.065	no	Putter	2007	[63]
13	1.13E-05	0.065	no	Putter	2007	[63]
14	1.13E-05	0.18	no	***Veldhoven***	2003	[66]
15	1.10E-05	0.13	yes	Mitteregger	2006	[50]
16	8.34E-06	0.09	yes	Breems	2007	[14]
17	8.20E-06	0.25	yes	Brewer	2005	[94]
18	8.08E-06	0.18	yes	Morrow	2005	[95]
19	7.66E-06	0.8	yes	Leung	1997	[80]
20	7.42E-06	0.35	yes	Yang	2003	[77]
21	7.14E-06	0.5	yes	Fujimori	2000	[79]
22	7.13E-06	0.18	yes	Yaghini	2005	[65]
23	6.31E-06	0.13	yes	Christen	2007	[96]
24	6.21E-06	0.065	no	***Veldhoven***	2009	[39]
25	5.66E-06	0.045	no	***Veldhoven***	2009	[39]
26	5.53E-06	0.065	no	***Veldhoven***	2008	[97]
27	5.49E-06	0.13	no	Yao	2005	[98]
28	5.47E-06	0.11	yes	Matsukawa	2009	[54]
29	5.25E-06	0.09	yes	Malla	2008	[59]
30	5.22E-06	0.18	no	Silva	2006	[89]

Table 8.4: *Top 30 of* $\textit{FOM}_{area}$ *benchmark*

8.5 Conclusions

To be able to benchmark modulator performance parameters with their cost, new FOMs are introduced: FOM$_{eq,th}$ for power consumption, FOM$_{HD3D}$ for third harmonic distortion, and FOM$_{area}$ for silicon area.

The power efficiency FOM, FOM$_{eq,th}$, unlike the traditional FOM$_{DR}$, includes the input signal to reflect the decreasing noise impedance required to achieve the same DR in a defined bandwidth at lower modulator input signals. It has been proven by [74] that it is not necessarily true that a small input signal will lead to a poor FOM$_{eq,th}$. However, the small input signal will cause a higher power consumption of the $\Sigma\Delta$ modulator as the its equivalent noise impedance has to be lower (quadratical relation). Therefore, according to FOM$_{DR}$ (Table 8.1) [74] is power inefficient, as the dynamic range bandwidth product compared to the consumed power is too low. As the choice of internal signal swings in the system have a major influence on the consumed power, the choice of signal swings should be an integral part of the system power efficiency optimization.

To come to a state-of-art $\Sigma\Delta$ modulator design, FOM$_{eq,th}$ and FOM$_{HD3D}$ with modulator (circuit) architecture, (and maybe even FOM$_{area}$) should be considered. The FOM$_{eq,th}$ and FOM$_{HD3D}$ plead for a low noise, highly linear input stage. The strong relation between FOM$_{eq,th}$ and FOM$_{HD3D}$ confirms that for a power efficient design, the majority of the current should be spent in the input stage, while reducing and power-optimizing overhead circuitry to reduce implementation loss. An RC integrator input stage combines the low noise and high linearity requirements at low power consumption, and therefore is the best solution for the modulator's input stage.

The FOM$_{area}$ predicts what kind of performance can be expected from a modulator per area. In the derivation of this FOM, a linear relation between modulator power and area is found.

The state-of-art FOMs are listed in the Table 8.5. In FOM$_{eq,th}$ Refs. [73] and [74] are excluded for reasons explained in this section.

FOM	State-of-art value	Dimension	Reference
FOM$_{DR}$	7.92E-14	[J/conversion]	[50]
FOM$_{eq,th}$	9.33E-03	[-]	[58]
FOM$_{HD3D}$	4.29E-02	[-]	[58]
FOM$_{area}$	1.38E-04	[W/mm^2]	[53]

Table 8.5: *State-of-art values for the different FOMs*

Chapter 9

ΣΔ Modulator Implementations and the Quality Indicators

The previous chapters have indicated that analog IP blocks can be judged by quality indicators. Furthermore, Chap. 2 argued that when the amount of digitization in an analog signal processing system is increased, the score on these quality indicators will be higher, as next generation technologies have more advantages for digital circuits compared to analog circuits. It also showed that this digitization can be done at different abstraction levels. In this chapter several implementation examples will be shown for each of these abstraction levels, which is schematically displayed in Fig. 9.1.

Section 9.1 will show $\Sigma\Delta$ modulator implementations that are suited for highly digitized receiver systems. Shifting more of the analog functionality into the digital domain yields a more robust and flexible receiver, but requires more performance and additional flexibility of the $\Sigma\Delta$ modulator. A few high dynamic range and high linearity modulator implementations will be shown, which can cope with the additional demand for flexibility, while using their resources (area, power) efficiently.

In Sect. 9.2 an implementation will be shown that is digitized at modulator architecture level. Analog circuit blocks are reduced and replaced by digital circuitry where possible.

In Sect. 9.3 modulators are shown which are designed using the digital design methodology of Sect. 6.1.2. Digitization is done at circuit and layout level to increase technology portability.

R.H.M. van Veldhoven, A.H.M. van Roermund, *Robust Sigma Delta Converters*, Analog Circuits and Signal Processing, DOI 10.1007/978-94-007-0644-6_9,

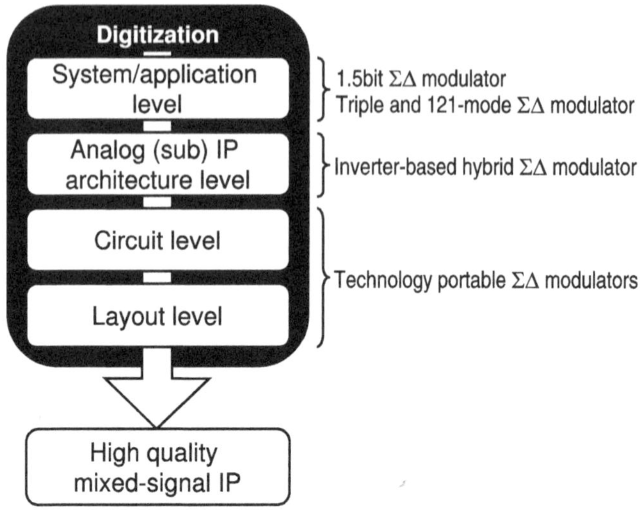

Figure 9.1: *System digitization at different abstraction levels*

At the end of this chapter the implemented $\Sigma\Delta$ modulators will be judged on the quality indicators of Chap. 2 and the FOMs of Chap. 8.

9.1 Digitization at System/Application Level: ΣΔ Modulators for Highly Digitized Receivers

As shown in the introduction of this book, the increasing number of wireless connectivity and cellular standards drives the need for flexible receiver systems. As flexibility is much easier to implement in the digital domain, the A/D converter is shifted closer to the antenna, digitizing the amount of analog AGC and filtering in front of the ADC, and replacing it by a digital equivalent. The reduction of filtering and AGC in front of the A/D converter translates into more stringent requirements on the ADC in terms of dynamic range and linearity.

In this section several $\Sigma\Delta$ modulator examples are shown that are suited for these flexible and highly digitized receiver systems. The first example is a high dynamic range ADC for a highly digitized receiver for UMTS. Furthermore, two $\Sigma\Delta$ modulator examples will be shown which are suited for a multi-mode receiver. This will put additional flexibility requirements on the ADC.

9.1.1 A 1.5-bit ΣΔ Modulator for UMTS

In this section a quadrature 4th-order, continuous-time, ΣΔ modulator with 1.5-bit quantizer and feedback DAC for a UMTS receiver is presented [18, 69]. It achieves a dynamic range of 70 dB in a 2 MHz bandwidth and the total harmonic distortion is −74 dB at full scale input. When used in an integrated receiver for UMTS, the dynamic range of the modulator substantially reduces the need for analogue AGC and its tolerance of large out-of-band interference also permits the use of only first order pre-filtering. The IC including an I and Q ΣΔ modulator, PLL, oscillator and bandgap dissipates 11.5 mW at 1.8 V. The active area is 0.41 mm^2 in a 0.18 μm, 1 poly, 5 metal, CMOS technology.

9.1.1.1 System Architecture

In Fig. 9.2, the zero-IF receiver architecture for UMTS is shown. The architecture comprises an RF front end, an ADC and a digital baseband processor. The front-end uses a quadrature down-converter to convert the RF channel to the zero IF. Both I and Q components are converted into the digital domain by a pair of ΣΔ modulators. The baseband processor subsequently provides all the necessary filtering of quantization noise and most of the receiver selectivity. The figure also

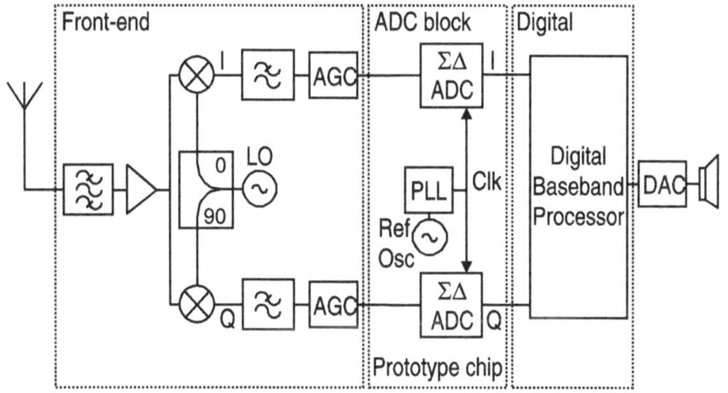

Figure 9.2: *Zero-IF receiver architecture for UMTS*

indicates the prototype chip, with the I and Q ADCs, a bandgap reference, a PLL and an oscillator on board.

9.1.1.2 Modulator Architecture

The functional block diagram of either the I or Q ΣΔ modulator is given in Fig. 9.3 showing the use of a feed-forward, 4th-order loop filter with a single

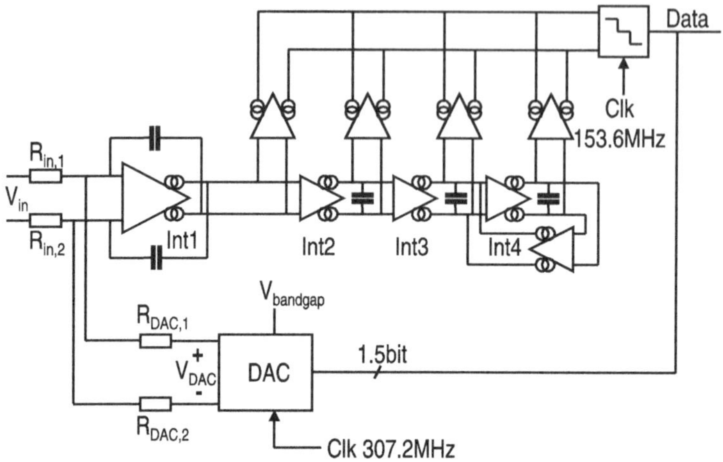

Figure 9.3: *4th order, 1.5-bit ΣΔ modulator architecture*

resonator, a 1.5-bit quantizer and feedback DAC. The 4th-order loop filter is built up out of an RC integrator followed by three gm-C integrators. To increase the spectral efficiency of the modulator (Sect. 5.1), an additional transconductor is connected head-to-tail to the last integrator stage. The four feed-forward coefficients are also implemented with transconductors. The dynamic range of the modulator depends on the following sources of noise: circuit noise emanating mainly from the input resistors, first integrator and feedback DAC, the quantization noise generated by the quantizer and the jitter noise originating from the clock. To minimize power consumption, the circuit noise is made dominant over the quantization and jitter noise which are designed 10 dB below the circuit noise. By using a 3 level rather than the more usual 2 level quantizer/DAC combination, maximum modulator input signal is increased by 1.6 dB (Eq. 5.8) and quantization noise introduced by the quantizer is reduced by a factor of two. Due to the smaller quantization noise step the in-band quantization noise-jitter noise products reduce accordingly. The clock frequency of the I and Q ΣΔ modulator is chosen a multiple of the UMTS chip rate to avoid fractional sample rate conversion. Because in UMTS the chip-rate is 3.84 MHz and a zero-IF architecture is used, the required conversion bandwidth is for both the I and Q modulator is 1.92 MHz. So the 153.6 MHz clock generated by a PLL, represents an oversampling ratio of 40 for both modulators. To have some margin, the I and Q modulators bandwidth is set to 2 MHz. When the modulator has a full scale signal at its input, the simulated SQNR for a single modulator is 80.3 dB in 2 MHz. Circuit noise simulations predict an SNR of 70 dB in the same bandwidth.

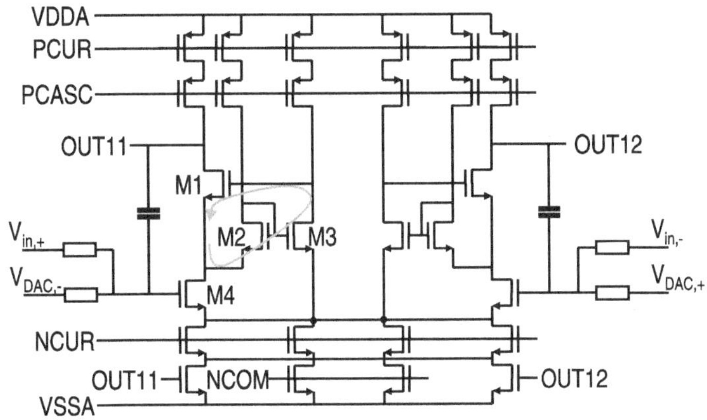

Figure 9.4: *Circuit diagram of the first integrator/input stage*

9.1.1.3 Circuit Design

In Fig. 9.4 the input stage of the loop filter is shown in detail. Input transistor M4 has a minimum channel length because of speed, and due to this small channel length its output resistance is very low. To obtain enough DC gain in the integrator stage a gain boosting technique is used. The supply to the gate of cascode transistor M1 is regulated via the level shift transistor M2 and amplifying transistor M3. The resulting DC gain is 80 dB and the maximum output swing is 0.8 Vpp differential. Figure 9.5 shows the schematic of the circuit used for the second, third, and fourth integrator stages, which also have regulated cascodes to

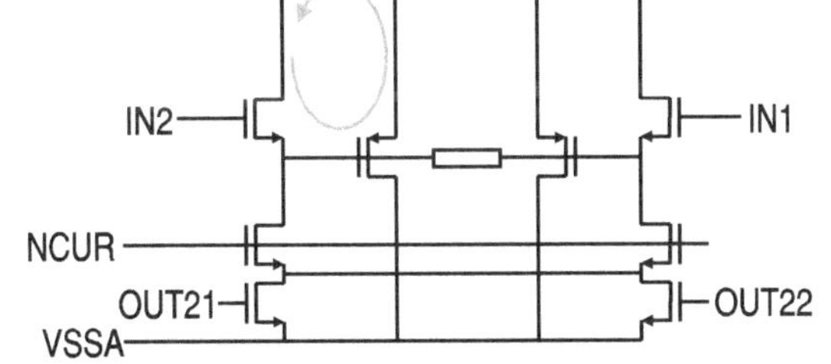

Figure 9.5: *Circuit diagram of the second to fourth integrator*

achieve 60 dB DC gain with minimum channel length input transistors. Their output swing is also 0.8 Vpp differential. The transconductor which creates the notch close to the edge of the bandwidth, is a scaled version of those used in the integrators to ensure good matching. Figure 9.6 shows the feed-forward coefficients and the 1.5-bit quantizer. The feed-forward transconductor is a gm scaled version of those used in the later integrator stages. The output currents of the feed-forward

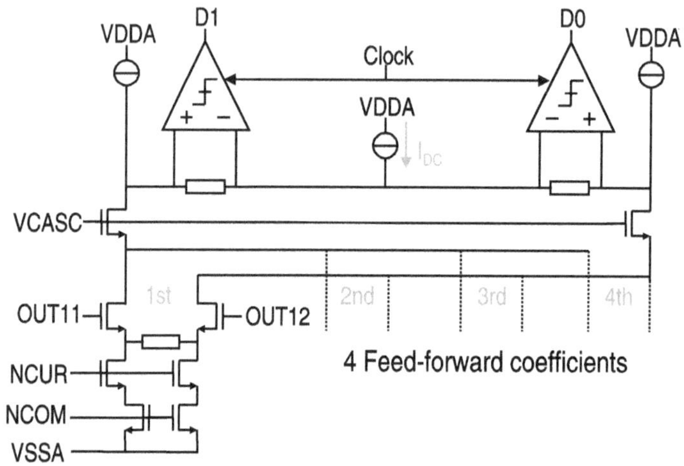

Figure 9.6: *Circuit diagram of the feed-forward transconductors and 1.5-bit quantizer*

coefficients are summed on a cascode and converted to voltages by two resistors. The current I_{DC} determines the separation of the comparator decision levels. The output bits D0 and D1 are fed to the DAC.

The 1.5-bit DAC has an RTZ period of 0.5 to reduce ISI. The schematic of the DAC is shown in Fig. 9.7. The input signal V_{in} is converted into a current by the input resistors $R_{in,1}$ and $R_{in,2}$. Dependent on the output data of the comparators, nodes n1 and n2 are switched to ground or to a bandgap reference voltage, through switches M1–M4. The data-dependent DAC output voltage is converted into a current by $R_{DAC1,2}$ resulting in a positive or negative feedback current, I_{DAC}. This feedback current is subtracted from the input current and the error signal is integrated on the capacitors of the input stage. By closing switch M5 and opening switches M1–M4 the RTZ level for the 1.5-bit DAC is set.

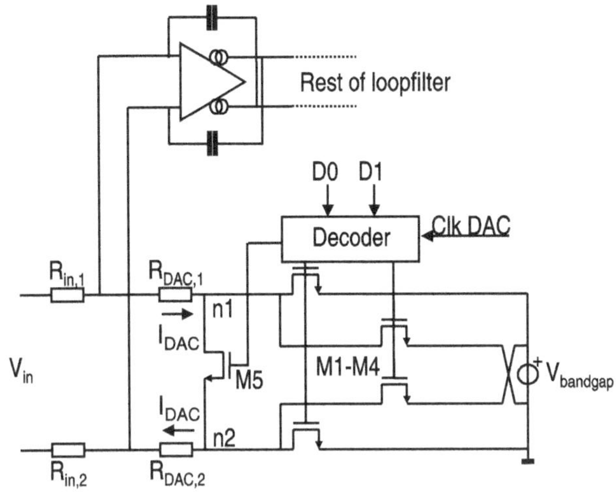

Figure 9.7: *Circuit diagram of the 1.5-bit DAC with RTZ coding*

9.1.1.4 Experimental Results

The prototype chip comprises 2 ADCs (I and Q), a reference oscillator and phase locked loop (PLL). The oscillator frequency was 30.72 MHz and the PLL output frequency 307.2 MHz, which is frequency-divided by two to produce the required 50 percent duty cycle sample clock. An on chip bandgap circuit provides all the necessary reference voltages and currents. During measurements, the signal generator is fed to the ADC via a highly-selective low-pass filter which removes the harmonic distortion of the generator. Figure 9.8 shows the measured output spectrum of a single ΣΔ modulator if a 1 MHz tone at full scale is applied to the

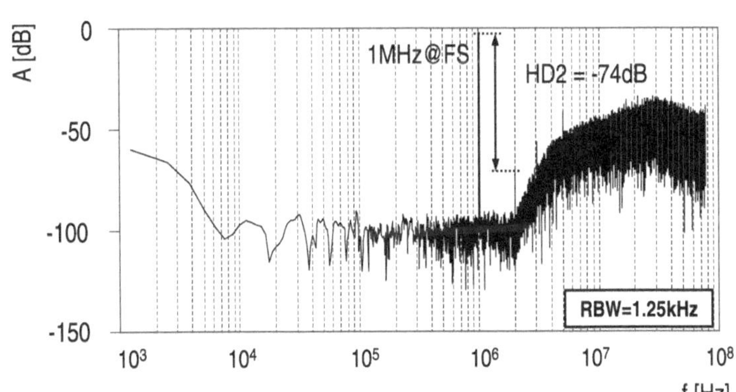

Figure 9.8: *Measurement at full scale input*

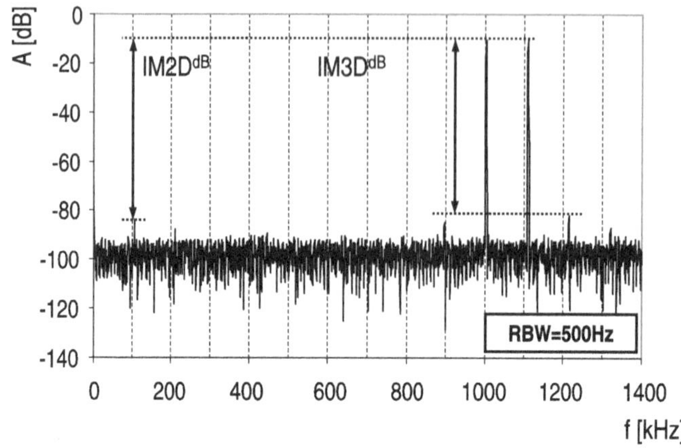

Figure 9.9: *Intermodulation measurement with −6 dBFS input signals at 1.003 and 1.11 MHz*

input. The measured dynamic range in a 2 MHz bandwidth is 70 dB, which is in good agreement with the earlier presented simulations. Second harmonic distortion is at −74 dB. In Fig. 9.9 an inter-modulation measurement is shown. The input frequencies applied to the input of a single ΣΔ modulator are 1.003 MHz and 1.11 MHz at −6 dBFS. The IM2 and IM3 distances are 76 and 74 dB respectively. Figure 9.8 shows the SNR and SNDR of the ADC as a function of the input power level of a 400 kHz tone. At high input powers, the 2nd and 3rd harmonics dominate the maximum SNDR figure of 68 dB. In the test setup for measuring the

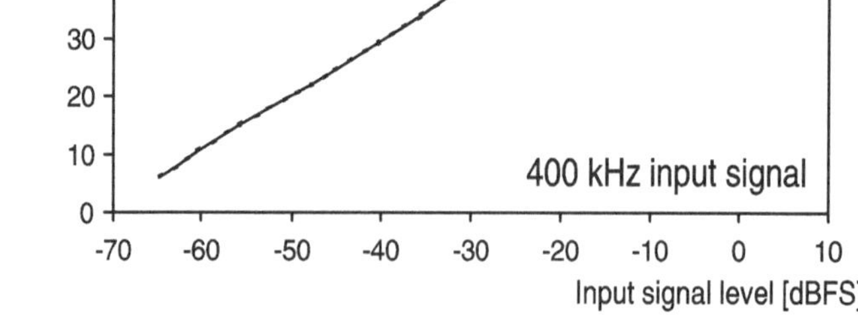

Figure 9.10: *Measured SNR and SNDR as function of input signal level*

complex output spectrum from both modulators (both I and Q ADCs active), the harmonic distortion filter was not connected. This was to avoid the effects of gain and phase mismatches in the two pre-filters, which would otherwise introduce a false image of the input tone and confuse the measured image rejection ratio of the two ADCs. Hence, the I and Q generator output signals were connected directly to the I and Q ADCs. In Fig. 9.11 a measurement of the complex output spectrum from the pair of ADCs is shown for an input tone of +500 kHz for a full-scale input signal. There is considerable distortion visible in the spectrum but investi-

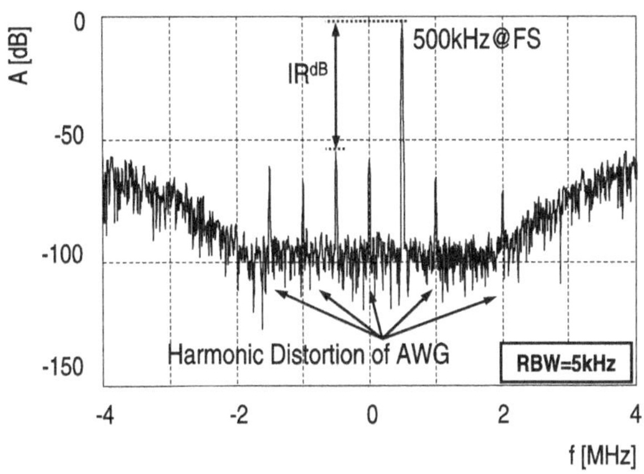

Figure 9.11: *Image rejection measurement*

gations have shown that this originates mainly in the signal generator. In complex networks of this kind, the third harmonic distortion appears only on the left side of the spectrum (Sect. 4.6.4). The measurement shows an image rejection of 53 dB while the dynamic range (excluding the power in the distortion products) is 70 dB in 4 MHz. Figure 9.12 summarizes performance and main design characteristics of the modulator. Figure 9.13 shows the die photograph of one modulator. The prototype IC was fabricated in a 5 metal, 1 poly, 0.18 μm, digital CMOS process. The input resistors, first integrator, 2nd–4th integrators, feed-forward coefficients, summing node, comparators and feedback resistors are indicated.

9.1.1.5 Conclusions

The design of a 4th order, 1.5-bit, continuous-time ΣΔ modulator has been presented. The quadrature modulator achieves a dynamic range of 70 dB in a +/−2 MHz bandwidth and an SNDR of 68 dB at full-scale input. All measurements where done clocked with the integrated PLL and oscillator. All reference voltages and currents are coming from the on-chip bandgap circuit. In these mod-

Conversion system	Zero-IF	
ΣΔ modulator	Continuous-Time, 4th order, 1.5bit	
Sampling rate	153.6MHz	
Signal bandwidth	2MHz (single modulator)	
Oversampling ratio	40	
Input voltage range	$0.5V_{rms}$ differential for a sinoid	
Dynamic range	70dB	
Total Harmonic Distortion	-74dB	
Image Rejection	>53 dB (only 6 samples measured)	
Process	1.8V, 1P, 5M, 0.18 μm CMOS	
Area and power consumption	Area [mm²]	Power@1.8V [mW]
I&Q ΣΔ modulator	2*0.12	2*3.3
PLL	0.14	3.6
Oscillator	0.029	0.72
Bandgap	0.02	0.54
Total :	0.43 mm²	11.5 mW

Figure 9.12: *Performance and main design characteristics summary of the 4th order, 1.5-bit modulator*

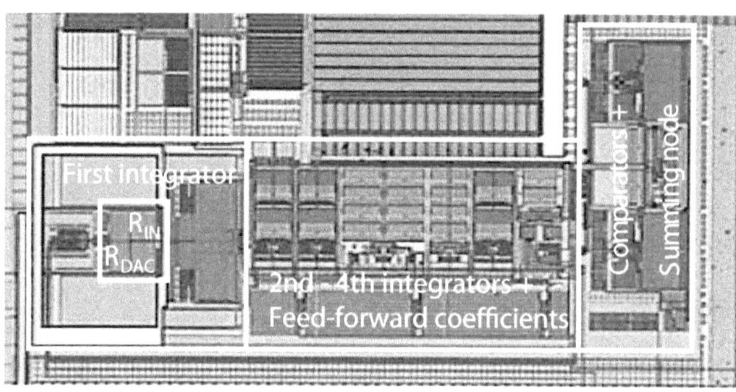

Figure 9.13: *Micrograph of a single modulator fabricated in a single poly, 5 metal layer 0.18 μm CMOS process*

ulators a 1.5-bit DAC is used to reduce the quantization noise, and at the same time reducing the influence of clock jitter on the achievable dynamic range. When used in a zero-IF UMTS receiver, the modulators provide enough dynamic range to substantially reduce the need for analogue pre-filters and AGC.

9.1.2 A Triple-Mode ΣΔ Modulator for GSM-EDGE, CDMA2000 and UMTS

In this section a high dynamic range, high linearity modulator is presented for a highly digitized multi-mode receiver [66, 66]. The I and Q CT 5th-order ΣΔ modulator can cover three system bandwidths for use in a GSM-EDGE, CDMA2000, and UMTS system and achieving a dynamic range of 92, 83, and 72 dB in a 200, 1228, and 3840 kHz bandwidth respectively. As the modulator is part of a receiver, an SC DAC rather than an SI DAC is used to reduce the effects of clock jitter on the dynamic of the modulator (Sect. 6.9). For linearity reasons the quantizer and DAC are both 1-bit. The measured intermodulation distances IM2 and IM3 are better than 87 dB in all modes. Both the I and Q modulator consumes a power of 3.8, 4.1, and 4.5 mW at 1.8 V. Processed in 0.18 μm CMOS, the 0.55 mm^2 active area includes a PLL, 2 oscillators and a bandgap.

9.1.2.1 System Architecture

The receiver architecture for which this modulator was designed is similar to the one of Sect. 9.1.1.1. The modulator has an additional mode input with which it can be switched to either GSM-EDGE, CDMA2000, or UMTS mode. The required filter bandwidths are scaled according to the program setting. Figure 9.14 summarizes the specifications on bandwidth and dynamic range in the different modes, from which the maximum allowed noise density can be derived. Because the modulator has to cope with the IF signals of 4 different systems, bandwidth and sample frequency has to be adapted to be able to achieve a high dynamic range at minimum power consumption. To assure re-usability of circuits, the maximum modulator input level provided by the front-end is set to $1V_{rms}$ differential in all modes. Furthermore, because GSM and EDGE bandwidths and dynamic ranges required are very close, it is decided to combine these two standards in one modulator mode. Figure 9.14 indicates that UMTS-mode has the highest bandwidth, which implies high bandwidth circuits, whilst GSM-mode requires the lowest noise density and determines power consumption due to the low impedances required to achieve this low noise density. Although the combination of the foregoing contradicts to the low power receiver solution needed in present-day telecom terminals, the few dB overhead in dynamic range in UMTS- and CDMA2000-mode can reduce the required analog pre-filtering in front of the modulator even more, and minimize cost. The AGC in front of the modulator has to be adopted accordingly, such that the dynamic range of signals at the output of the RF front-end exactly fits the input dynamic range of the modulator (Chap. 4).

9.1.2.2 ΣΔ Modulator Architecture

Figure 9.15 shows the block diagram of the ΣΔ modulator. A 5th order feed-forward loop filter is implemented with two resonators to increase noise shaping efficiency. A 1-bit quantizer is used together with a 1-bit inherently lin-

Mode	UMTS	CDMA2000	EDGE	GSM
Receiver topology	ZIF	ZIF	LIF	LIF
Channel bandwidth [kHz]	3840	1228	270	190
SNR Specification [dB]	70	80	87	90
Noise Density [nV/√Hz] ($V_{in,max}$=1V_{rms})	228	128	85	72
ADC Sample Frequency [MHz]	153.6	76.8	26	26
Calculated SNR Thermal [dB]	78	85	88	90
Simulated SNR Quantization [dB]	82	103	102	105
Simulated SNR Total [dB]	76	85	88	90

▒ ADC target specification

Figure 9.14: *ADC requirements in the different modes*

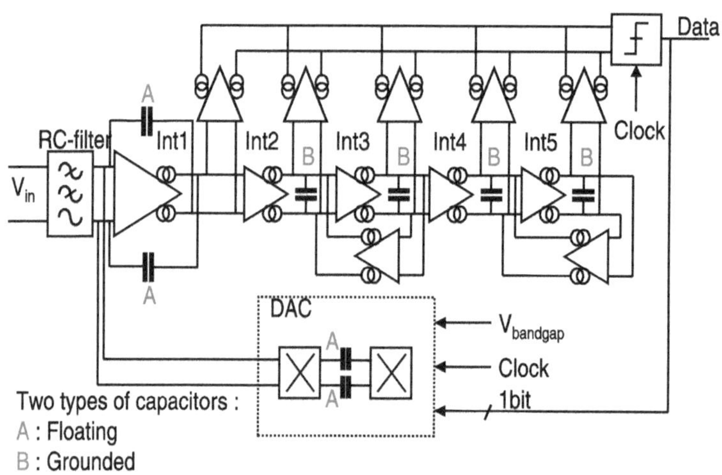

Figure 9.15: *Block diagram of the 1-bit, 5th order, feed-forward ΣΔ modulator*

ear, switched-capacitor DAC. This way, the advantages of low-jitter sensitivity of switched capacitor ΣΔ modulators (Sect. 6.9.2, the TPJE jitter model) and high anti-alias suppression of continuous-time ΣΔ modulators are combined. Figure 9.14 shows the maximum achievable SQNR of the 5th order modulator in all modes. The sample frequencies in GSM, CDMA2000, and UMTS mode are 26, 76.8, and 153.6 MHz respectively. Without clock jitter, the theoretical SQNR in GSM, CDMA2000, and UMTS mode is 102, 103, and 82 dB. Because the SQNR in all modes is at least 10 dB better than required, thermal noise is dominant which results in the lowest power consumption. Including the thermal noise the simulated SNR in GSM, CDMA2000, and UMTS mode is 90, 85, and 76 dB respectively.

9.1.2.3 Circuit Design

All capacitors of the modulator are implemented as NMOS in N-well devices, which have a well defined absolute value and show good matching. Because the NMOS in N-well capacitors are normally-on devices, these capacitors show good linearity at low bias voltages. A disadvantage of this type of capacitor is the large parasitic capacitance from N-well to substrate. Capacitor type A (Fig. 9.16) uses two capacitors of which the gates are connected together and via a diode to the analog supply (VDDA), to create a floating feedback capacitor. At start-up the

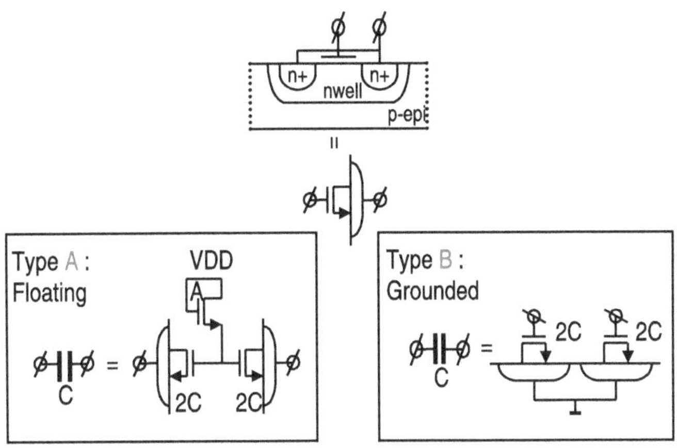

Figure 9.16: *Capacitors implemented with NMOS in N-well devices*

diode pulls up the capacitor gates to the VDDA, to bias the capacitors in their linear region. When the gates are charged to VDDA, the diode has a V_{gs} of zero volt and presents a high impedance. The N-well terminals are the terminals of the capacitor. The parasitic capacitors of non-floating capacitor type B are shorted by substrate contacts and (external) ground connections.

The circuit of the amplifier used in the first integrator shown in Fig. 9.17, uses a regulated NMOS cascode which compensates for the low output impedance of the input transistors, which have minimum channel length to achieve high speed. The gain-boost amplifier is biased with a resistor to avoid the need for an additional common-mode circuit. The integrator stage achieves a DC gain of 80 dB. The second to fifth integrator and feed-forward coefficients are implemented with

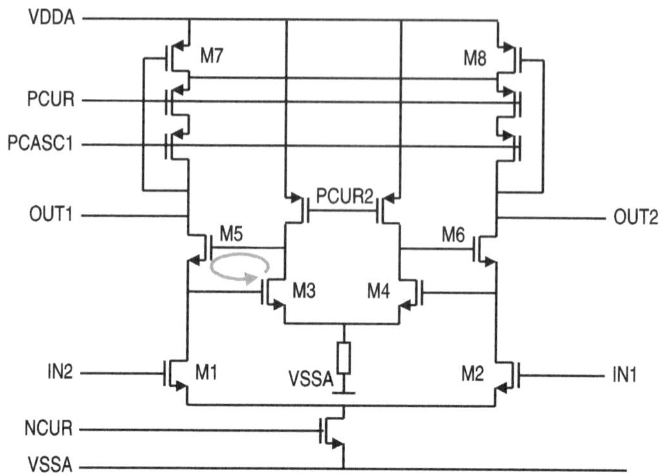

Figure 9.17: *Circuit diagram of the OTA used in the first integrator*

scaled gm-C stages, similar to the one presented in Fig. 9.5. The interface circuit between loop filter and comparator is straightforward compared to the one of the modulator presented in Sect. 9.1.1 because only one instead of two comparators is used in this design, and therefore is not shown.

The scaling of the loop filter is shown in Fig. 9.18. In UMTS mode all switches are open and the loop filter has the largest bandwidth. In CDMA2000 mode the switches numbered 2 are closed and additional capacitance is added to the out-

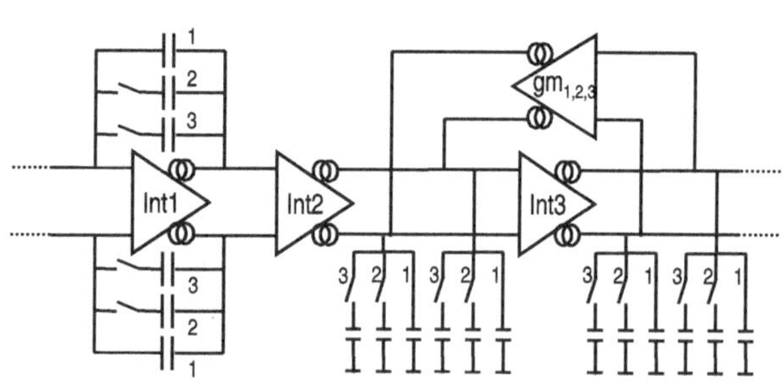

Mode	Switch 3	Switch 2	Switch 3
UMTS	X	X	X
CDMA2000		X	X
GSM-EDGE			X

Figure 9.18: *Scaling of the loop filter*

put of the integrator. In GSM-EDGE mode the switches 2 and 3 are closed and again additional capacitance is switched to the integrator outputs to increase the integrator time-constants further to adapt the unity gain frequencies to the lower clock frequency. The local feedback transconductors are also scaled to move the resonators to the wanted frequencies, maximizing noise shaping efficiency in each mode.

To reduce the negative effect of jitter on the dynamic range of the converter, a switched capacitor feedback DAC is used. In Fig. 9.19, the switched capacitor DAC circuit is shown in detail. The capacitors used in the DAC are of the semi-

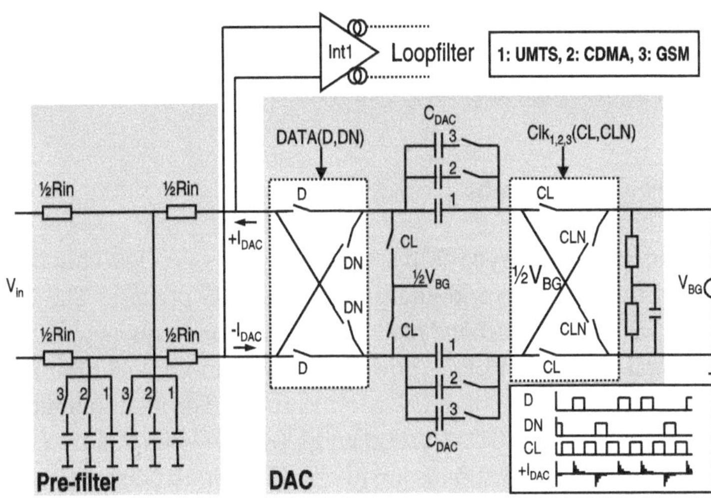

Figure 9.19: *Input filter and switched capacitor feedback DAC*

floating type A. In the first clock phase, the capacitors are charged to half the bandgap voltage by closing switches CL (switches CLN are open). In the second clock phase switches CLN are closed (switches CL are open) and the capacitors are discharged in a data-dependent way by closing switches D or DN depending on the output of the quantizer. The DAC output current is subtracted from the input current and integrated on the capacitors of the first integrator. In GSM and CDMA2000 additional capacitors are switched on to keep a constant ratio between the input resistance and the effective DAC feedback resistance, which changes proportional to $1/(f_s C_{DAC})$. This is to have the correct modulator input-output gain. In each mode, the cut-off frequency of the pre-filters (Fig. 9.19) is adapted to the lowest value possible to achieve the highest possible attenuation of out-of-band interferers.

Mode	f_{osc}[MHz]
UMTS	30.72(X1)
CDMA2000	30.72 (X1)
GSM-EDGE	52 (X2)

Mode	f_{PLL}[MHz]
UMTS	307.2
CDMA2000	153.6
GSM-EDGE	n.a.

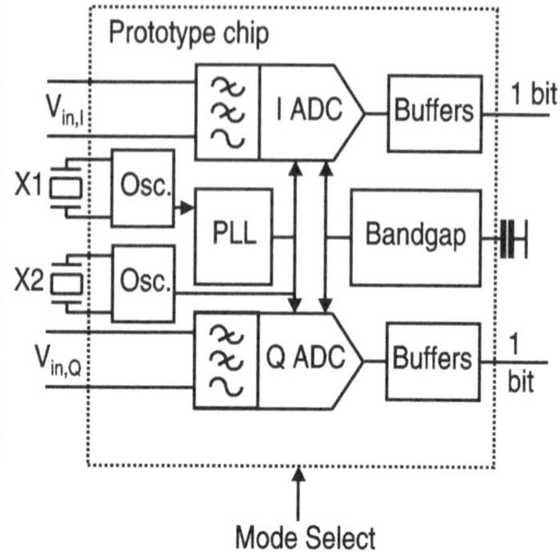

Figure 9.20: *Block diagram of the prototype chip*

9.1.2.4 Experimental Results

A block diagram of the prototype chip is shown in Fig. 9.20. The chip is fabricated in a single poly, five metal layer digital 0.18 μm CMOS process. The IC includes two oscillators, a PLL and a bandgap. The oscillator frequency is 52 MHz in GSM-EDGE mode and the PLL is powered down. In CDMA2000 and UMTS mode an oscillator frequency of 30.72 MHz is used. The PLL output frequency is 153.6 and 307.2 MHz for CDMA2000 and UMTS mode respectively. The 52, 153.6, and 307.2 MHz output signals are divided by two to create 50% duty-cycle sample clocks of 26, 76.8, and 153.6 MHz respectively.

The micrograph of the chip containing the I and Q modulator is shown in Fig. 9.21. Next to the I and Q modulator, the oscillators, PLL and digital multi-mode channel filters are indicated. A zoom-in on the chip photo reveals a single triple-mode ΣΔ modulator, shown in Fig. 9.22. The input filter, the integrator stages including loop filter capacitor bank, comparator and SC DAC are indicated in the micrograph.

In Fig. 9.23 the SNR as a function of the input signal level is plotted. At full-scale input signals with frequencies of 150, 530, and 1700 kHz for GSM-EDGE, CDMA2000, and UMTS mode, the peak SNR is 92, 83, and 72 dB in bandwidths of 200, 1228, and 3840 kHz. The differential input swing is $1V_{rms}$ in all measurements. Over a power supply range of 1.6–2.9 V the peak SNR only varies +/−1 dB. The measured I and Q output spectra are shown on the left in Figs. 9.24,

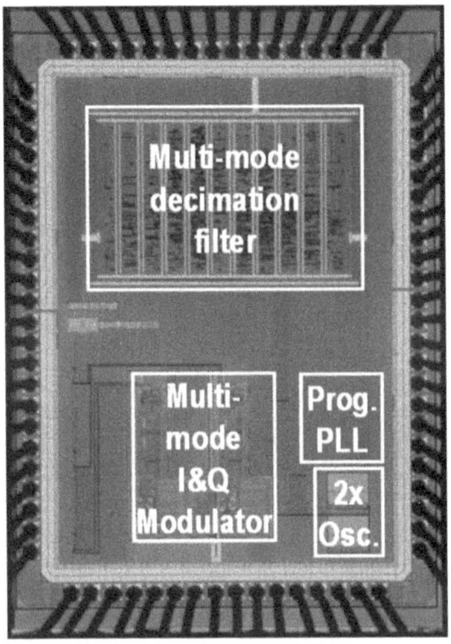

Figure 9.21: *Micrograph of the chip containing the I and Q modulator fabricated in a single poly, 5 metal layer 0.18 μm CMOS process*

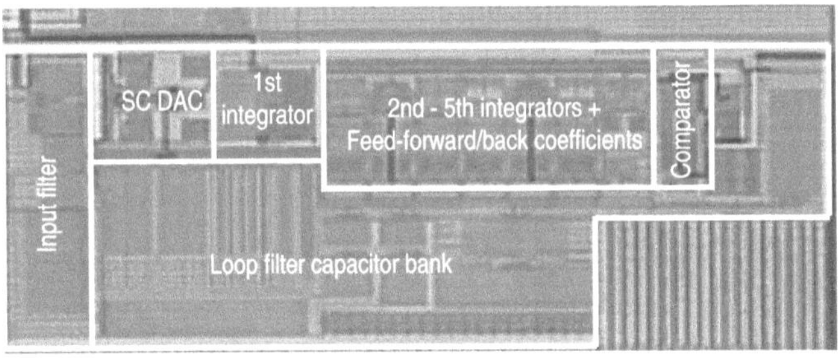

Figure 9.22: *Micrograph of a single modulator*

9.25 and 9.26. In UMTS, CDMA2000, EDGE and GSM mode, the peak SNR is 72, 83, 90, and 92 dB in a 3840, 1228, 271, and 200 kHz bandwidth respectively. In all three modes the IR is 50 dB, which is limited by the amplitude and phase mismatch of the quadrature input signals generated by the AWG. The harmonic distortion visible also originates from the AWG. This is confirmed by the intermodulation IM measurements displayed at the right of Figs. 9.24, 9.25 and 9.26. Two separate channels of the AWG generate two differential tones at a level of −6 dBFS, which are combined with a resistive 50 Ω network. This way, no IM components are produced by the AWG. The measured IM2 and IM3 distances are is better than 87 dB in all modes (the exact measured values are presented in Fig. 9.27).

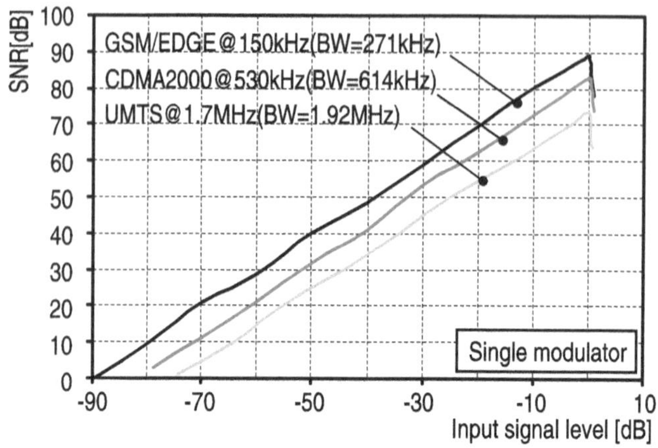

Figure 9.23: *SNR as function of input level*

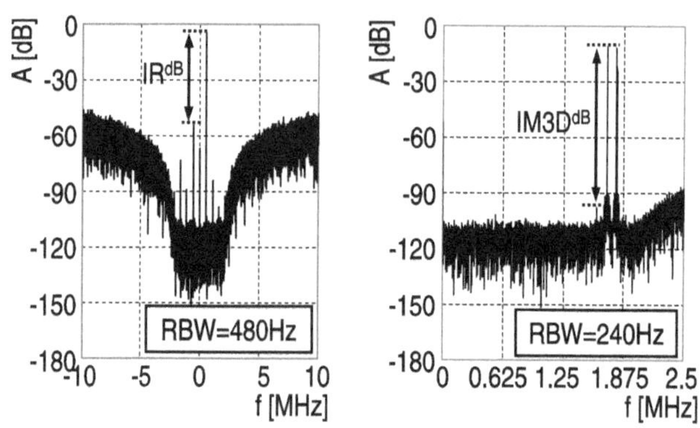

Figure 9.24: *Image rejection and intermodulation in UMTS mode*

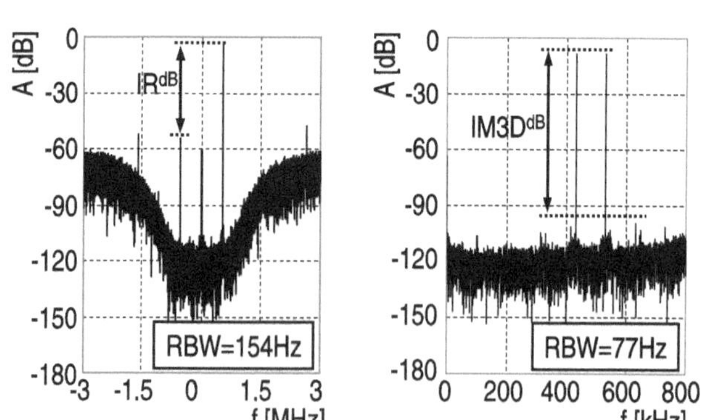

Figure 9.25: *Image rejection and intermodulation in CDMA2000 mode*

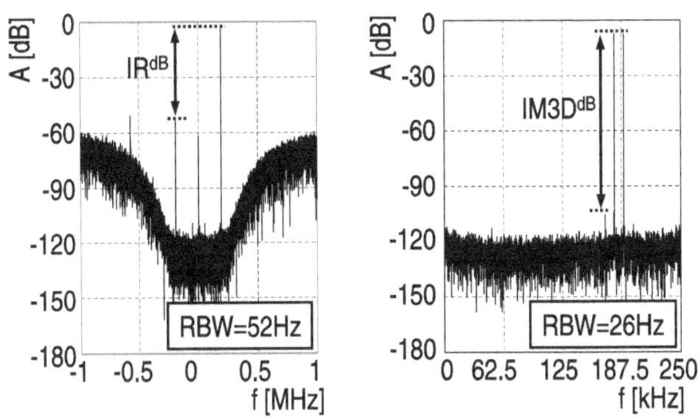

Figure 9.26: *Image rejection and intermodulation in GSM mode*

Process	1P, 5M, standard 0.18 μm CMOS			
Supply voltage	1.6 – 2.9V (+/- 1dB DNR performance deviation)			
ΣΔ modulator	5th order CT, feedforward, 1 bit with SC DAC			
Input voltage range	1 V rms, differential			
Mode	**UMTS**	**CDMA2000**	**GSM (EDGE)**	
IF	Zero-IF	Zero-IF	Low-IF	
Sampling rate	153.6 MHz	76.8MHz	26MHz	
Signal bandwidth	3.84 MHz	1.228MHz	200kHz *(271kHz)*	
Oversampling ratio	40	64	65 (48)	
Dynamic range	74dB (f_s=250MHz)	83dB	92dB *(90dB)*	
Intermodulation distances	IM2>110dB*	IM2>98dB	IM2>110dB*	
	IM3>87dB	IM3>91dB	IM3>97dB	
Image Rejection	>50dB**	>50dB**	>50dB**	
	**** Not visible because of noise floor***			
	***** Measurement limited by test setup***			
Area and power	**Power@1.8V [mW]**			**Area [mm²]**
I&Q ΣΔ modulator	2*4.5	2*4.1	2*3.8	2*0.18
PLL	3.8	3.6	--	0.14
Oscillator	0.8	0.8	1	0.03
Bandgap	0.5	0.5	0.5	0.02
Total :	14.1	13.1	9.1	0.55

Figure 9.27: *Performance summary and main design characteristics of the triple-mode ΣΔ modulator*

The performance summary and main design characteristics of the modulator are presented in Fig. 9.27. Power consumption of both the I and Q modulator is 3.8 mW in GSM-EDGE, 4.1 mW in CDMA2000 and 4.5 mW in UMTS mode at 1.8 V supply.

9.1.2.5 Conclusions

A quadrature 5th-order 1-bit modulator has been presented that combines the advantages of low-jitter sensitivity of switched capacitor ΣΔ modulators and high anti-alias suppression of continuous-time ΣΔ modulators. The I and Q modulators together achieve a dynamic range of 92, 83, and 72 dB in a bandwidth of 200, 1228, and 3840 kHz. The measured IM2 and IM3 distances are better than 87 dB and the IR performance is limited to 50 dB by the measurement setup. This low-power high-resolution triple-mode modulator reduces the amount of pre-filtering and AGC required in front of the ADC and thus enables a low-cost, highly integrated and multi-mode receiver for telecom applications.

9.1.3 An Extremely Scalable ΣΔ Modulator for Cellular and Wireless Applications

The implementation example of the previous section exploited the ΣΔ modulator scaling theory presented in Sects. 5.1 and 7.4 to enable a modulator with three optimized bandwidth and dynamic range settings for different telecom standards.

In this section, an extensively re-configurable continuous-time, 5th-order, 1-bit ΣΔ modulator is presented which can cover the complete range of bandwidths presented in Fig. 7.1, and is designed for a highly digitized and extremely flexible receiver [85]. The modulator was designed using the scaling theory presented in Sects. 5.1, 7.4 and in [28]. The dynamic range and bandwidth is programmable from 85 dB@100 kHz to 52 dB@10 MHz in 121 steps. Processed in 90 nm CMOS, the 0.36 mm^2 IC includes two ΣΔ modulators, a bandgap reference and a decimation filter. The power consumption of a single modulator in different modi ranges from 1.44 mW to 6.6 mW at a 1.2 V supply.

9.1.3.1 System Architecture

The receiver architecture the modulator was designed for is basically the one presented in Fig. 1.4. It concerns the extremely flexible receive pipe which can cope with any telecom standard thinkable. This puts huge demands on the ADC as it has to adapt to the right bandwidth at a good power-performance ratio in each mode. This section shows the design of a ΣΔ modulator that fits such a flexible receive pipe.

9.1.3.2 ΣΔ Modulator Architecture

Figure 9.28 shows the block diagram of the ΣΔ modulator. A 5th order CT feed-forward loop filter is implemented with two programmable notches that suppress the quantization noise at the edge of the signal band. A 1-bit quantizer is used together with a 1-bit inherently linear switched-capacitor (SC) feedback DAC for

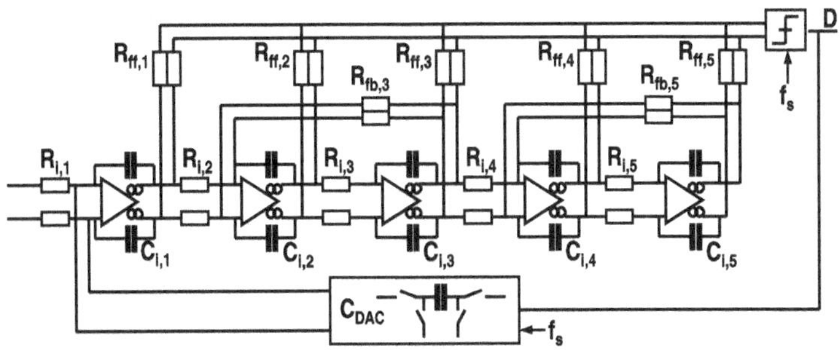

Figure 9.28: *5th order 1-bit ΣΔ modulator architecture*

reasons explained earlier. The loop filter consists of five RC integrator stages that use identical OTA circuits. The feed-forward coefficients are implemented with resistors that are combined on a virtual ground node (not shown). The virtual ground node is implemented with the same OTA circuit as used in the integrators.

For the ADC clock frequency, the strategy of Sect. 7.2.2 is used. The modulator clock frequency is optimized to the bandwidth the modulator is expected to convert. The sample frequency of the modulator can be programmed from 20 MHz to 400 MHz. To limit the amount of programmability required in the modulators loop filter, the clock frequency range is split into sub-ranges using the method presented in Sect. 7.4. For each clock frequency range $[f_{s,1}, f_{s,2}]$, a single unity gain frequency for the loop filter integrators is calculated (Eqs. 5.3 and 5.4). The unity gain frequency of this range is based on the lowest sample frequency within that range $f_{s,1}$, to avoid overload of the internal signal swings in the loop filter and instability of the modulator. At clock frequencies higher than $f_{s,1}$ the signal swing on the integrator outputs will get lower, keeping the integrators in their linear region. The upper boundary of the clock frequency range $f_{s,2}$ is required, to avoid unnecessarily low integrator unity gains in high bandwidth modes. For higher clock frequencies the unity gain of the integrators are increased to reduce the thermal noise contributions of the later integrator stages. To have a maximum signal swing variation on the integrators of about 30%, 11 sub-sets of clock frequency ranges are required (Eq. 7.3), to cover the sample frequency range of 20 MHz to 400 MHz, which means 11 settings for the unity gain frequencies in the modulators loop filter. The unity gain programmability is shown in Fig. 9.29a. As proposed in Sect. 7.1, the modulators noise impedance is chosen to be fixed which means that the integrator input resistor values $R_{i,n}$ are taken to be fixed. This means in order to scale the unity gain frequency of the modulator, the integrator capacitors are required to change. The capacitor values can be calculated

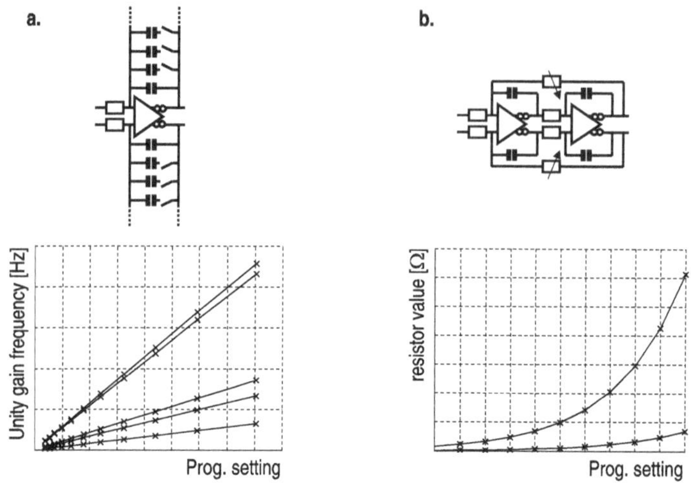

Figure 9.29: *Unity gain and local feedback coefficient programming*

using Eq. 7.4. In each clock frequency range the capacitors required for the higher clock frequency ranges are re-used to minimize the modulators area.

To optimize the modulators bandwidth in each clock frequency range, the local feedback coefficients b are also made programmable. The OSR can be programmed from 10x to 64x within each clock frequency range. To reduce the amount of resistors an OSR programming accuracy of 20% is chosen. According to Eq. 7.5, this means 11 resistor values. The same 11 resistor values are used for each sample frequency sub-range, as the b-coefficients are only dependent on the OSR (Sect. 7.4). This means that in each sample frequency sub-range, you can program the same OSR values. Figure 9.29b shows the b-coefficient programming.

Figure 9.30 gives an overview of the programming of the unity gain frequencies and the local feedback coefficients. The spectrum shown in the figure is stretched by the modulators clock frequency (and its accordingly scaling unity gains), and the modulators bandwidth (OSR) is set by the local feedback coefficients which position the resonators such that the noise shaping efficiency of the modulator is optimized.

The capacitors in the feedback DAC also need programming for different sample frequencies as the DAC impedance to input resistor ration has to be constant, to eliminate modulator input-output gain changes. In this design only one DAC ca-

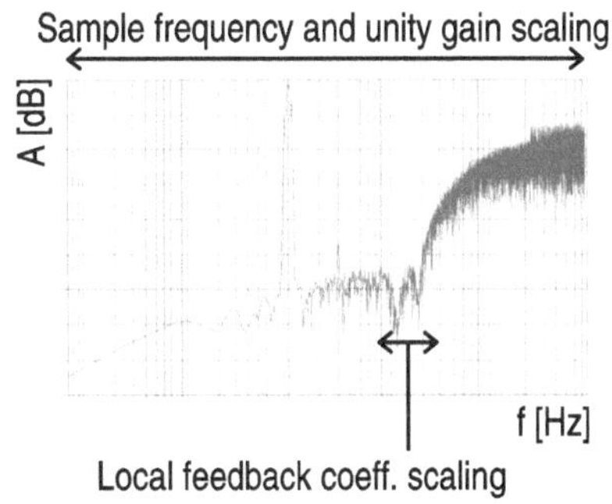

Figure 9.30: *Unity gain and local feedback coefficient programming*

pacitor is implemented for each clock frequency range. As the clock frequency can change by 30% in each range, the modulator input-output gain variation is limited to 30%, as $R_{DAC} = 1/(f_s C_{DAC})$. The feed-forward coefficients are fixed as they only scale with the maximum integrator output signal swing which is fixed. As there is a 30% integrator signal swings variation within each sample frequency range, the output signal of the feed-forward summing OTA will also vary by 30%. In a 1-bit modulator, the quantizer gain is automatically adapted. The bias reference current of each of the circuits used in the $\Sigma\Delta$ modulator is programmable with 4-bit resolution. In this way, the minimum required power consumption can be set in each mode depending on bandwidth and noise requirements.

The combination of local feedback coefficients and unity gains in the loop filter give the 121 different bandwidth settings, which are displayed in Fig. 9.31. In the figure each dot represents a simulation of the modulator including thermal noise for all possible combinations of sample frequencies and bandwidths. As can be seen from the figure, at low oversampling ratios, the quantization noise is the dominant noise contributor, while at high oversampling ratios the circuit noise is dominant.

9.1.3.3 Experimental Results

A block diagram and layout of the prototype chip are shown in Fig. 9.32. The IC is fabricated in a single poly, 6 metal layer, 90 nm CMOS process. It includes two ADCs, a bandgap reference, decimation filter and a serial interface for programming. The modulator operates from 1.1 V–1.3 V with only 3 dB SNR variation and uses between 1.44 mW and 6.6 mW from the native 1.2 V supply. Figure 9.33 shows measurements for both GSM (left) and UMTS (right) mode. In GSM mode

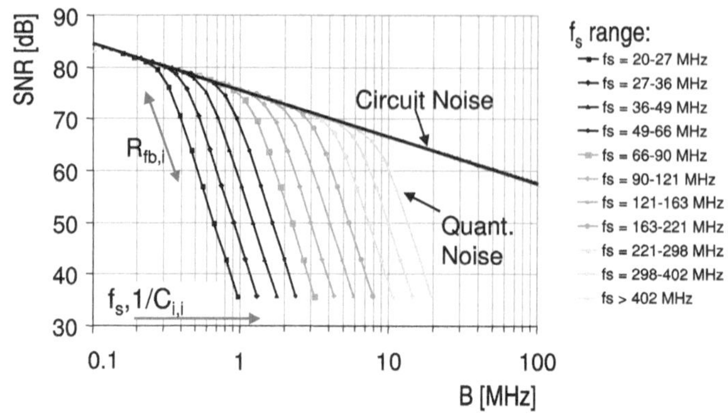

Figure 9.31: *Simulation of the ΣΔ modulator including thermal noise for all possible combinations of sample frequency and bandwidths*

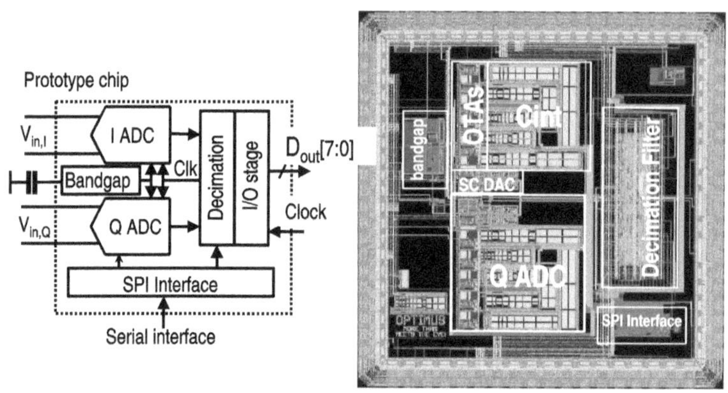

One ΣΔ Modulator: **600μm x 500μm = 0.3mm²**
Prototype chip: **1.6mm x 1.6mm = 2.56mm²**

Figure 9.32: *Micrograph of the test chip fabricated in a single poly, 6 metal layer, 90 nm CMOS process*

the modulator achieves a peak SNR of 82 dB in 200 kHz and sampled at 26 MHz. In UMTS mode the modulator achieves 71 dB in a bandwidth of 3.84 MHz. Sample frequency in this mode is 312 MHz. In both cases IR is better than 50 dB. Figure 9.34 shows the same measurements for Bluetooth (left) and WLAN (right) modes. Measured peak SNR is 75 and 52 dB in 1 and 20 MHz bandwidth respectively. Sample frequencies are 200 MHz and 400 MHz. In both modes the IR is better than 50 dB. In Fig. 9.35 the measured dynamic range (crosses) in a

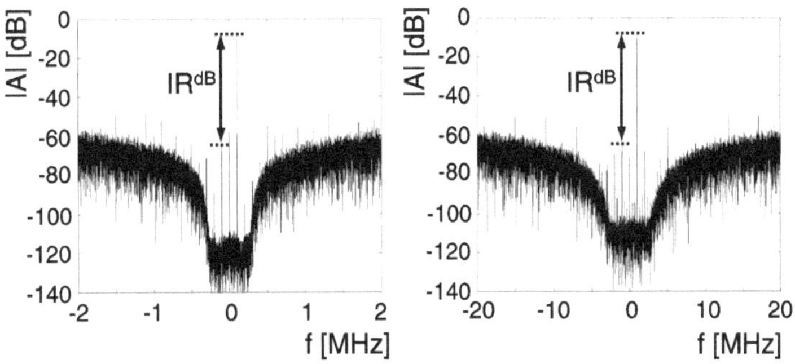

Figure 9.33: *I and Q measurement in GSM and UMTS mode*

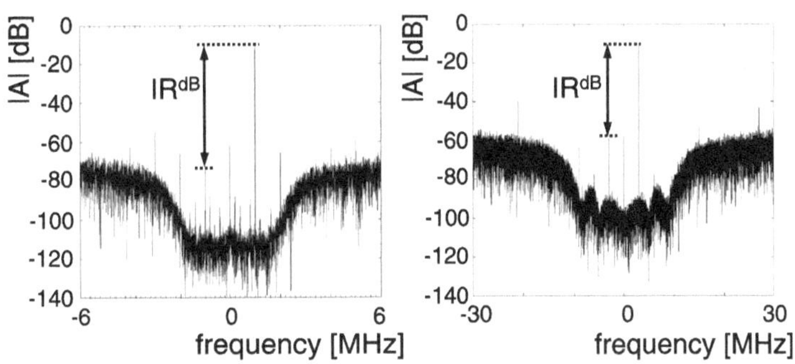

Figure 9.34: *I and Q measurement in Bluetooth and WLAN mode*

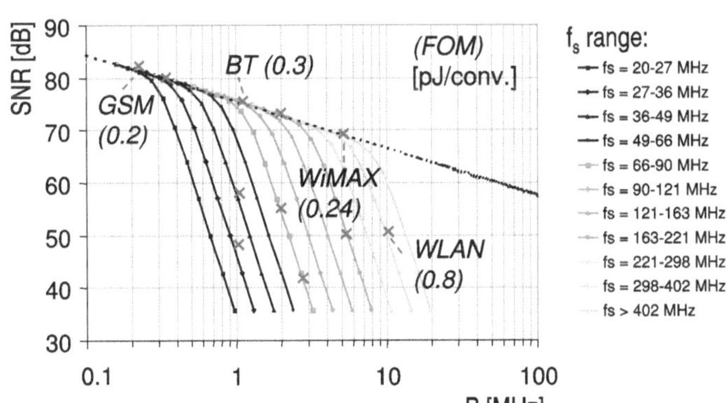

Figure 9.35: *Comparison of SNR simulations and measurements*

number of relevant modes is compared with the calculated (lines). The simulated and measured values match within 2 dB. The power efficiency is indicated by the FOM given for some characteristic modes. A performance summary is given in Fig. 9.36.

Process	1P, 6M, standard 90 nm CMOS			
Supply voltage	1.1V – 1.3V (+/- 3dB DR performance deviation)			
ΣΔ modulator	5th-order CT, feed forward, 1-bit with SC DAC			
Input voltage range	0.45Vrms,differential			
Modes	121	GSM	BT	WLAN
Sampling rate	20MHz - 400MHz	26MHz	200MHz	400MHz
Signal bandwidth	100kHz - 10MHz	270kHz	1MHz	10MHz
SNR_{max}	52dB - 82dB	82dB	75dB	52dB
Dynamic range	54dB - 84dB	84dB	76dB	54dB
Intermod. distances	IM2D > 70dB IM3D > 75dB			
Image Rejection	> 50dB			
Power@1.2V, one ΣΔ modulator	1.44mW - 6.6mW	1.44mW	3.36mW	6.6mW
FOM	0.21pJ/conv. - 0.81pJ/conv.	0.21pJ/conv.	0.33pJ/conv.	0.81pJ/conv.

Figure 9.36: *Performance summary and main design characteristics of the 121-mode ΣΔ modulator*

9.1.3.4 Conclusions

A 121-mode CT ΣΔ modulator that achieves a competitive Figure of Merit (from 0.2 pJ to 0.8 pJ/conversion). The modulator is programmable in bandwidth (over two decades: 0.1 MHz–10 MHz), SNR (85 dB–52 dB) and power (1.44 mW–6.6 mW). It combines the advantages of low-jitter sensitivity of SC ΣΔ modulators and high anti-alias suppression of CT ΣΔ modulators. This low power, high resolution, high linearity multi-mode modulator enables a low-cost, multi-mode, highly integrated receiver for almost all mobile and connectivity standards. Furthermore, due to the re-usable nature of the modulator receiver systems can be put faster on the market, as the same A/D converter can be used for several different receiver applications.

9.1.4 Multi-mode Modulator Clock Strategy

In the previous sections three modulator implementations for highly digitized receivers with an increasing number of modes were described. In Sect. 9.1.3 it has been explained that to cover the clock frequency range of 20 MHz to 400 MHz, a large amount of capacitors will be required. Section 7.2.3 proposed to always use the highest clock frequency for the modulator, thus also in low bandwidth modes. In this case only one set of capacitors is required in the loop filter, and only the local feedback coefficients are used to scale modulator bandwidth, reducing the number of modulator modes from 121 to 11. Figure 9.37 shows 11 simulations for the modulator of Sect. 9.1.3 sampled only at its highest clock frequency. In each simulation different values for the local feedback coefficients are programmed, like in Fig. 7.3. For each local feedback coefficient setting the total in-band noise is integrated from 1 kHz to the required modulator bandwidth B. For example, if the modulator is programmed to feedback coefficient 9, and the required bandwidth B is 4 MHz, the expected peak SNR of the modulator is 55 dB, while in program setting 7, it would have been 65 dB. As Sect. 7.2.3 described, one of

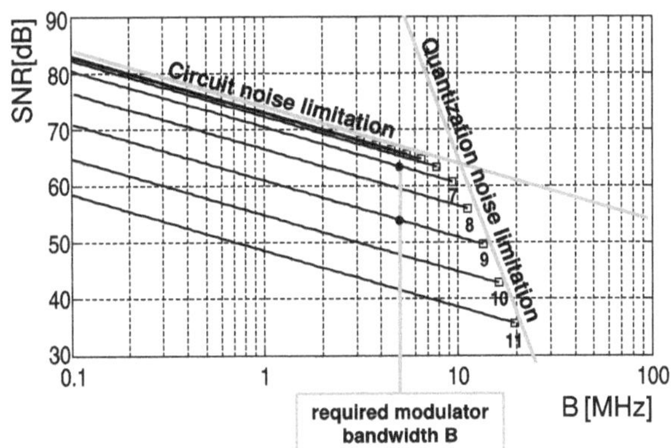

Figure 9.37: *Modulator SNR as a function of the local feedback coefficients and modulator bandwidth*

the advantages of the fixed high clock frequency is that the capacitors in the loop filter can be reduced to the capacitors required for the highest clock frequency which are small. For the 121-mode modulator this would mean an 80% smaller modulator area (Fig. 9.32). Because of the higher clock frequency a higher decimation factor will be required in low bandwidth modes which means a larger decimation filter. As the first decimation stages will have very relaxed transition band requirements because of the high OSR, the area required to implement the additional stages will be limited. The increase in the decimators complexity can easily be implemented within the 80% saved area, as it represents 100 k to 200 k gates (2 input NAND) in the 90 nm technology the modulator was designed in.

The high clock frequency strategy will lead to additional power spending in the modulator and digital part. For the digital part this power increase is expected to be limited as the additionally required complexity is limited, and the efficiency of digital circuits becomes higher in each next technology node. For the modulator there will be additional power spending as more bandwidth is required from its circuits. For the 121-mode implementation, Fig. 9.36 revealed that the difference in modulator power in the lowest and highest bandwidth modes is a factor of four. One of the main reasons is the increased bandwidth requirement for the feed-forward summing node and quantizer to maintain proper operation of the modulator at higher sample frequencies. This means higher bias currents, and thus higher modulator power consumption. However, with the use of the loop delay compensation techniques presented in Sect. 6.8, additional power spending at higher sample frequencies can be avoided. The excess phase of the feed-forward summing node can be compensated, and/or the decision time for quantizer can be increased, by a deliberately implemented delay between quantizer decision and DAC feedback clocking, giving the quantizer more time to make a proper decision. Unfortunately, at the time of design of this modulator the delay compensation technique theory was not available, and therefore was not applied to this modulator to save power.

Using the high modulator clock frequency strategy increases the digitization of the receiver system, as more programmability is shifted into the digital domain, at the same time exploiting the area, power and speed advantages of deep sub-micron technologies making this clock strategy future proof. A thorough analysis of the interchange of complexity, area and power between modulator and decimator are out of the scope of this book. Furthermore, the impact at receiver system level, like the possible requirement of fractional sample-rate conversion, should be further investigated.

9.2 Digitization at Analog IP Architecture Level: A Hybrid, Inverter-Based ΣΔ Modulator

In this section an example is shown of a modulator which is digitized at architecture and circuit level [97]. The number of analog blocks is reduced replacing them by digital circuitry, in order to speed up technology porting of analog IP. To digitize the IP the design methodology presented in Sect. 6.1.2 is used.

The presented modulator, with a first order analog filter, 5 bit quantizer, second order digital filter, 1-bit quantizer and 1-bit DAC, achieves a peak SNR of 77 dB in 200 kHz. The active circuitry is implemented solely with inverter circuits and standard digital cells in a 65 nm CMOS technology. Power consumption is 950 μW at 1.2 V and the area is only 0.03 mm^2.

9.2.1 ΣΔ Modulator Architecture

The modulator architecture is based on the architecture of Sect. 5.4 (Fig. 5.6). The advantages of lower quantization noise using a multi-bit quantizer and the linearity of a 1-bit feedback path are combined. As design technology for this modulator is 65 nm CMOS, a high sample frequency (150 MHz) in combination with a first order analog loop filter and a 5 bit quantizer is chosen, to reduce the in-band quantization noise. Using this architecture, only a very limited amount of analog circuits is required which are the integrator stage of the first order loop filter, a 5 bit quantizer and a 1-bit SC DAC. Furthermore a digital filter and digital 1-bit quantizer is required. For the digital filter, a second order IIR filter is used to sufficiently push the in-band 1-bit quantization noise below the 5-bit quantization noise. This way the modulator exploits the multi-bit quantizer, while an inherently linear 1-bit feedback DAC can still be used. Furthermore, as the modulator's loop order is increased, the digital filter is used to de-correlate the signal and the quantization noise at the output Y.

In Fig. 9.38a the transfer function of the first order analog filter and the second order digital filter are shown. In Fig. 9.38b a simulation of the modulator is shown.

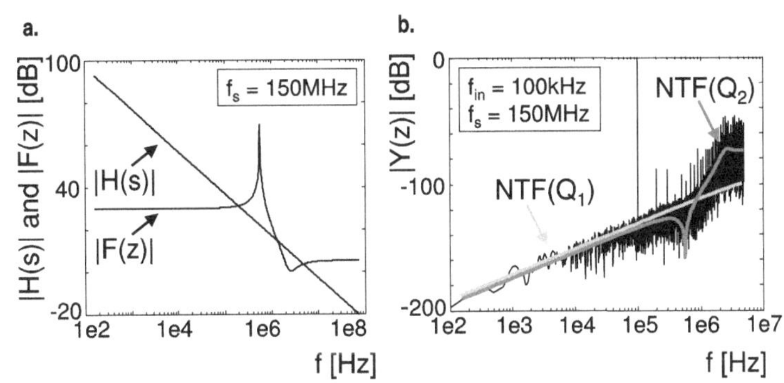

Figure 9.38: *Analog and digital filter transfer function and modulator simulation*

The noise transfer functions of the quantization noise sources are indicated. From the figure it can be seen that there still is correlation between input signal and the quantization noise coming out of the modulator. This can be better observed with an input signal of lower frequency. Figure 9.39a shows a simulation of the modulator with an input frequency of 10 kHz. The correlation between input signal and quantization noise can clearly be seen from the spectrum. In Fig. 9.39b the expected thermal noise is added to the system. The thermal noise greatly

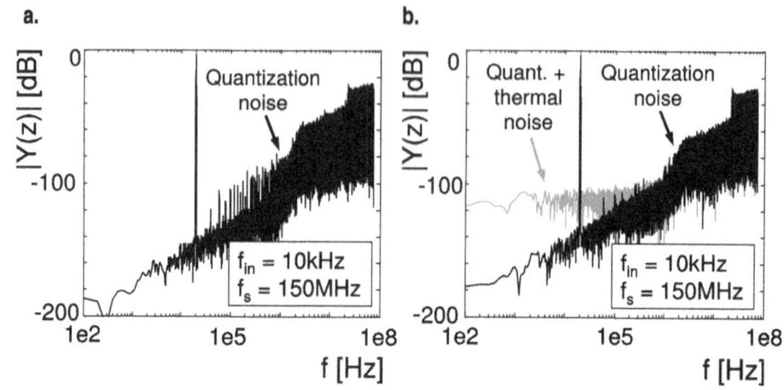

Figure 9.39: *Reduction of correlation between input signal and quantization noise due to inclusion of circuit noise in the simulation*

reduces the correlation between input signal and quantization noise, as can be seen from the quantization noise plot. The simulated SQNR is 90 dB in 200 kHz at a clock frequency of 150 MHz.

9.2.2 Circuit Design

Figure 9.40 shows the block diagram (a) and transfer function (b) of the digital filter. It is a second order IIR filter with a resonator. The filter architecture only requires 1 addition of two numbers in the direct forward path, which reduces loop delay. Because f_s is much smaller than the f_T of the 65 nm technology, the worst-case filter delay is 400 ps which is much smaller than the sampling time and modulator stability is not compromised. The 1-bit quantization is implemented by taking the MSB of the output word of the digital filter. The filter includes bubble correction at its input V to correct for quantizer errors.

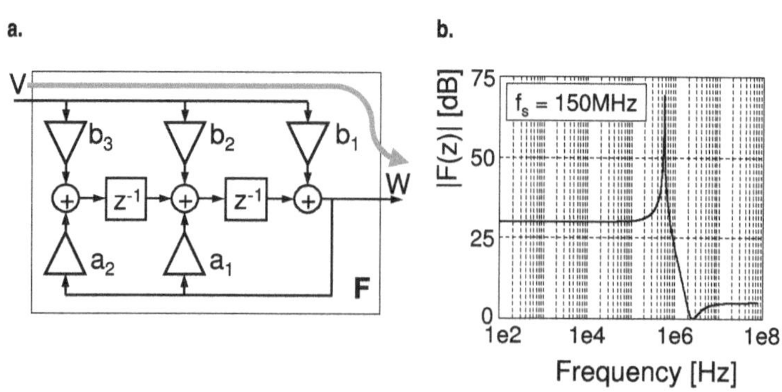

Figure 9.40: *Digital filter architecture and transfer function*

The digitization process is carried through to all circuits of the modulator. The quantizer (Fig. 9.41a) consists of a resistor ladder and $N = 2^{B-1}$ identical comparators. The ladder outputs N reference voltages $V_{c,x}$ which are a combination

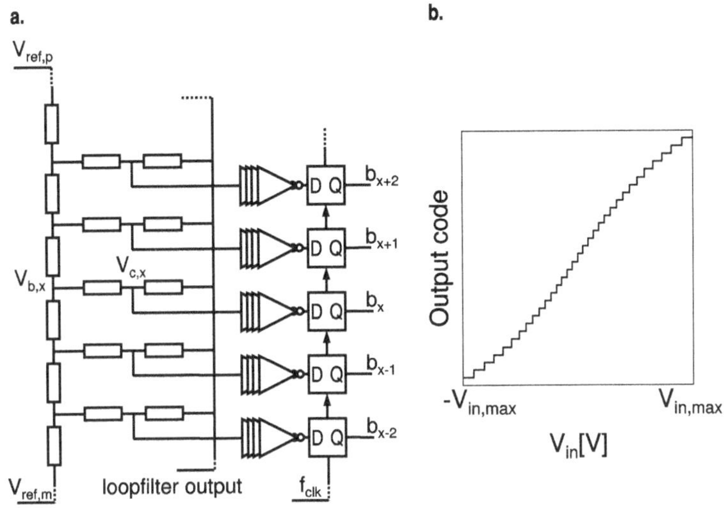

Figure 9.41: *5-bit quantizer implementation and transfer function*

of the $V_{b,x}$ voltages and V_{in}. Each of the comparators consists of a cascade of the analog inverters of Fig. 6.3 and standard digital inverters followed by a flip-flop. The cascade ensures that the input signal of the flip-flop is rail to rail before it is converted to a logic 0 or 1. Each comparator has its decision level at $(V_{refp} + V_{refm})/2$ which reduces the design complexity of the comparators. The quantizer will output a thermometer code depending on V_{in}. Because the output impedance of the resistor ladder is not the same for each output tap, the comparator decision levels will not be linearly distributed over the input signal range (Fig. 9.41b), unless the resistor values are scaled accordingly. However, when the resistors are not scaled, the non-linearity errors of the quantizer are suppressed by the analog loop filter gain, and in this case have very limited impact on the modulators performance. The trip levels of the comparators are subject to transistor mismatch, which causes shifting of the inverter trip levels. The consecutive comparator trip levels and their offset distribution are shown in Fig. 9.42. The shifting of decision levels can cause that some comparators can fall out of the used quantizer input range and the order of trip levels might change. The levels that shift out of range are lost, but the ones that change order are corrected by a bubble correction, to preserve the quantizers monotonic behavior. The LSB of the quantizer is 20 mV, and its simulated comparator offset σ value is 3.3 mV. A simulation of the

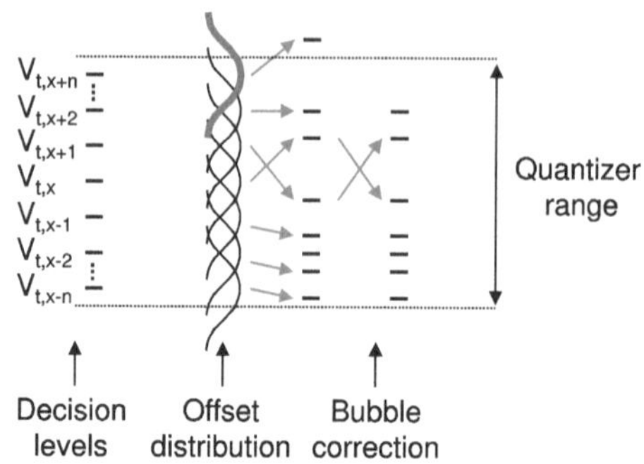

Figure 9.42: *Comparator offset distribution and bubble correction*

modulator model with different quantizer offset σ values is shown in Fig. 9.43. The figure shows the simulated SQNR and SNR for the modulator on the Y-axis

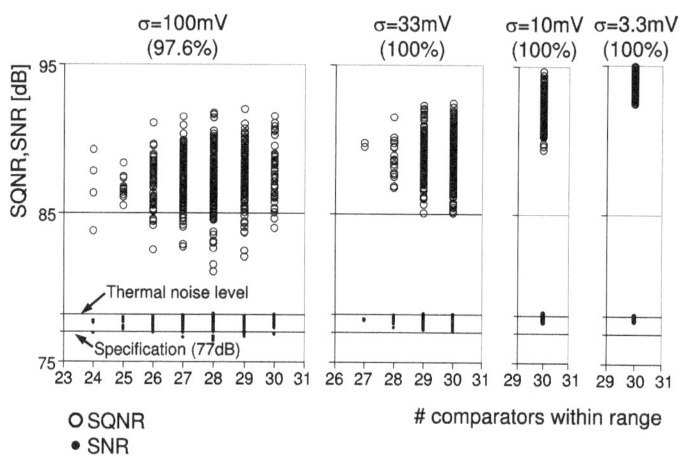

Figure 9.43: *Comparator offset spread impact on modulator SQNR for 4 different σ values*

for different σ values. The thermal noise level is indicated in the figure and is at 76 dB in 200 kHz bandwidth. The X-axis displays the number of comparator levels within the quantizer input range for 4 different σ values. For each σ value 1000 simulations are done. For σ values of 3.3 and 10 mV, all comparators are within the quantizer input range. At σ values of 33 mV comparators start to drop out of the quantizer input range for various trials. For a σ of 100 mV, the SNR starts to degrade below the target specification of 77 dB in 200 kHz (indicated in the figure) for two reasons. More and more comparator levels drop out of the quantizer input range, and the distribution of comparator levels over the quantizer input range becomes more non-linear. The percentage of simulations within specification is mentioned in the figure below each σ value. Even for the σ value of 100 mV, the percentage of drop-outs is only 2.4%.

The first order analog loop filter is implemented with an RC integrator stage. The OTA is completely made out of unit cell inverters and will be shown in Sect. 9.3 (Fig. 9.54). The feedback DAC is similar to the one used in the modulator presented in Sect. 9.1.2.3, and only uses switches, capacitors and a few standard digital cells.

9.2.3 Experimental Results

The block diagram of the prototype chip is shown in Fig. 9.44. At an f_s of 150 MHz, the thermal noise of the input stage dominates all other noise contributions. The output signal of the analog loop filter is converted to a single

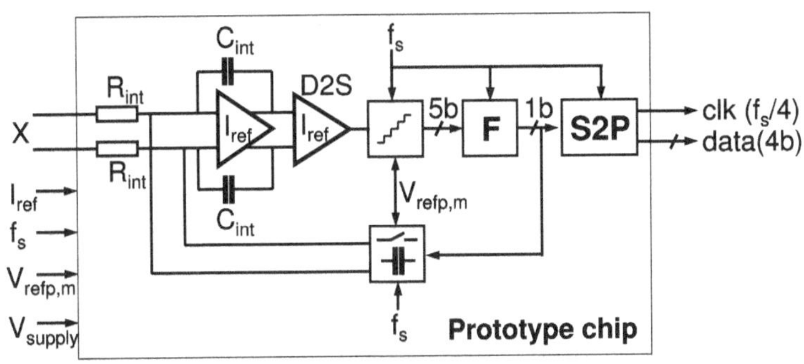

Figure 9.44: *Block diagram of the implemented hybrid modulator*

ended signal to drive the 5-bit quantizer which output is fed to the digital filter F. The digital 1-bit output code is clocked by a flip-flop before it is fed to the 1-bit DAC.

Figure 9.45 shows the measured modulator output spectrum when a 100 kHz full scale signal is applied to the modulator input. The left figure shows a zoom-in of the spectrum on the right. The measured peak SNR is 77 dB in 200 kHz, THD is −79 dB. The left of Fig. 9.46 shows a zoom-in on the intermodulation measurement presented on the right. The IM2 and IM3 distances are 86 and 82 dB respectively, when a 200 kHz and 250 kHz signal is applied to the modulator, both with an amplitude of −6 dBFS. Figure 9.47a shows the SNR as function of the input signal amplitude for 3 different bias currents ($I_{ref,nom} = 25$ µA). Peak SNR

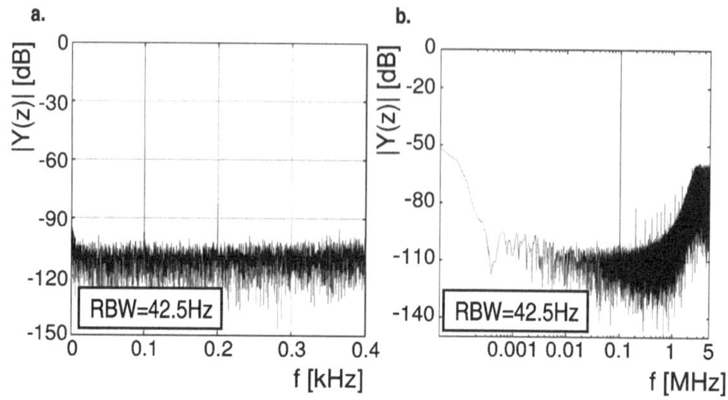

Figure 9.45: *Full scale measurement to determine modulators SNR and THD*

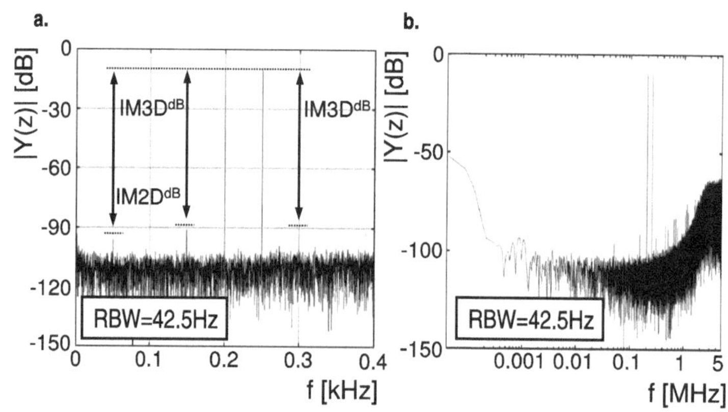

Figure 9.46: *Second and third order intermodulation measurement*

is 77 dB in 200 kHz in all cases. There is almost no current dependency as the design is limited by the thermal noise of the input resistors. Figure 9.47b shows the peak SNR of 36 measured modulators. The SNR of 11 samples measure 76 dB, 25 measure 77 dB.

Figure 9.48 shows the peak SNR, as a function of f_s. The modulator achieves 77 dB in 200 kHz over a $f_s = 75$–300 MHz range. At $f_s < 75$ MHz the quantization noise starts to dominate the noise contributions, as the digital filter transfer function scales with f_s. Figure 9.48 illustrates the potential of this modulator architecture to become multi-mode, as the bandwidth and power of the digital filter scale with clock frequency.

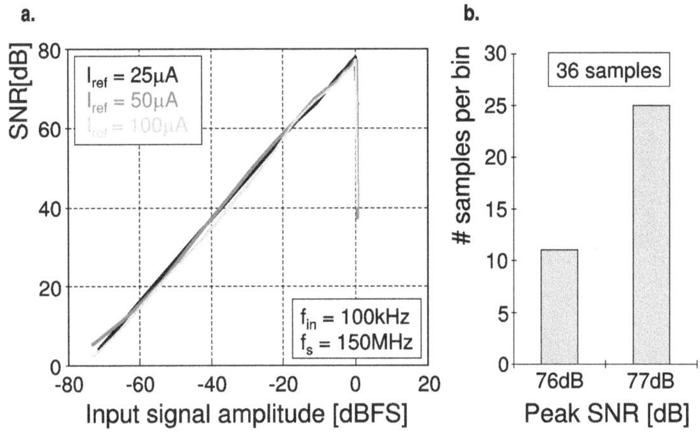

Figure 9.47: *SNR as a function of input signal amplitude (a) and measured peak SNR over 36 samples (b)*

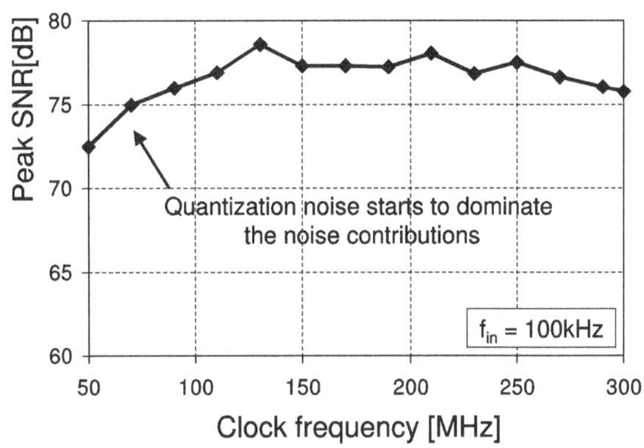

Figure 9.48: *Measured modulator peak SNR as a function of f_s*

Figure 9.49 shows a benchmark of all low-pass ΣΔ modulators presented at the ISSCC over the past 11 years. On the Y-axis the modulators active area is shown, on the X-axis the technology feature size is shown. The modulator presented in this section is two times smaller than the smallest published. The grey circles indicate modulators, which have a similar bandwidth and dynamic range as the modulator presented. Their bandwidth is in the range of 100–400 kHz. Their dynamic range is 77 dB plus or minus 3 dB range calculated in 200 kHz. If only these modulators are considered the presented converter is even 4 times smaller. The prototype chip shown on the left in Fig. 9.50, is fabricated in a single poly,

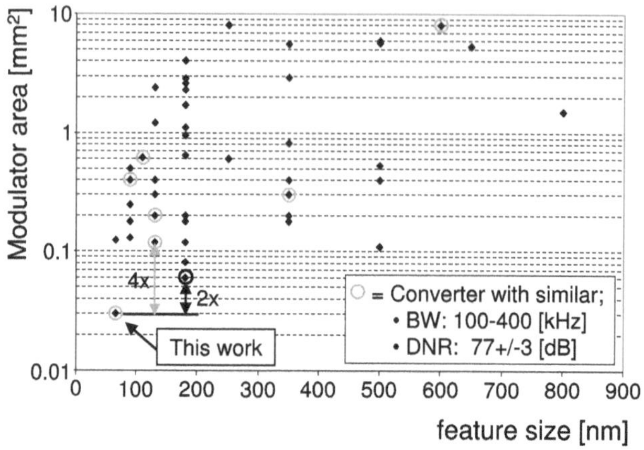

Figure 9.49: *Area benchmark of the presented modulator with other low-pass modulators presented in literature*

7 metal layer, 65 nm CMOS process. The sub-blocks are indicated in the layout plot on the right of Fig. 9.50. The main design characteristics and the performance summary of the presented modulator are shown in Fig. 9.51.

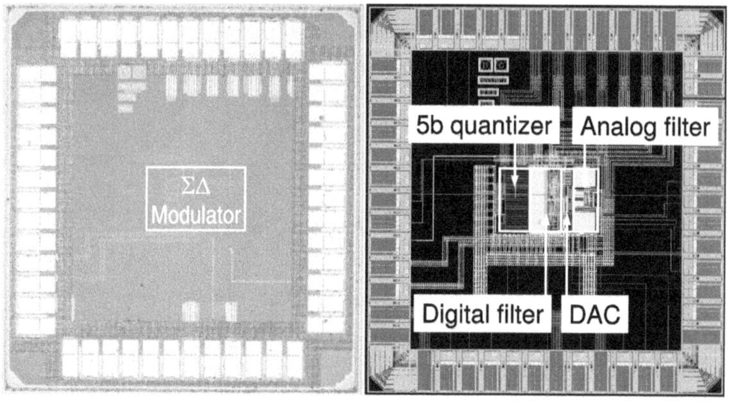

Figure 9.50: *Micrograph of the 1-5-2-1 hybrid modulator fabricated in a single poly, 7 metal layer, 65 nm CMOS process*

9.2.4 Conclusions

The ΣΔ modulator presented uses a first order analog filter, a 5-bit quantizer, a 2nd order digital filter and a 1-bit quantizer, forming a 1-5-2-1 inverter-based hybrid ΣΔ modulator. The ΣΔ modulator exploits the area scaling advantages of

Technology	1P, 7M, standard 65nm CMOS
ΣΔ Modulator	1st order analog filter / 5-bit quantizer / 2nd order digital filter / 1-bit quantizer / 1-bit SC feedback DAC or **1-5-2-1** hybrid ΣΔ modulator
Input voltage range	0.45Vrms, differential
Sampling range	50-300MHz
Signal bandwidth	200kHz
Dynamic range	77dB (I_{ref}=25-100µA, f_s=150MHz)
IM2D, IM3D	86dB, 82dB
Power consumption	**950µW (I_{ref}=25µA, f_s=150MHz)**
FoM	**0.3pJ/conversion**
Active area	0.03mm^2

Figure 9.51: *Main design characteristics and performance summary of the 1-5-2-1 hybrid ΣΔ modulator*

digital circuits in deep-submicron technologies, by shifting analog functionality into the digital domain as much as possible. At architecture level, a complex analog filter is replaced by a first order filter and a low requirement quantizer and digital filter. The advantage of a modulator with multi-bit quantizer is combined with the advantage of a 1-bit inherently linear feedback path. At circuit level, the remaining analog circuits are implemented using inverter unit cells. The modulator achieves a peak SNR of 77 dB in 200 kHz and the IM2 and IM3 measure −82 and −86 dB respectively. With a power consumption of 950 µW, the figure of merit is 0.3 pJ/conversion. Due to the digitization of the modulator, its active area is only 0.03 mm^2. Due to the limited amount of analog blocks in this analog IP, portability to future nm-technologies has been greatly improved.

The large clock frequency range over which this modulator can be scaled makes the hybrid modulator architecture extremely suitable for implementing a scalable (multi-mode) modulator for highly digitized receivers.

9.3 Digitization at Circuit and Layout Level: Technology Portable ΣΔ Modulators

In this section several examples of an analog IP block will be shown that were designed using the digitized circuit design methodology of Sect. 6.1.2. It concerns three 5th order, 1-bit, ΣΔ modulators which achieve a peak SNR of 60 dB in

15 MHz at a power consumption of 9 mW, each measuring only 0.04 mm^2 [39]. A 65 nm test chip contains a feedback modulator, for which a sample frequency of 1 GHz was chosen to test the technology's speed limitations. Furthermore, to prove that the digitized circuit design approach increases the portability of analog IP, the feedback modulator is ported from a 65 nm to a 45 nm technology, without major changes to the design.

This section also will show an example on how to make a modulator more robust to circuit imperfections by making changes at ΣΔ modulator architecture level. In the example, the excess delay in a feed-forward modulator designed in a 45 nm technology, will be compensated using the delay compensation method of Sect. 6.8.

9.3.1 ΣΔ Modulator Architecure

Figure 9.52a shows the modulator architecture. For the loop filter architecture two options are available: feed-forward and feedback. Feed-forward ΣΔ modulators require an additional summing node for the feed-forward coefficients (grey dotted components), which, due to limited bandwidth of the summing node, introduces an additional pole in the loop (Fig. 9.52a). The feedback ΣΔ modulator does not need the additional summing node which makes this architecture more suitable for high sample rate modulators. In return, the feedback modulator needs L DACs. Although a feed-forward modulator is preferred (Sect. 6.3), the parasitic pole in its summing node has to be compensated for at these high speed

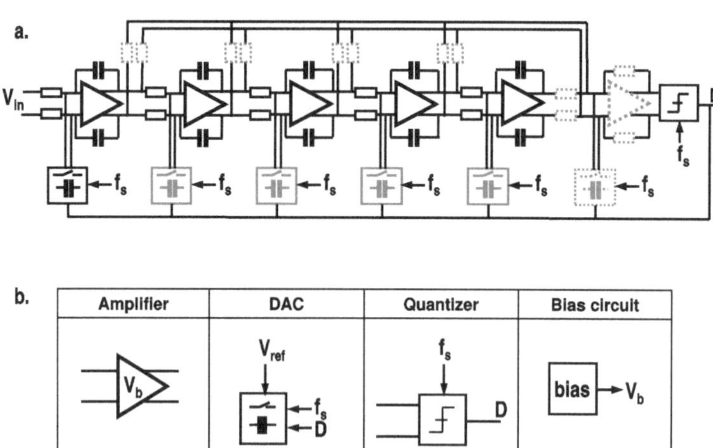

Figure 9.52: *Feedback or feed-forward ΣΔ modulator architecture*

clocks to avoid excessive power spending in the summing node, or even worse modulator instability. The excess phase compensation of the feed-forward modulator can be implemented using the theory presented Sect. 6.8.2. To be able to apply the compensation method, a delay compensation DAC (the right most DAC in Fig. 9.52a which is grey dotted) is required. Figure 9.53 shows the simulated SQNR for a modulator uncompensated and compensated parasitic pole, relative to the ideal modulator without parasitic pole (and compensation). Obviously, the ideal modulator has no SQNR loss, as there is no parasitic pole in the loop. The uncompensated modulator with parasitic pole in its loop loses performance when the parasitic pole is shifted down in frequency. If the pole drops below $0.3f_s$ the maximum achievable SQNR of the modulator drops very rapidly. After compensation, the SQNR is restored to its original value, independent of pole frequency.

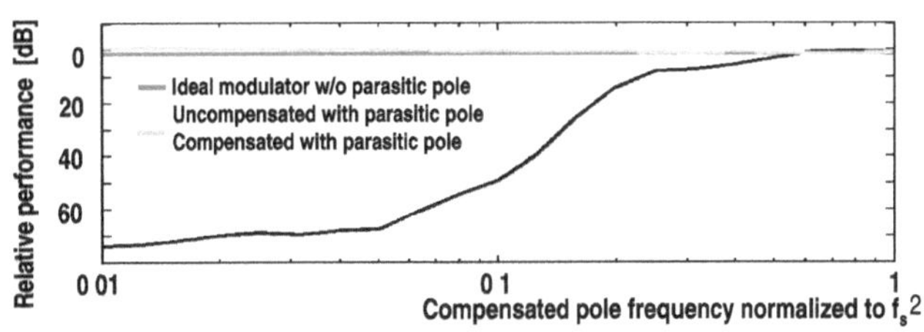

Figure 9.53: *Simulated SQNR of uncompensated and compensated modulator with parasitic pole as a function of normalized parasitic pole frequency and relative to an ideal modulator without parasitic pole (and compensation)*

9.3.2 Circuit Design and Layout

For the circuit and the layout, the digitized design methodology of Sect. 6.1.2 is used. The ΣΔ modulator (either feed-forward or feedback) is split in only a limited number of circuits, which are an amplifier (OTA), a switched capacitor DAC, a quantizer and a bias circuit. The modulator block diagrams and the separate blocks are indicated in Fig. 9.52b. After the split in circuit blocks, a common divisor is searched for the active circuitry. For the loop filter integrator, a top-down bottom-up (Sect. 6.1.2) search for a unit cell based integrator leads to a three stage inverter OTA, used as an RC integrator (Fig. 9.54). All the inverters in the OTA consist of one or multiple instantiations of an analog unit cell inverter in parallel. A single analog unit cell inverter is displayed on the bottom right of Fig. 9.54. The W/L of the PMOS and NMOS transistor in the inverter unit cell is fixed for

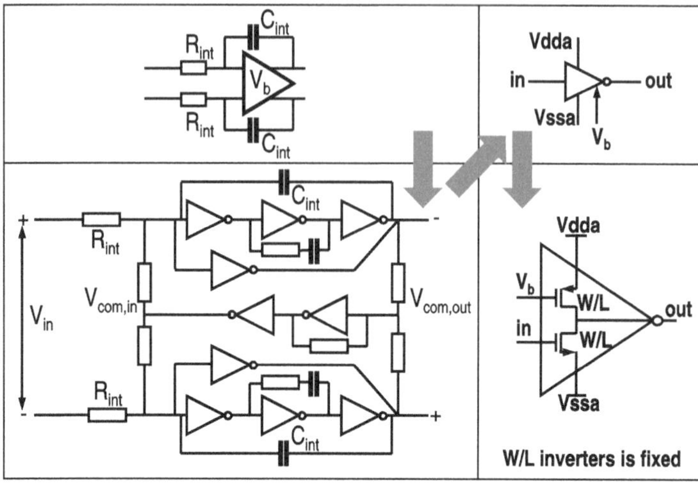

Figure 9.54: *Breaking up circuits: integrator*

all used unit cell inverters. The PMOS and NMOS gate are not connected together unlike in a digital inverter, to be able to connect the PMOS and NMOS transistor in different ways. Note that the unit cell inverter is also used for the common-mode arrangement in the OTA.

Figure 9.55 presents the bias circuit for the inverters used in the OTA. The NMOS is switched off by connecting its gate to ground and the gate-drain connected

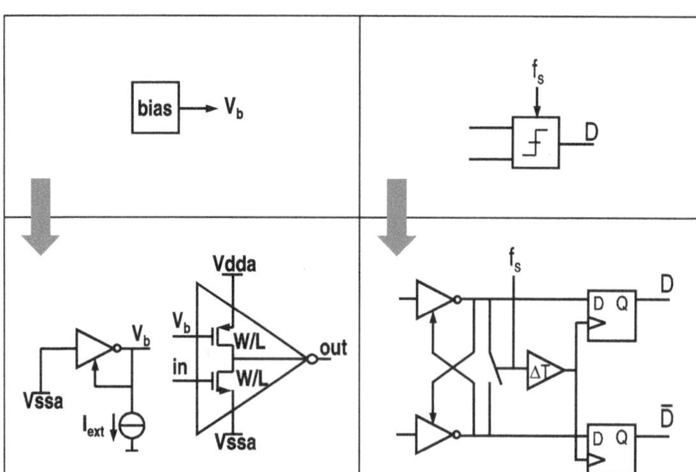

Figure 9.55: *Breaking up circuits: OTA bias circuit and quantizer circuit*

PMOS transistor is used to bias the PMOS current sources in the inverter. The quantizer is shown on the right of Fig. 9.55. It uses two cross coupled inverter unit cells with a switch in between, to reset the quantizer latch. After a delay circuit made out of standard digital cells, the data coming from the latch is re-clocked by a standard digital master-slave flipflop. The SC DAC is similar to the one presented in Sect. 9.1.2.3.

The feedback ΣΔ modulator uses 5 times exactly the same OTA, a quantizer and 5 times the same DAC. The feed-forward modulator uses 6 times exactly the same OTA, a quantizer, and 2 times the same DAC. All the analog blocks are built out of the analog unit cell inverter stage, switches, resistors, capacitors, and standard digital cells only. This makes the porting of such a modulator very simple, as only a very limited amount of circuits have to be re-simulated and re-layouted.

Figure 9.56 shows an example of a digital inverter plus its p-cell layout, together with its analog equivalent used to implement the modulator. The analog inverter

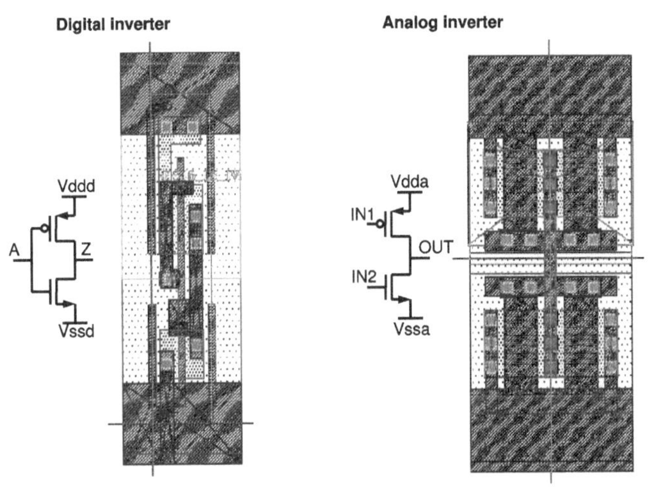

Figure 9.56: *Digital and analog inverter circuit and p-cell layout*

has the same layout properties as the digital inverter (for example the supply lines and a placement boundary box) in order to be able to layout the analog circuits with the automatic layout tools normally used for digital circuits. For the other unit cells (resistors, capacitors, switches, etc.) a similar p-cell layout is made. For the analog inverter only one layout is made which is re-used in the OTAs, quantizer and bias circuit.

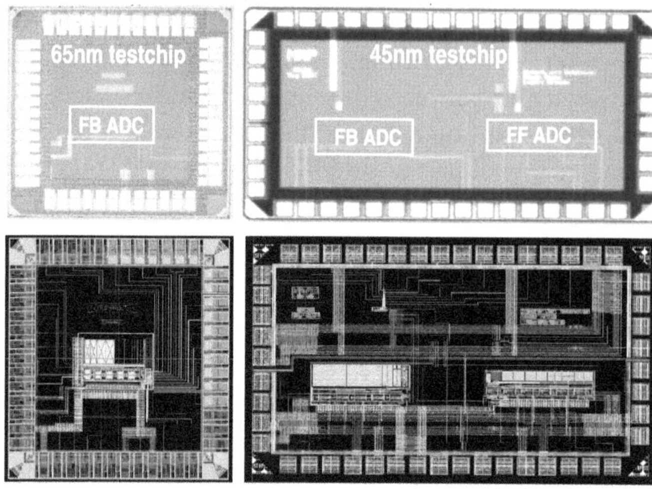

Figure 9.57: *Chip micrographs of the 45 nm and 65 nm test chip and their layout plots*

9.3.2.1 Experimental Results

The chip micro graphs and layout plots of the 45 nm and 65 nm test chip are shown in Fig. 9.57. Figure 9.58 shows the full-scale input signal measurement of the 65 nm and 45 nm test chip. Both achieve a peak SNR of about 60 dB. The clock frequency of the 65 nm test chip is 1 GHz, the clock frequency of the 45 nm test chip is 1.5 GHz. As both the 45 nm and 65 nm feedback modulator have

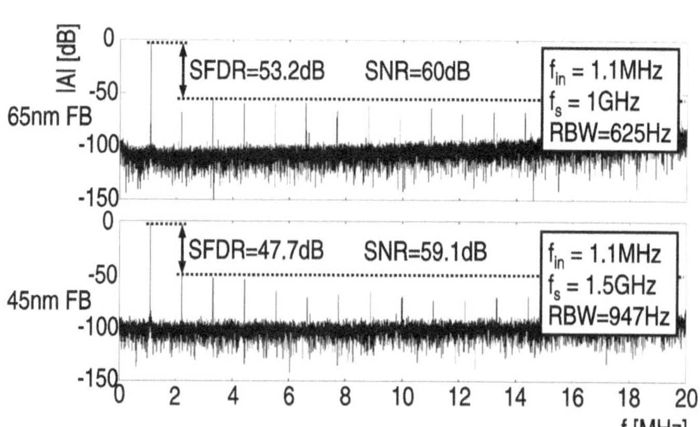

Figure 9.58: *Full scale input signal measurement to determine the peak SNR and SFDR of the 65 nm and 45 nm feedback modulator*

comparable performance, the technology porting is completed successfully. The port from 65 to 45 nm was done in one month only, including re-simulation of the unit-cells, re-generation of all circuit views and layout generation.

The chip micrograph and layout plot of the 45 nm feed-forward modulator are also shown in Fig. 9.57. Figure 9.59 shows the full-scale input signal measurement of the 45 nm feed-forward modulator. It achieves a peak SNR of about 60 dB at a clock frequency of 1.5 GHz. Figure 9.60 compares the presented mod-

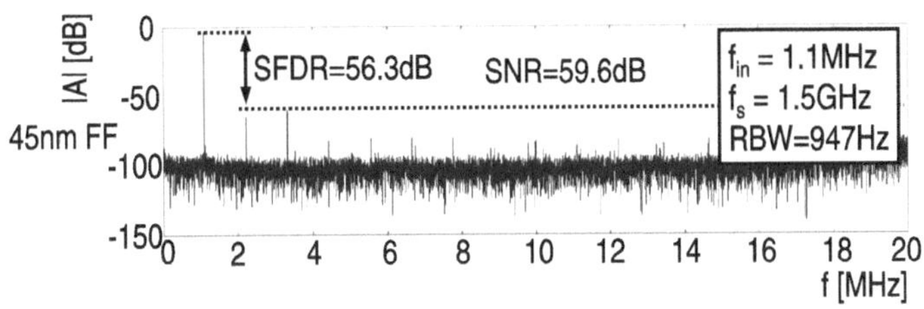

Figure 9.59: *Full scale measurement to determine the peak SNR and SFDR of the 45 nm feed-forward modulator*

ulators of this section with published modulators with comparable specifications (B = 10–20 MHz and SNR = 60–67 dB) on area, FOM_{DR} and $\text{FOM}_{eq,th}$. The

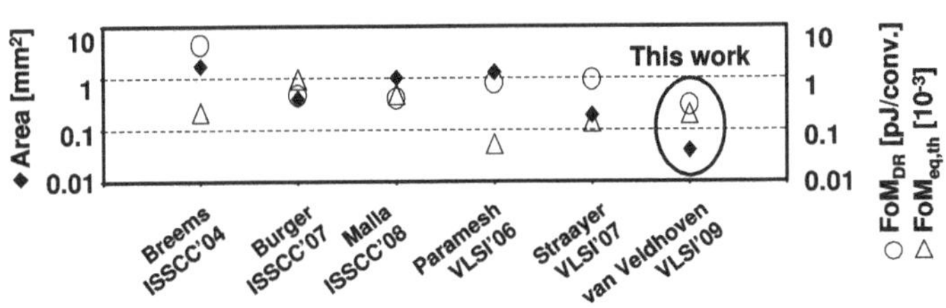

Figure 9.60: ΣΔ *modulator area, FOM_{DR} and $FOM_{eq,th}$ benchmark*

modulators presented in this section are at least 5 times smaller (left Y-axis) and achieve a state-of-art FOM_{DR} (right Y-axis) compared to the other modulators in this benchmark. The $\text{FOM}_{eq,th}$ will give a more realistic view on power efficiency as explained in Chap. 8. This FOM is four to five times worse than state-of-art. This can be explained by the fact that power consumption/performance ratio of the modulators presented in this section was not optimized to keep the modulators as simple and straightforward as possible. For example all five three-stage integrators presented in Fig. 9.54 use the same amount of current as they are exact copies. As the later integrators are far less critical in terms of noise compared to the first integrator, the later integrators can be scaled by using fewer inverters in parallel in each stage of the integrator, while keeping stage ratios the same. This way the power efficiency of these modulators can be increased. However, the power efficiency increase is limited by the stability criteria of the modulator. In the feedback modulator the last integrator determines the phase shift of the loop filter at high frequencies, and to avoid modulator instability the power consumption of this integrator can not be reduced only based on noise requirements. In the feed-forward modulator the first integrator and the summing node determine the high frequency phase shift. The bias current of the first integrator should be high because of its large contribution to the noise (which is also the case in the feedback modulator), which will have a positive impact on its bandwidth. In the summing node the power consumption can be reduced when the excess loop delay compensation technique is applied.

Figure 9.58 gives the main design characteristics and performance summary for the feedback and feed-forward modulator.

ΣΔ modulator architecture	5th order, 1-bit quantizer, 1 bit SC feedback DAC(s)		
Loopfilter architecture	feedback	feedback	feed-forward
Number of required feedback DACs	5	5	2
Technology (standard CMOS)	**1.2V, 1P, 7M, 65nm**	**1.1V,1P, 7M, 45nm**	
Differential input voltage range	$0.45V_{rms}$	$0.45V_{rms}$	$0.45V_{rms}$
Differential input impedance	2 · 25kΩ	2 · 25kΩ	2 · 25kΩ
Sampling frequency	1GHz	1.5GHz	1.5GHz
Over-sampling ratio	33	50	50
Signal bandwidth	15MHz	15MHz	15MHz
Power consumption	9mW	9mW	9mW
Peak SNR	60dB	59.1dB	59.6dB
SFDR	53.2dB	47.7dB	56.3dB
FoM	0.30pJ/conv	0.33pJ/conv	0.31pJ/conv
Active area	**0.040mm²**	**0.044mm²**	**0.035mm²**

Figure 9.61: *Main design characteristics and performance summary of the technology portable ΣΔ modulators*

9.3.3 Conclusions

Several modulators have been made using a digitized design methodology. The modulators are built up out of only a limited number of analog unit cells. Using the unit cell design approach, a feedback modulator has been successfully ported from a 65 nm to a 45 nm without major changes to the circuits. The port was done in an extremely short time-frame of 1 month (circuit design and layout), proving the portability increase due to the unit cell design methodology.

Furthermore, a 45 nm feedback modulator has been successfully transformed into an excess phase compensated feed-forward modulator without performance loss, proving the excess phase compensation theory presented in this book.

The digitally designed modulators have a very low area, proving that this design methodology does not necessarily come at the cost of additional area. The power efficiency FOMs of the modulators are competitive, but there is room for improvement by further optimizing the power consumption of the integrator stages for the stages which are not critical in terms of noise or modulator stability.

9.4 Implementations Judged on the FOMs and Quality Indicators

In Chap. 6 the 1-bit, high order, CT feed-forward $\Sigma\Delta$ modulator was selected as the preferred modulator architecture, by using the quality indicators. In this chapter several implementations of this and other $\Sigma\Delta$ modulator architectures have been demonstrated. In Fig. 9.62 the FOMs of Chap. 8 for the implemented $\Sigma\Delta$ modulators are presented. Figure 9.63 presents a matrix with the quality indica-

FOM	1.5-bit modulator	Triple-mode modulator	121-mode modulator	Hybrid modulator	GHz-rate modulators
FOM_{DR} [pJ/conv.]	0.26	GSM: 0.24	GSM: 0.29	0.33	65nm FB: 0.3
		CDMA: 0.24	BT: 0.3		45nm FB: 0.33
		UMTS: 0.22	WLAN: 0.88		45nm FF: 0.31
$FOM_{eq,th}$ [-]	$1.3\cdot10^{-3}$	GSM: $4.5\cdot10^{-3}$	GSM: $2.6\cdot10^{-3}$	$1.2\cdot10^{-3}$	65nm FB: $2\cdot10^{-4}$
		CDMA: $1.6\cdot10^{-3}$	BT: $1.1\cdot10^{-3}$		45nm FB: $1.3\cdot10^{-4}$
		UMTS: $0.6\cdot10^{-3}$	WLAN: $2.7\cdot10^{-5}$		45nm FF: $1.5\cdot10^{-4}$
FOM_{HD3D} [-]	$5\cdot10^{-7}$	GSM: $8.7\cdot10^{-4}$	GSM: $8.7\cdot10^{-6}$	$2.14\cdot10^{-6}$	65nm FB: $4.3\cdot10^{-10}$
		CDMA: $2\cdot10^{-5}$	BT: $6.5\cdot10^{-7}$		45nm FB: $0.95\cdot10^{-10}$
		UMTS: $6.6\cdot10^{-7}$	WLAN: $9.4\cdot10^{-12}$		45nm FF: $3.6\cdot10^{-10}$
FOM_{area} [W/mm²]	$2.8\cdot10^{-6}$	GSM: $2.9\cdot10^{-5}$	GSM: $1.3\cdot10^{-6}$	$5.5\cdot10^{-5}$	65nm FB: $6.2\cdot10^{-6}$
		CDMA: $1.1\cdot10^{-5}$	BT: $1.3\cdot10^{-6}$		45nm FB: $5.1\cdot10^{-6}$
		UMTS: $0.5\cdot10^{-5}$	WLAN: $6.6\cdot10^{-8}$		45nm FF: $5.7\cdot10^{-6}$

Figure 9.62: *Implementations judged on the FOMs*

tors and the different designs presented in this chapter. A brief explanation on the judgments based on the FOMs and quality indicators is given in the summary below:

Quality indicator		1.5-bit modulator	Triple-mode modulator	121-mode modulator	Hybrid modulator	GHz-rate modulators
Algorithmic accuracy	Overall	☺	☺	☺	☺	☺
Robustness	Overall	😐	☺	☺	☺	😐
Flexibility	Overall	😐	☺	☺	☺	☺
Efficiency	Power	☺	☺	☺	☺	😐
	Area	☺	☺	☹	☺	☺
	Time-to-market	☹	☹	☹/☺	☺	☺
Emission	Not investigated	X	X	X	X	X

Figure 9.63: *Implementations judged on the quality indicators*

Algorithmic accuracy:

- For the signal bandwidths currently required in cellular and wireless terminals, a high order, CT feed-forward ΣΔ modulator with 2 or 3 level DAC provides sufficient SQNR in all cases. However, in high bandwidth modes the sample frequency is in the giga-hertz range, and technology speed limitations will be increasingly dominant in the design of such modulators, and hence higher power consumption is expected. To reduce this technology dominance, the most state-of-art technology and/or a more complex loop architecture (like the one in Sect. 5.5) should be selected.

Robustness:

- The FOMs of the implemented modulators prove that the RC integrator stage combines linearity and low noise leading to high performance and power efficient modulators.

- The 1.5-bit modulator requires accurate loop gain. Furthermore, because of its 3 level DAC the 1.5-bit modulator has a lower FOM_{HD3D}. However, improvement is possible by using the data chopping technique presented in Sect. 6.6.4.3.

- As the presented multi-mode modulators use optimized high gain circuits in combination with a 1-bit DAC, the FOM_{HD3D} of these modulators is very competitive in the low-bandwidth modes, and slightly less competitive in the high bandwidth modes, as in-band loop filter gain decreases due to the lower oversampling ratio.

- The GHz-rate modulators have a limited FOM_{HD3D} because of the deliberately non-optimized bias current distribution. The feed-forward GHz modulator indeed shows better linearity compared to the feedback GHz modulator as predicted by Sect. 6.3.

- The delay compensation technique makes a modulator more robust at architecture level. It has been successfully applied to the GHz-rate feed-forward $\Sigma\Delta$ modulator.
- All modulators use either a 1.5-bit SI or a 1-bit SC DAC, which both improve the robustness to clock jitter. A 1.5-bit benefits from the lower out-of-band quantization noise, and therefore, lower in-band jitter-quantization noise convolution products. The modulator with SC DAC benefits from the jitter shaping of such DAC due to the amplitude modulation described by the TPJE model.
- The 1.5-bit and multi-mode designs use custom designed circuits to achieve high performance at lowest possible power. Therefore these designs are less portable, as changing technology requires thorough simulation of all circuits. The hybrid and GHz-rate modulators as much as possible were made out of digital circuits and analog unit cells. Circuits were not optimized; instead common divisors in circuits were searched, to reduce the designs into a few analog unit cells and therefore increase the designs' portability.

Flexibility:

- The multi-mode modulators demonstrated that it is possible to cover a large range of bandwidths with a single modulator, by using the scaling theory presented in Sect. 7.4.
- The hybrid modulator architecture demonstrated its potential for a flexible bandwidth modulator, as the bandwidth of the digital filter scales with clock frequency.
- The digitally designed GHz modulators are very flexible in terms of technology.

Efficiency:

- Power: All presented modulators achieve a comparable FOM_{DR} of about 0.3 pJ per conversion. The FOM_{DR} of WLAN mode of the 121-mode modulator is worse because this mode is dominated by quantization noise (clearly visible in the right of Fig. 9.34). The $\text{FOM}_{eq,th}$ predicts that the 1.5-bit and multi-mode modulators are more power efficient, compared to the hybrid and GHz-rate modulators. This proves that the FOM$_{eq,th}$ is better than the FOM$_{DR}$: in the design of the 1.5-bit and multi-mode modulators a lot of effort is put in the power efficiency of the modulators' circuits. The hybrid and technology portable modulators were not optimized for power, but for portability. Most certain this holds for the GHz-modulators; all the integrator stages use the same amount of power to keep the circuit as simple as possible, which does not lead to the most power efficient design.

- Area: Because of its huge clock frequency range, the 121-mode modulator has area overhead due to the large amount of capacitors and switches required to adapt the ADC bandwidth, but its area can be greatly reduced, when a high clock frequency is used in all modes, at the cost of a slightly increased power consumption. The triple-mode modulator has the best FOM$_{area}$ in GSM mode. In CDMA and UMTS mode FOM$_{area}$ gradually decreases due to the capacitor overhead of the lower bandwidth modes. The same holds for the 121-mode modulator; at larger bandwidths, the FOM$_{area}$ becomes worse because of capacitor area overhead. The GHz-rate modulators, which were designed following the digitized design methodology are area efficient, despite the unit cell approach.

- Time-to-market: the 121-mode modulator, re-configurable over two decades of bandwidths, can help to achieve a very early market introduction of receivers as it can be re-used in receivers with different bandwidth requirements, but its flexibility in terms of technology portability is reasonably low. The hybrid modulator is very technology portable because of digitization at architecture, circuit and layout level. The GHz-rate modulators are very portable because of digitization at circuit and layout level.

The outcome of the digitization process of a $\Sigma\Delta$ modulator at different abstraction levels is summarized in Fig. 9.64.

9.5 Conclusions

This section presented several different types of modulators for highly digitized receivers for wireless and cellular communication. Key aspect is digitization at all levels: system (modulator application) level, modulator architecture level, and modulator circuit and layout level. Digitization at all these levels creates more robust, flexible, efficient and portable $\Sigma\Delta$ modulators which still can have competitive figure-of-merits.

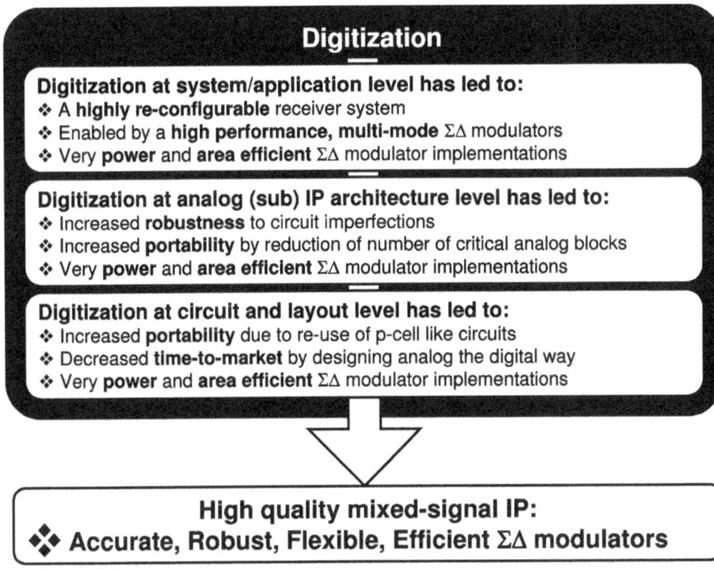

Figure 9.64: *Outcome of multi abstraction level digitization*

The combination of a high sample frequency and a high degree of digitization leads to very small modulator areas, pushing more of the analog area into the digital domain. This way the advantages of technology scaling are exploited.

In the design of the presented modulators, the categorization of modulator design properties in the quality indicators of Fig. 2.4 has played a crucial role, as they have identified critical design aspects (like noise and distortion) and related them to cost (like power and area). This way careful design trade-offs can be made, to come to competitive modulator implementations.

The conclusion-only reader is encouraged to also read Sect. 9.4.

Chapter 10

Conclusions

System properties can be categorized in the quality indicators accuracy, robustness, flexibility, efficiency and emission. Applying these quality indicators to a $\Sigma\Delta$ modulator as an ADC architecture has allowed identification of important modulator performance parameters, design parameter relations, and performance-cost relations.

The quality indicators predict, that to exploit the advantages of modern digitally optimized IC technologies, digitization should be carried through to all IC design abstraction levels as far as system optimization allows. For a system on a chip these abstraction levels are: system/application level, analog IP architecture level, circuit topology level and layout level.

The generalized system property categorization in quality indicators, and the digitization at different levels of system design, is named the digital design methodology. In this book this methodology is applied to $\Sigma\Delta$ modulators, leading to high quality, mixed-signal $\Sigma\Delta$ modulator implementations, which are more accurate, more robust, more flexible and/or more efficient.

For the ADC in a highly digitized, multi-mode (N)ZIF receiver architecture, a 1-bit $\Sigma\Delta$ modulator with a high order, continuous-time, feed-forward loop filter with a high over-sampling ratio is preferred, as it can supply sufficiently high algorithmic accuracy (SQNR) and in combination with RC integrator stage and 1-bit SC feedback DAC is the most robust $\Sigma\Delta$ modulator implementation compared to multi-bit or cascaded $\Sigma\Delta$ modulator architectures as these architectures are require more complex analog circuitry which more sensitive to technology impairments.

R.H.M. van Veldhoven, A.H.M. van Roermund, *Robust Sigma Delta Converters*,
Analog Circuits and Signal Processing, DOI 10.1007/978-94-007-0644-6_10,

A 1-bit CT modulator can be made flexible over a large range of bandwidths either by scaling all its circuit blocks according to the receiver-requested ADC bandwidth, or by creating an open-pipe ADC of which the clock frequency is as high as its largest conversion bandwidth demands. Choosing the latter approach, almost all the flexibility is shifted in the digital domain where it is comparably easy to implement.

A 1-bit CT modulator has proven to be an efficient ADC implementation, as it combines a high dynamic range and a high linearity with low power consumption and area.

$\Sigma\Delta$ modulator implementations are presented, that are designed according to the digital design methodology. The presented modulators have increased reconfigurability and portability, are more robust against circuit imperfections, are efficient in terms of area and power, and can be put on the market more quickly. Therefore, the modulators score high on the quality indicators.

For a more in depth description of the conclusions the author refers to the conclusion at the end of each chapter.

Appendix A

Harmonic and Intermodulation Distortion in an I&Q System

This appendix shows the impact of distortion on two complementary blocks in an I&Q system. First the second and third order distortion of a complex signal is reviewed, next the distortion of two complementary I&Q blocks is analyzed, in this case the I&Q ADCs.

A.1 Double Sided Spectrum of Second and Third Order Distortion of a Complex Signal

If Eq. 4.19 is reused, but now with a complex signal as input signal:

$$x_{in}(t) = A \cdot e^{j\omega_w t} \tag{A.1}$$

the output signal yields:

$$y_{in}(t) = A \cdot e^{j\omega_w t} + aA^2 \cdot e^{j2\omega_w t} + bA^3 \cdot e^{j3\omega_w t} \tag{A.2}$$

The output spectrum is: Fig. A.1 shows that the fundamental input signal and its harmonic components are on the same side of the spectrum. If the input signal would have been at $-\omega_w$ its second and third order harmonics would have been at $-2\omega_w$ and $-3\omega_w$.

R.H.M. van Veldhoven, A.H.M. van Roermund, *Robust Sigma Delta Converters*, Analog Circuits and Signal Processing, DOI 10.1007/978-94-007-0644-6,

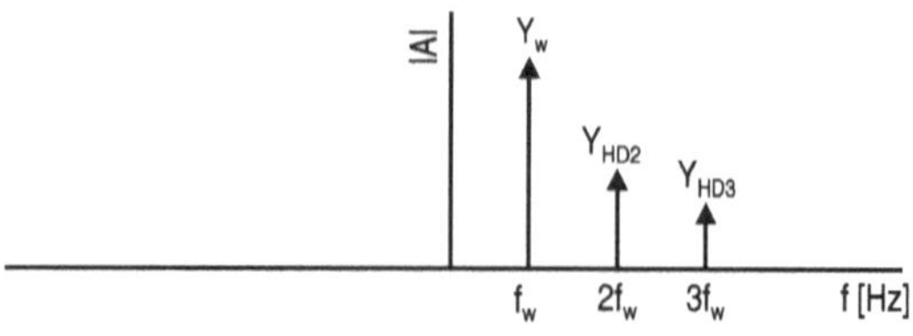

Figure A.1: *Double sided output spectrum*

A.2 Double Sided Spectrum of Second and Third Order Distortion in a Complex System

In Sect. 3.1 Fig. 3.2, the highly but not completely digitized radio receiver was chosen as the most promising receiver architecture in terms of ADC power consumption. It uses I&Q down conversion to be able to distinguish the positive from the negative half of the frequency spectrum. The non-linearity in the complementary I&Q blocks can be correlated (depending on their nature), which will have an impact on the combined I&Q output spectrum Y_{out}. If it is assumed that the I path has HD2 and HD3 coefficients a and b, and the Q path has HD2 and HD3 coefficients $a + \Delta a$ and $b + \Delta b$, it can be calculated that:

$$y_{DC} = \frac{1+j}{2}A^2 a + \frac{j}{2}A^2 \Delta a \tag{A.3}$$

$$y_w(t) = \left(A + \frac{3}{4}A^3 b + \frac{3}{8}A^3 \Delta b\right) e^{j\omega_w t} + \frac{3}{8}A^3 \Delta b e^{-j\omega_w t} \tag{A.4}$$

$$y_{HD2}(t) = \left(\frac{1-j}{4}A^2 a + \frac{1}{4}A^2 \Delta a\right)\left(e^{j2\omega_w t} + e^{-j2\omega_w t}\right) \tag{A.5}$$

$$y_{HD3}(t) = \left(\frac{1}{8}A^3 \Delta b\right) e^{j3\omega_w t} + \left(\frac{1}{4}A^3 b + \frac{1}{8}A^3 \Delta b\right) e^{-j3\omega_w t} \tag{A.6}$$

In case of $\Delta a = \Delta b = 0$, which means the distortion in the I&Q ADCs is fully correlated, the output spectrum of Fig. A.2a shows that the second order harmonic distortion is both on the positive and negative side of the frequency spectrum. The third order distortion is only visible in the opposite frequency side compared to the input signal. Input offset differences in the differential pairs of the I and of the Q ADC, leads to a shift of the HD2 and HD3 coefficients, and $\Delta a = \Delta b \neq 0$. The result is shown in Fig. A.2b. The fundamental input signal now leaks into the opposite half of the frequency spectrum, and the HD3 now is also visible at the same side as the input signal.

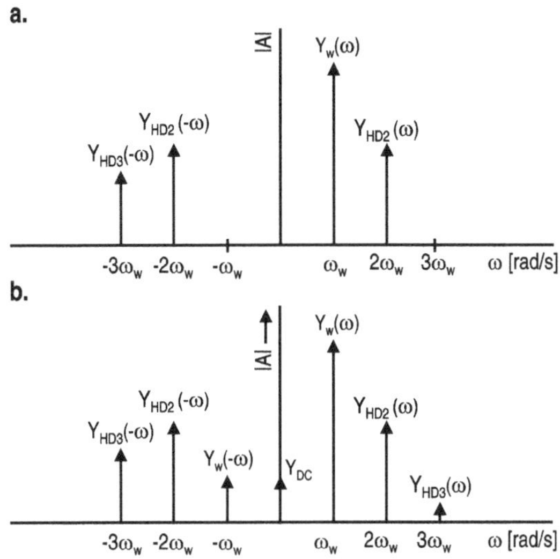

Figure A.2: *Double sided output spectrum*

Appendix B

Distortion of a Differential Input Transistor Pair Biased in Weak Inversion

For weak inversion the drain current exponentially relates to the gate-source voltage:

$$I_D = I_{D0} \cdot \frac{W}{L} \cdot e^{\frac{V_{gs} \cdot q}{n \cdot k \cdot T}} \text{ [A]} \tag{B.1}$$

For the output current of the differential pair with differential input signal $\hat{V}_{in}$ one can write:

$$I_{D,diff} = I_{D0} \cdot \frac{W}{L} \cdot \left(e^{\frac{\frac{1}{2}\hat{V}_{in} \cdot q}{n \cdot k \cdot T}} - e^{\frac{-\frac{1}{2}\hat{V}_{in} \cdot q}{n \cdot k \cdot T}}\right) \text{ [A]} \tag{B.2}$$

Using Taylor expansion for an exponential given by

$$e^x = 1 + x + \frac{x^2}{2!} + \frac{x^3}{3!} + \frac{x^n}{n!} + \mathbf{O}(x^{n+1}) \tag{B.3}$$

yields:

$$I_{D,diff} = I_{D0} \cdot \frac{W}{L} \cdot \left(c \cdot \hat{V}_{in} + \frac{1}{24} c^3 \hat{V}_{in}^3\right) \text{ [A]} \tag{B.4}$$

In which the even order terms cancel because of the differential nature of the circuit, the odd 5th and higher order components are neglected, and $c = \frac{q}{n \cdot k \cdot T} = gm/I_D$.

The HD3 distance for a weak inversion biased differential input pair excited with a sine wave now becomes:

$$\text{HD3D} = \frac{96 I_D^2}{gm^2 \hat{V}_{in}^2} \text{ [-]} \tag{B.5}$$

R.H.M. van Veldhoven, A.H.M. van Roermund, *Robust Sigma Delta Converters*, Analog Circuits and Signal Processing, DOI 10.1007/978-94-007-0644-6,

Appendix C

Fourier Series Expansion and Return-to-Zero

The Fourier series is a well known representation for repetitive waveforms. Below the Fourier series are given for the square wave from Fig. C.1.

$$a_n = \frac{2A}{n\pi} \sin(n\pi D) \tag{C.1}$$

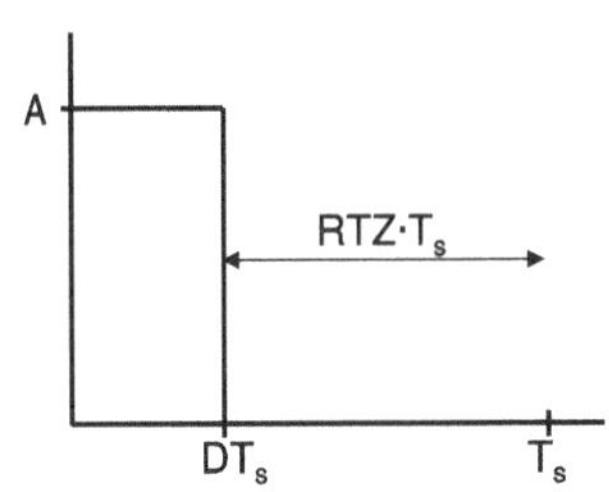

Figure C.1: *Square wave with duty cycle D and return-to-zero period RTZ*

Equation C.1 can also be written as a function of the RTZ period, which leads to:

$$a_n = \frac{2A}{n\pi} \sin(n\pi(1 - \text{RTZ})) \tag{C.2}$$

R.H.M. van Veldhoven, A.H.M. van Roermund, *Robust Sigma Delta Converters*, Analog Circuits and Signal Processing, DOI 10.1007/978-94-007-0644-6,

Appendix D

Clock Jitter in an I&Q System According to the TPJE Clock Jitter Model

This appendix extends the TPJE model for an I&Q ADC. A modulator with SI DAC is taken as an example here; for a modulator with SC DAC a similar analysis can be done.

Using Eqs. 6.72 and 6.73, the I and Q modulator output signals can be written in the form:

$$y_{\tilde{I}}(t) = \sin(\omega_i t) - \frac{1}{\Delta_i}\sin((\omega_i + \omega_m)t) + \frac{1}{\Delta_i}\sin((\omega_i - \omega_m)t) \tag{D.1}$$

and

$$y_{\tilde{Q}}(t) = \cos(\omega_i t) - \frac{1}{\Delta_i}\cos((\omega_i + \omega_m)t) + \frac{1}{\Delta_i}\cos((\omega_i - \omega_m)t) \tag{D.2}$$

Combining $y_{\widetilde{I\&Q}}(t) = y_{\tilde{I}}(t) + j \cdot y_{\tilde{Q}}(t)$ with Eqs. D.1 and D.2 yields:

$$y_{\widetilde{I\&Q}}(t) = e^{j\omega_i t} - \frac{1}{\Delta_i}e^{j(\omega_i+\omega_m)t} + \frac{1}{\Delta_i}e^{j(\omega_i-\omega_m)t} \tag{D.3}$$

The sine wave induced jitter components and the input signal component are on the same side of the spectrum ($\omega_m < \omega_i$). Due to I&Q path gain and phase mismatch the input signal and jitter components can leak to the opposite side of the spectrum as well. The amplitude distance between the image and its jitter components will be the same as the distance between the wanted signal and its jitter

R.H.M. van Veldhoven, A.H.M. van Roermund, *Robust Sigma Delta Converters*, Analog Circuits and Signal Processing, DOI 10.1007/978-94-007-0644-6,

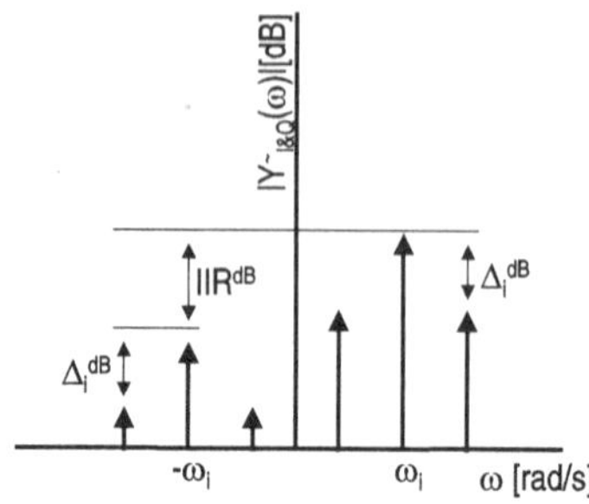

Figure D.1: *Sine wave induced clock jitter in an I and Q modulator*

components. The amplitude difference between image and wanted signal is defined by the image rejection ratio. The above is illustrated in Fig. D.1. Note that if the I and Q wanted signals are correlated, and the I and Q ADC is clocked with the same jittered clock, the jitter noise belonging to the wanted signal will also be correlated in the combined I and Q output spectrum. Therefore, the distance between signal and jitter noise will be the same for the I and Q combined output spectrum of both ADCs, and the spectrum at the outputs of the I and Q ADC separately.

If the input signals of the I and Q ADCs are uncorrelated (for example noise), the jitter in the I and Q combined output spectrum related to these uncorrelated input signals, will also be uncorrelated. In case of wide-band jitter, the jitter of the I ADC might overlap with the jitter of the Q ADC, and the total expected jitter noise in the output spectrum will be 3 dB higher.

References

[1] G.E. Moore, "Cramming more components onto integrated circuits," *Electronics*, vol. 38, pp. 114–117, 1965.

[2] G.E. Moore, "Progress in digital integrated electronics," *International Electron Devices Meeting*, vol. 21, pp. 11–13, 1975.

[3] M. Vertregt, P. Scholtens, "Scalable high speed analog circuit design," in *Analog Circuit Design* (A.v.R.J. Huijsing, M. Steyaert, ed.), pp. 3–21, Kluwer Academic Publishers, 2004.

[4] H. Veendrick, "From basics to ASICs," in *Nanometer CMOS ICs*, pp. 695–701, Springer, 2008.

[5] C.F. Hill, "Noise margin and noise immunity of logic circuits," *Microelectronics*, vol. 1, pp. 16–21, 1968.

[6] J. Kenny, L.R. Carley, "CLANS: A high-level synthesis tool for high resolution data converters," *Proceedings of the International Conference on Computer Aided Design*, vol. 1, pp. 496–499, 1988.

[7] J.C. Lee, "Fully automated layout generators for high-performance analog VLSI modules," *TENCON*, vol. 4, pp. 893–896, 1989.

[8] D. Janokis, K. Folkesson, J. Dabrowski, P. Eriksson, C. Svensson, "A 2.4-GHz RF sampling receiver front-end in 0.18μm CMOS," *IEEE J. Solid-State Circuits*, vol. 40, no. 6, pp. 1265–1277, 2005.

[9] B. Razavi, *RF Microelectronics*. Prentice-Hall, 1998.

[10] C.E. Shannon, "A mathematical theory of communications," *Bell Systems Technical Journal*, vol. 27, pp. 379–423 and 623–656, 1948.

[11] T. Rappaport, *Wireless Communications*. Prentice-Hall, 1996.

R.H.M. van Veldhoven, A.H.M. van Roermund, *Robust Sigma Delta Converters*,
Analog Circuits and Signal Processing, DOI 10.1007/978-94-007-0644-6,

[12] GSM specification, Draft prETS 300 910 (GSM 05.05, version 5.10.0, release 1996, November 1998).

[13] A. Marques, V. Peluso, M. Steyaert, W. Sansen, "Optimal parameters for $\Delta\Sigma$ topologies," *ISSCC Digest of Technical Papers*, vol. 33, pp. 1232–1241, 1998.

[14] L.J. Breems, R. Rutten, R.H.M. van Veldhoven, G. van der Weide, H. Termeer, "A 56mW CT quadrature cascaded $\Sigma\Delta$ modulator with 77dB DR in a near zero-IF 20MHz band," *ISSCC Digest of Technical Papers*, vol. 42, pp. 238–599, 2007.

[15] S.H. Ardalan, J.J. Paulos, "An analysis of nonlinear behaviour in delta-sigma modulators," *IEEE Transactions on Circuits and Systems*, no. 6, pp. 593–603, 1987.

[16] L.J. Breems, *Continuous-Time Sigma-Delta Modulation for A/D Conversion in Radio Receivers*. Delft University Press Science, 2001.

[17] S.R. Norsworthy, R. Schreier, G.C. Temes, *Delta-Sigma Data Converters: Theory, Design and Simulation*. IEEE Computer Society Press, 1996.

[18] R.H.M. van Veldhoven, K. Philips, B.J. Minnis, "A 3.3 mW $\Sigma\Delta$ modulator for UMTS in 0.18μm CMOS with 70 dB dynamic range in 2 MHz bandwidth," *IEEE J. Solid-State Circuits*, vol. 37, no. 12, pp. 1645–1652, 2002.

[19] J. Kenny, L.R. Carley, "Design of multi-bit noise-shaping data converters," *Analog Int. Circuits Signal Proceedings Journal*, vol. 3, pp. 259–272, 1993.

[20] L.R. Carley, "An oversampling analog-to-digital converter topology for high resolution data acquisition systems," *IEEE Transactions on Circuits and Systems*, no. 1, pp. 83–91, 1987.

[21] R.H.M. van Veldhoven, "A multi-bit Sigma Delta modulator with 1 bit inherently linear feedback," *Patent*, no. Patent filed, 2006.

[22] R.H.M. van Veldhoven, L.J. Breems, R. Rutten, "Using a noise-shaping coder in a Sigma Delta modulator," *Patent*, no. Patent filed, 2007.

[23] L.J. Breems, R. Rutten, G. Wetzker, "A cascaded continuous-time $\Sigma\Delta$ modulator with 67dB dynamic range in 10MHz bandwidth," *IEEE J. Solid-State Circuits*, vol. 39, no. 12, pp. 2152–2160, 2004.

[24] L.J. Breems, R. Rutten, R.H.M. van Veldhoven, G. van der Weide, "A 56 mW continuous-time quadrature cascaded $\Sigma\Delta$ modulator with 77 dB DR in a near zero-IF 20 MHz band," *IEEE J. Solid-State Circuits*, vol. 42, no. 12, pp. 2696–2705, 2007.

[25] W.S.J.A. Cherry, *Continuous-Time Delta-Sigma Modulators for High-Speed A/D Conversion*. Kluwer Academic Publishers, 2000.

[26] R. Rutten, L.J. Breems, G. Wetzker, "Digital calibration of a continuous-time cascaded $\Sigma\Delta$ modulator based on variance derivative estimation," *European Solid-State Circuits Conference*, vol. 32, pp. 199–202, 2006.

[27] C.A. Gobet, A. Knob, "Noise analysis of switched capacitor networks," *IEEE Transactions on Circuits and Systems*, no. 1, pp. 37–43, 1983.

[28] R.H.M. van Veldhoven, "$\Sigma\Delta$ modulators for future multi-standard wireless communication receivers," Tech. Note 2005/0663, Philips Research, 2005.

[29] P.G.R. Silva, *High-Resolution IF-to-Baseband $\Sigma\Delta$ ADC for AM/FM Car Radios*. PrintPartners Ipskamp, Enschede, 2001.

[30] E.J. van der Zwan, E.C. Dijkmans, "A 0.2mW CMOS $\Sigma\Delta$ modulator for speech coding with 80dB dynamic range," *IEEE J. Solid-State Circuits*, vol. 31, no. 12, pp. 1873–1880, 1996.

[31] P. Wambacq, W. Sansen, *Distortion Analysis of Analog Integrated Circuits*. Kluwer Academic Publishers, 2001.

[32] R.H.M. van Veldhoven, "Circuit with a digital to analog converter," *Patent*, no. US7042378B2, 2006.

[33] R.H.M. van Veldhoven, L.J. Breems, E.C. Dijkmans, "Electronic circuit with a sigma delta A/D converter," *Patent*, no. US7113119B2, 2006.

[34] H. Nyquist, "Certain topics in telegraph transmission theory," *Transactions of the American Institute of Electrical Engineers*, no. 2, pp. 617–644, 1928.

[35] R.W. Hickman, F.V. Hunt, "On electronic voltage stabilizers," *Review of Scientific Instruments*, vol. 10, pp. 6–21, 1939.

[36] E. Sackinger, W. Guggenbuhl, "A high-swing, high-impedance MOS cascode circuit," *IEEE J. Solid-State Circuits*, vol. 25, no. 1, pp. 289–298, 1990.

[37] K. Bult, G. Geelen, "A fast-settling CMOS opamp with 90dB DC-gain and 116Mhz unity-gain frequency," *ISSCC Digest of Technical Papers*, vol. 25, pp. 108–109, 1990.

[38] J.A. Cherry, W.M. Snelgrove, "Excess loop delay in continuous-time delta-sigma modulators," *IEEE Transactions on Circuits and Systems II*, no. 4, pp. 376–389, 1999.

[39] R.H.M. van Veldhoven, N. Nizza, L.J. Breems, "Technology portable, 0.04mm^2, GHz-rate $\Sigma\Delta$ modulators in 65nm and 45nm CMOS," *VLSI Digest of Technical Papers*, vol. 29, pp. 72–73, 2009.

[40] F. Gerfers, M. Ortmanns, *Continuous-Time Sigma-Delta A/D Conversion: Fundamentals, Performance Limits and Robust Implementations*. Springer, 2006.

[41] R.H.M. van Veldhoven, "A triple-mode continuous-time $\Sigma\Delta$ modulator with switched-capacitor feedback DAC for a GSM-EDGE/CDMA2K/UMTS receiver," *IEEE J. Solid-State Circuits*, vol. 38, no. 12, pp. 2069–2076, 2003.

[42] P. Wambacq, W. Sansen, *Principles of Communication Systems*. McGraw-Hill, New York, 1971.

[43] S. Harris, "The effects of sampling clock jitter on Nyquist sampling analog-to-digital converters, and on oversampling $\Delta\Sigma$ ADCs," *IEEE J. Audio Engineering Society*, no. 7/8, pp. 537–542, 1990.

[44] P. Nuijten, *The effects of phase lock loop jitter on the performance of A/D and D/A converters*. Philips Research Tech. Note, 1993.

[45] R.H.M. van Veldhoven, P.A.C.M.Nuijten, P.T.M. van Zeijl, "The effect of clock jitter on the DR of $\Sigma\Delta$ modulators," *Proceedings of International Symposium on Circuits and Systems*, pp. 2009–2012, 2006.

[46] P.T.M. van Zeijl, R.H.M. van Veldhoven, P.A.C.M.Nuijten, "Sigma-Delta ADC clock jitter in digitally implemented receiver architectures," *European Conference on Wireless Technology*, vol. 9, 2006.

[47] K. Philips, "A 4.4mW 76dB complex $\Sigma\Delta$ ADC for Bluetooth receivers," *ISSCC Digest of Technical Papers*, vol. 38, pp. 64–65/478, 2003.

[48] H.J.M. Veendrick, "The behaviour of flip-flops used as synchronizers and prediction of their failure rate," *IEEE J. Solid-State Circuits*, vol. 15, no. 2, pp. 169–176, 1980.

[49] B. Murmann, *ADC performance survey 1997–2010, Available at http://www.stanford.edu/~murmann/adcsurvey.html.*

[50] G. Mitteregger, C. Ebner, S. Mechnig, T. Blon, C. Holuigue, E. Romani, A. Melodia, V. Melini, "A 14b 20mW 640MHz CMOS CT $\Delta\Sigma$ ADC with 20MHz signal bandwidth and 12b ENOB," *ISSCC Digest of Technical Papers*, vol. 41, pp. 131–132/140, 2006.

[51] Y. Chae, G. Han, "Low power $\Sigma\Delta$ modulator using class-C inverter," *VLSI Digest of Technical Papers*, vol. 27, pp. 240–241, 2007.

[52] S. Kwon, F. Maloberti, "A 14mW multi-bit $\Delta\Sigma$ modulator with 82dB SNR and 86dB DR for ADSL2+," *ISSCC Digest of Technical Papers*, vol. 41, pp. 161–162/170, 2006.

[53] L. Bos, G. Vandersteen, J. Ryckaert, P. Rombouts, Y. Rolain, G. van der Plas, "A multirate 3.4 to 6.8mW 85 to 66dB DR GSM/Bluetooth/UMTS cascade DT $\Delta\Sigma$ modulator in 90nm digital CMOS," *ISSCC Digest of Technical Papers*, vol. 44, pp. 176–177, 2009.

[54] K. Matsukawa, Y. Mitani, M. Takayama, K. Obata, S. Dosho, A. Matsuzawa, "A fifth order delta-sigma modulator with single-opamp resonator," *VLSI Digest of Technical Papers*, vol. 29, pp. 68–69, 2009.

[55] S. Huang, Y. Lin, "A 1.2V 2MHz BW 0.084mm^2 CT $\Delta\Sigma$ ADC with -97.7dBc THD and 80dB DR using low-latency DEM," *ISSCC Digest of Technical Papers*, vol. 44, pp. 172–173, 2009.

[56] P. Crombez, G. van der Plas, M. Steyaert, J. Craninckx, "A 500kHz–10MHz multimode power-performance scalable 83-to-67dB DR CT $\Sigma\Delta$ in 90 nm digital CMOS with flexible analog core circuitry," *VLSI Digest of Technical Papers*, vol. 29, pp. 70–71, 2009.

[57] J. Koh, Y. Choi, G. Gomez, "A 66dB DR 1.2V 1.2mW single-amplifier double-sampling 2nd-order $\Delta\Sigma$ ADC for WCDMA in 90nm CMOS," *ISSCC Digest of Technical Papers*, vol. 40, pp. 170–171/591, 2005.

[58] H. Park, K.Y. Nam, D.K. Su, K. Vleugels, B.A. Wooley, "A 0.7V 100dB 870μW digital audio $\Sigma\Delta$ modulator," *VLSI Digest of Technical Papers*, vol. 28, pp. 178–179, 2008.

[59] P. Malla, H. Lakdawala, K. Kornegay, K. Soumyanath, "A 28mW spectrum-sensing reconfigurable 20MHz 72dB-SNR 70dB-SNDR DT $\Delta\Sigma$

ADC for 802.11n/WiMAX receivers," *ISSCC Digest of Technical Papers*, vol. 43, pp. 496–497/631, 2008.

[60] L. Doerrer, F. Kuttner, P. Greco, S. Derksen, "A 3mW 74dB SNR 2MHz CT $\Delta\Sigma$ ADC with a tracking-ADC-quantizer in 0.13μm CMOS," *ISSCC Digest of Technical Papers*, vol. 40, pp. 492–493/612, 2005.

[61] M. Park, M. Perrott, "A 0.13μm CMOS 78dB SNDR 87mW 20MHz BW CT $\Delta\Sigma$ ADC with VCO-based integrator and quantizer," *ISSCC Digest of Technical Papers*, vol. 44, pp. 170–171, 2009.

[62] T. Burger, Q. Huang, "A 13.5mW, 185MSample/s $\Delta\Sigma$ modulator for UMTS/GSM dual-standard IF reception," *ISSCC Digest of Technical Papers*, vol. 37, pp. 44–45/427, 2002.

[63] B. Putter, "A 5th-order CT/DT multi-mode $\Sigma\Delta$ modulator," *ISSCC Digest of Technical Papers*, vol. 42, pp. 244–245, 2007.

[64] V. Dhanasekaran, M. Gambhir, M.M. Elsayed, E. Sánchez-Sinencio, J. Silva-Martinez, C. Mishra, L. Chen, E. Pankratz, "A 20MHz BW 68dB DR CT $\Delta\Sigma$ ADC based on a multi-bit time-domain quantizer and feedback element," *ISSCC Digest of Technical Papers*, vol. 44, pp. 174–175, 2009.

[65] N. Yaghini, D. Johns, "A 43mW CT complex $\Delta\Sigma$ ADC with 23MHz of signal bandwidth and 68.8dB SNDR," *ISSCC Digest of Technical Papers*, vol. 40, pp. 502–503/613, 2005.

[66] R.H.M. van Veldhoven, "A triple-mode continuous-time $\Sigma\Delta$ modulator with switched-capacitor feedback DAC for a GSM-EDGE/CDMA2K/UMTS receiver," *ISSCC Digest of Technical Papers*, vol. 38, pp. 60–61/477, 2003.

[67] L. Yao, M. Steyaert, W. Sansen, "A 1V 88dB 20kHz $\Sigma\Delta$ modulator in 90nm CMOS," *ISSCC Digest of Technical Papers*, vol. 39, pp. 80–81/514, 2004.

[68] K. Lee, S. Kwon, F. Maloberti, "A 5.4mW 2-channel time-interleaved multi-bit $\Delta\Sigma$ modulator with 80dB SNR and 85dB DR for ADSL," *ISSCC Digest of Technical Papers*, vol. 41, pp. 171–172/180, 2006.

[69] R.H.M. van Veldhoven, K. Philips, B.J. Minnis, "A 3.3 mW $\Sigma\Delta$ modulator for UMTS in 0.18μm CMOS with 70 dB dynamic range in 2 MHz bandwidth," *ISSCC Digest of Technical Papers*, vol. 37, pp. 176–459, 2002.

[70] K. Lee, J. Chae, M. Aniya, K. Hamashita, K. Takasukal, S. Takeuchi, G.C. Temes, "A noise-coupled time-interleaved $\Delta\Sigma$ ADC with 4.2MHz BW, −98dB THD, and 79dB SNDR," *ISSCC Digest of Technical Papers*, vol. 43, pp. 494–495/631, 2008.

[71] Y. Shu, B. Song, K. Bacrania, "A 65nm CMOS CT $\Delta\Sigma$ modulator with 81dB DR and 8MHz BW auto-tuned by pulse injection," *ISSCC Digest of Technical Papers*, vol. 43, pp. 500–501/631, 2008.

[72] A. Das, R. Hezar, R. Byrd, G. Gomez, B. Haroun, "A 4th-order 86dB CT $\Sigma\Delta$ ADC with two amplifiers in 90nm CMOS," *ISSCC Digest of Technical Papers*, vol. 40, pp. 496–497/612, 2005.

[73] P. Fontaine, A.N. Mohieldin, A. Bellaouar, "A low-noise low-voltage CT $\Delta\Sigma$ modulator with digital compensation of excess loop delay," *ISSCC Digest of Technical Papers*, vol. 40, pp. 498–499/613, 2005.

[74] H. Neuteboom, B.M.J. Kup, Mark Janssens, "A DSP-based hearing instrument IC," *IEEE J. Solid-State Circuits*, vol. 32, no. 11, pp. 1790–1806, 1997.

[75] B. Nauta, "A CMOS transconductance-C filter technique for very high frequencies," *IEEE J. Solid-State Circuits*, vol. 27, no. 2, pp. 142–153, 1992.

[76] E.J. van der Zwan, "A 2.3 mW CMOS $\Sigma\Delta$ modulator for audio applications," *ISSCC Digest of Technical Papers*, vol. 31, 1996.

[77] Y. Yang, A. Chokhawala, M. Alexander, J. Melanson, D. Heser, "A 114dB 68mW chopper-stabilized stereo multi-bit audio ADC in 5.62mm^2," *IEEE J. Solid-State Circuits*, vol. 38, no. 12, pp. 2061–2068, 2003.

[78] B. Putter, "$\Sigma\Delta$ adc with finite impulse response feedback dac," *ISSCC Digest of Technical Papers*, vol. 39, pp. 76–77, 2004.

[79] I. Fujimori, L. Longo, A. Hairapetian, K. Seiyama, S. Kosic, C. Jun, C. Shu-Lap, "A 90dB SNR 2.5MHz output-rate ADC using cascaded multibit $\Delta\Sigma$ modulation at 8x oversampling ratio," *IEEE J. Solid-State Circuits*, vol. 35, no. 12, pp. 1820–1828, 2000.

[80] K.Y. Leung, E.J. Swanson, K. Leung, S.S. Zhu, "A 5V, 118dB $\Delta\Sigma$ analog-to-digital converter for wideband digital audio," *ISSCC Digest of Technical Papers*, vol. 32, pp. 218–219/460, 1997.

[81] J. Sauerbrey, T. Tille, D. Schmitt-Landsiedel, R. Thewes, "A 0.7V MOSFET-only switched-opamp $\Sigma\Delta$ modulator," *ISSCC Digest of Technical Papers*, vol. 37, pp. 310–311/469, 2002.

[82] O. Oliaei, P. Clement, P. Gorisse, "A 5mW $\Sigma\Delta$ modulator with 84dB dynamic range for GSM/EDGE," *IEEE J. Solid-State Circuits*, vol. 37, no. 1, pp. 2–10, 2002.

[83] Y. Geerts, M. Steyaert, W. Sansen, "A 2.5MSample/s multi-Bit $\Delta\Sigma$ CMOS ADC with 95dB SNR," *ISSCC Digest of Technical Papers*, vol. 35, pp. 336–337/468, 2000.

[84] R. Jiang, T.S. Fiez, "A 1.8V 14b $\Delta\Sigma$ A/D converter with 4MSamples/s conversion," *ISSCC Digest of Technical Papers*, vol. 37, pp. 220–461, 2002.

[85] S. Ouzounov, R.H.M. van Veldhoven, C. Bastiaansen, K. Vongehr, R. van Wegberg, G. Geelen, L.J. Breems, A. van Roermund, "A 1.2V 121-Mode CT $\Sigma\Delta$ modulator for wireless receivers in 90nm CMOS," *ISSCC Digest of Technical Papers*, vol. 42, pp. 242–600, 2007.

[86] M.G. Kim; G.C. Ahn, P.K. Hanumolu, Sang-Hyeon Lee, Sang-Ho Kim, Seung-Bin You, Jae-Whui Kim, G.C. Temes, Un-Ku Moon "A 0.9V 92dB double-sampled switched-RC $\Sigma\Delta$ audio ADC," *VLSI Digest of Technical Papers*, vol. 26, pp. 160–161, 2006.

[87] G. Gomez, B. Haroun, "A 1.5V 2.4/2.9mW 79/50dB DR $\Sigma\Delta$ modulator for GSM/WCDMA in a 0.13μm digital process," *ISSCC Digest of Technical Papers*, vol. 37, pp. 242–243/490, 2002.

[88] K. Nguyen, B. Adams, K. Sweetland, H. Chen, K. McLaughlin, "A 106dB SNR hybrid oversampling ADC for digital audio," *ISSCC Digest of Technical Papers*, vol. 40, pp. 176–177/591, 2005.

[89] P.G.R. Silva, L.J. Breems, K.A.A. Makinwa, R. Roovers, J.H. Huijsing, "A 110dB dynamic range continuous-time IF-to-baseband $\Sigma\Delta$ modulator for AM/FM/IBOC receivers," *ISSCC Digest of Technical Papers*, vol. 41, pp. 151–160, 2006.

[90] C.W. Tsang, Y. Chiu, B. Nikolic, "A 1.2V, 10.8mW, 500kHz $\Sigma\Delta$ modulator with 84dB SNDR and 96dB SFDR," *VLSI Digest of Technical Papers*, pp. 162–163, 2006.

[91] Y. Chae, I. Lee, G. Han, "A 0.7V 36μW 85dB-DR audio $\Sigma\Delta$ modulator using class-C inverter," *ISSCC Digest of Technical Papers*, vol. 43, pp. 490–491/630, 2008.

[92] P. Balmelli, Q. Huang, "A 25MS/s 14b 200mW $\Sigma\Delta$ modulator in 0.18μm CMOS," *ISSCC Digest of Technical Papers*, vol. 39, pp. 74–75/514, 2004.

[93] W. Yang, W. Schofield, H. Shibata, S. Korrapati, A. Shaikh, N. Abaskharoun, D. Ribner, "A 100mW 10MHz-BW CT $\Delta\Sigma$ modulator with 87dB DR and 9ldBc IMD," *ISSCC Digest of Technical Papers*, vol. 43, pp. 498–499/631, 2008.

[94] R. Brewer, J. Gorbold, P. Hurrell, C. Lyden, R. Maurino, M. Vickery, "A 100dB SNR 2.5MS/s output data rate $\Delta\Sigma$ ADC," *ISSCC Digest of Technical Papers*, vol. 40, pp. 172–173/591, 2005.

[95] P. Morrow, M. Chamarro, C. Lyden, P. Ventura, A. Abo, A. Matamura, M. Keane, R. O'Brien, P. Minogue, J. Mansson, N. McGuinness, M. McGranaghan, I. Ryan, "A 0.18μm 102dB-SNR mixed CT SC audio-band $\Delta\Sigma$ ADC," *ISSCC Digest of Technical Papers*, vol. 40, pp. 178–179/592, 2005.

[96] T. Christen, T. Burger, Q. Huang, "A 0.13μm CMOS EDGE/UMTS/ WLAN Tri-Mode $\Delta\Sigma$ ADC with −92dB THD," *ISSCC Digest of Technical Papers*, vol. 39, pp. 240–241/599, 2007.

[97] R.H.M. van Veldhoven, R. Rutten, L.J. Breems, "An inverter based hybrid $\Sigma\Delta$ modulator," *ISSCC Digest of Technical Papers*, vol. 43, pp. 492–493/630, 2008.

[98] L. Yao, M. Steyaert, W. Sansen, "A 1-V, 1-MS/s, 88-dB sigma-delta modulator in 0.13μm digital CMOS technology," *VLSI Digest of Technical Papers*, vol. 25, pp. 180–183, 2005.

[99] A. Namdar, B.H. Leung, "A 400MHz 12b 18mW IF digitizer with mixer inside a $\Sigma\Delta$ modulator loop," *IEEE J. Solid-State Circuits*, vol. 34, no. 12, pp. 1765–1776, 1999.

[100] S. Jantzi, K. Martin, A. Sedra, "A quadrature bandpass $\Sigma\Delta$ modulator for digital radio," *ISSCC Digest of Technical Papers*, vol. 32, pp. 216–217/460, 1997.

[101] L.J. Breems, E.J. van der Zwan, E.C. Dijkmans, J.H. Huijsing, "A 1.8mW CMOS ΣΔ modulator with integrated mixer for A/D conversion of IF signals," *ISSCC Digest of Technical Papers*, vol. 34, pp. 52–53, 1999.

[102] F. Henkel, U. Langmann, A. Hanke, S. Heinen, E. Wagner, "A 1-MHz-bandwidth second-order continuous-time quadrature bandpass sigma-delta modulator for low-IF radio receivers," *IEEE J. Solid-State Circuits*, vol. 37, no. 12, pp. 1628–1635, 2002.

[103] T. Nagai, H. Satou, H. Yamazaki, Y. Watanabe, "A 1.2V 3.5mW ΔΣ modulator with a passive current summing network and a variable gain function," *ISSCC Digest of Technical Papers*, vol. 40, pp. 494–495/612, 2005.

[104] E.J. van der Zwan, K. Philips, C. Bastiaansen, "A 10.7MHz IF-to-baseband ΣΔ A/D conversion system for AM/FM radio receivers," *ISSCC Digest of Technical Papers*, vol. 35, pp. 340–341/469, 2000.

[105] J.A.E.P. van Engelen, R.J. van de Plassche, E. Stikvoort, A.G. Venes, "A sixth-order continuous-time bandpass sigma-delta modulator for digital radio IF," *IEEE J. Solid-State Circuits*, vol. 34, no. 12, pp. 1753–1764, 1999.

[106] D. Senderowicz, G. Nicollini, S. Pernici, A. Nagari, P. Confalonieri, C. Dallavalle, "Low-voltage double-sampled ΣΔ converters," *ISSCC Digest of Technical Papers*, vol. 32, pp. 210–211/458, 1997.

[107] F. Chen, S. Ramaswamyand , B. Bakkaloglu, "A 1.5V 1mA 80dB passive ΣΔ ADC in 0.13μm digital CMOS process," *ISSCC Digest of Technical Papers*, vol. 38, pp. 54–55, 2003.

[108] J. Goes, B. Vaz, R. Monteiro, N. Paulino, "A 0.9V ΣΔ modulator with 80dB SNDR and 83dB DR using a single-phase technique," *ISSCC Digest of Technical Papers*, vol. 41, pp. 74–75, 2006.

[109] V. Peluso, P. Vancorenland, A. Marques, M. Steyaert, W. Sansen, "A 900mV 40μW switched opamp ΣΔ modulator with 77dB DR," *ISSCC Digest of Technical Papers*, vol. 33, pp. 68–69/414, 1998.

[110] A.K. Ong, B.A. Wooley, "A two-path bandpass sigma-delta modulator for digital IF extraction at 10MHz," *IEEE J. Solid-State Circuits*, vol. 32, no. 12, pp. 1920–1934, 1997.

[111] A.L. Coban, P.E. Allen, "A 1.5V 1.0mW audio $\Delta\Sigma$ modulator with 98dB dynamic range," *ISSCC Digest of Technical Papers*, vol. 34, pp. 50–51, 1999.

[112] F. Esfahani, P. Basedau, R. Ryter, R. Becker, "A fourth order continuous-time complex $\Sigma\Delta$ ADC for low-IF GSM and EDGE receivers," *VLSI Digest of Technical Papers*, vol. 23, pp. 75–78, 2003.

[113] V.S.L. Cheung, H.C. Luong, "A 3.3-V 240-MS/s CMOS bandpass $\Sigma\Delta$ modulator using a fast-settling double-sampling SC filter," *VLSI Digest of Technical Papers*, vol. 24, pp. 84–87, 2004.

[114] K.S. Lee, F. Maloberti, "A 1.8 V, 1 MS/s, 85 dB SNR 2+2 mash $\Sigma\Delta$ modulator with ± 0.9 V reference voltage," *VLSI Digest of Technical Papers*, vol. 23, pp. 71–74, 2003.

[115] S. Yan, E. Sanchez-Sinencio, "A continuous-time $\Sigma\Delta$ modulator with 88dB dynamic range and 1.1MHz signal bandwidth," *ISSCC Digest of Technical Papers*, vol. 38, pp. 62–63, 2003.

[116] T. Ueno, T. Itakura, "A 0.9V 1.5mW continuous-time $\Delta\Sigma$ modulator for WCDMA," *ISSCC Digest of Technical Papers*, vol. 39, pp. 78–79/514, 2004.

[117] T.L. Brooks, D.H. Robertson, D.F. Kelly, A. del Muro, S.W. Harston, "A 16b $\Sigma\Delta$ pipeline ADC with 2.5MHz output data-rate," *ISSCC Digest of Technical Papers*, vol. 32, pp. 208–209/458, 1997.

[118] A. Tabatabaei, K. Kaviani, B. Wooley, "A two-path bandpass $\Sigma\Delta$ modulator with extended noise shaping," *ISSCC Digest of Technical Papers*, vol. 35, pp. 342–343/468, 2000.

[119] K. Vleugels, S. Rabii, B. Wooley, "A 2.5V $\Sigma\Delta$ modulator for broadband communications applications," *IEEE J. Solid-State Circuits*, vol. 36, no. 12, pp. 1887–1899, 2001.

[120] K. Vleugels, S. Rabii, B. Wooley, "A 2.5V $\Sigma\Delta$ modulator for broadband communications applications," *IEEE J. Solid-State Circuits*, vol. 36, no. 12, pp. 1887–1899, 2003.

[121] T. Salo, S. Lindfors, K.A.I. Halonen, "A 80MHz bandpass $\Delta\Sigma$ modulator for a 100MHz IF receiver," *IEEE J. Solid-State Circuits*, vol. 37, no. 7, pp. 798–808, 2002.

[122] S. Gupta, V. Fong, "A 64MHz clock-rate ΣΔ ADC with 88dB SNDR and −105dB IM_3 distortion at a 1.5MHz signal frequency," *IEEE J. Solid-State Circuits*, vol. 37, no. 12, pp. 1653–1661, 2002.

[123] A. Dezzani, E. Andre, "A 1.2-V dual-mode WCDMA/GPRS ΣΔ modulator," *ISSCC Digest of Technical Papers*, vol. 38, pp. 58–59, 2003.

[124] A. Tabatabaei, K. Onodera, M. Zargari, H. Samavati, D. Su, "A dual channel ΣΔ ADC with 40MHz aggregate signal bandwidth," *ISSCC Digest of Technical Papers*, vol. 38, pp. 66–67/478, 2003.

[125] R. Gaggl, A. Wiesbauer, G. Fritz, C. Schranz, P. Pessl, "A 85dB dynamic range multibit delta-sigma ADC for ADSL-CO applications in 0.18μm CMOS," *IEEE J. Solid-State Circuits*, vol. 38, no. 7, pp. 1105–1114, 2003.

[126] F. Ying, F. Maloberti, "A mirror image free two-path bandpass ΣΔ modulator with 72dB SNR and 86dB SFDR," *ISSCC Digest of Technical Papers*, vol. 39, 2004.

[127] J. Yu, F. Maloberti, "A low-power multi-bit ΔΣ modulator in 90nm digital CMOS without DEM," *ISSCC Digest of Technical Papers*, vol. 40, pp. 168–169/591, 2005.

[128] A. Bosi, A. Panigada, G. Cesura, R. Castello, "An 80MHz 4x oversampled cascaded ΔΣ-Pipelined ADC with 75dB DR and 87dB SFDR," *ISSCC Digest of Technical Papers*, vol. 40, pp. 174–175/591, 2005.

[129] Y. Fujimoto, Y. Kanazawa, P. Lore, M. Miyamoto, "An 80/100MS/s 76.3/70.1dB SNDR ΔΣ ADC for digital TV receivers," *ISSCC Digest of Technical Papers*, vol. 41, pp. 201–210, 2006.

[130] L. Doerrer, F. Kuttner, A. Santner, C. Kropf, T. Puaschitz, T. Hartig, M. Punzenberger, "A continuous time AM ADC for voice coding with 92dB DR in 45nm CMOS," *ISSCC Digest of Technical Papers*, vol. 43, pp. 502–503/631, 2008.

[131] S.D. Kulchycki, R. Trofin, K. Vleugels, B.A. Wooley, "A 1.2V 77dB 7.5MHz CT/DT cascaded ΣΔ modulator," *VLSI Digest of Technical Papers*, vol. 27, pp. 238–239, 2007.

[132] M. Vadipour, C. Chen, A. Yazdi, M. Nariman, T. Li, P. Kilcoyne, H. Darabi, "A 2.1mW/3.2mW delay-compensated GSM/WCDMA ΣΔ ADC," *VLSI Digest of Technical Papers*, vol. 28, pp. 180–181, 2008.

[133] O. Rajaee, T. Musah, S. Takeuchi, M. Aniya, K. Hamashita, P. Hanumolu, U. Moon, "A 79dB 80 MHz 8x-OSR hybrid $\Delta\Sigma$/pipeline ADC," *VLSI Digest of Technical Papers*, vol. 29, pp. 74–75, 2009.

[134] P. Balmelli, H. Qiuting, F. Piazza, "A 50-mW 14-bit 2.5-MS/s $\Sigma\Delta$ modulator in a 0.25μm digital CMOS technology," *VLSI Digest of Technical Papers*, vol. 20, pp. 142–143, 2000.

[135] Jae Hoon Shim, Beomsup Kim, "A third-order $\Sigma\Delta$ modulator in 0.18μm CMOS with calibrated mixed-mode integrators," *VLSI Digest of Technical Papers*, vol. 24, pp. 78–81, 2004.

[136] J. Paramesh, R. Bishop, K. Soumyanath, D. Allstot, "An 11-bit 330MHz 8X OSR $\Sigma\Delta$ modulator for next-generation WLAN," *VLSI Digest of Technical Papers*, vol. 26, pp. 166–167, 2006.

[137] Z. Zhang, J. Steensgaard, G.C. Temes, J. Wu, "A split 2-0 MASH with dual digital error correction," *VLSI Digest of Technical Papers*, vol. 27, pp. 242–243, 2007.

Index

1-bit DAC, 69, 73, 80, 101, 113, 150, 240, 245, 258
1/f noise, 33, 34, 36–39, 41, 43, 45, 62, 63, 68, 70, 106, 107, 170, 205
1.5-bit DAC, 73, 80, 114, 115, 218, 219, 222

A
Accuracy, 7, 13, 15, 16, 20, 21, 24, 25, 68, 205, 234
Adaptability, xvii, 5, 6, 16
ADC, xiii, xix, xx, 1, 6, 11, 12, 27–31, 33–38, 41–50, 52–56, 59–70, 86, 87, 95, 175–184, 186, 187, 191–196, 263, 264
ADC clock, 177–180, 182, 186, 233
Additive error-feedback loop, 73, 82, 85, 88, 89, 102–104
Algorithmic accuracy, v, 69, 71–73, 88, 91, 113, 258
Aliasing, 28, 31, 69, 71, 91, 97–100, 116, 118–120, 122–124, 169, 171, 173
Anti-alias distance (AAD), xiii, xix, 118, 119, 121, 122, 127, 171
Architecture, 9, 28, 30, 31, 35, 37, 38, 67, 74, 82, 88, 89, 177–179, 181, 199, 215, 223, 232, 240, 241
Area benchmark, 209, 210, 248
Area benchmarking, 209
Automatic gain control (AGC), xiii, 50, 65–69, 177, 214, 215, 222, 223, 232

B
b-bit DAC, 73, 78, 101, 113, 114
Benchmark, 13, 17, 19, 91, 189, 190, 193, 194, 198, 199, 202, 247, 255
Bit error rate (BER), xiii, xix, 50–52, 61, 66, 166, 186
Bit-rate, xx, 50, 61, 175, 178, 181
Bubble correction, 242–244

C
Carrier to noise ratio (CNR), xiii, xix, 152, 154, 167
Cascaded, 73, 87, 88, 94, 104, 263
Cascode, 217, 218, 225
Cellular terminal, v, 1, 2
Circuit noise, 20, 91, 105, 107, 158, 159, 199, 216, 235, 242
Clock jitter, 91, 98, 132, 134–136, 139–141, 160, 162–164, 166–169, 171–173, 222–224, 259, 273
Clock strategy, 182, 186, 238, 240
Co-channel interference, xx, 36, 37, 62
Coefficient, xix, xx, 69, 73–78, 83, 84, 86, 87, 98–100, 125, 127–130, 171, 179–181, 183–186, 216, 218, 221, 233–235, 239

R.H.M. van Veldhoven, A.H.M. van Roermund, *Robust Sigma Delta Converters*, Analog Circuits and Signal Processing, DOI 10.1007/978-94-007-0644-6,

Co-existence, xvii, 3
Co-habitation, xvii
Comparator offset, 243, 244
Complexity, 2–5, 37, 73, 95, 104, 112, 114, 171, 176, 177, 208, 239, 240, 243
Connectivity terminals, v, 1, 2
Continuous-time (CT), xiii, 9–11, 69, 71–75, 87, 97–99, 104, 116, 132, 134, 136, 173–175, 190, 238, 257, 258, 263, 264
Crest factor, 33, 48, 49, 63, 64
Cross-modulation, xiii, 58, 59, 153

D
DAC, 72–74, 78–83, 86–88, 98–108, 112–116, 118–120, 122–124, 127, 128, 131–133, 136–138, 143–145, 162–164, 170–173, 183, 215, 216, 218, 219
DC gain, 217, 218, 225
DC offset, 33, 34, 36–39, 56, 62, 70
Differential pair, 59, 109–111, 195, 266, 269
Digitization, 12, 24, 25, 46, 94, 95, 174, 213, 214, 240, 249, 260, 263
Digitized, 6, 19, 28–31, 46, 87, 95, 100, 213, 214, 240, 250, 251, 256, 260, 266
Discrete time (DT), xiii, 87, 91, 97–99, 128, 190
Distortion, 20, 23, 31, 54, 59, 62, 71, 100, 105, 108, 110, 111, 189, 192, 193, 200, 215, 265, 266
Distortion benchmark, 193, 211
Distortion benchmarking, 202
Durability, xvii, 16
Dynamic range, xiii, xix, xxiii, 29, 30, 47, 48, 63, 65, 100, 132, 133, 162, 163, 165, 192, 214–216, 221–223, 232, 247

E
Efficiency, v, xvii, xix, 4, 5, 7, 13, 15–17, 22, 24, 50, 51, 69, 73, 93, 94, 181, 187, 189
Emission, v, 7, 13, 16, 17, 23, 24, 69, 263
Excess loop delay, 91, 127
Excess phase delay compensation, 129, 130, 251
Excess time delay compensation, 128

F
Feedback, 73, 88, 99, 100, 110–113, 118, 119, 122, 130, 131, 134–136, 183, 218, 250, 251, 253, 256
Feed-forward, xx, 71, 73–76, 88, 99–101, 128–130, 171, 173–175, 215, 216, 218, 223–225, 232, 233, 235, 240, 250, 251, 255–259
Flexibility, v, xvii, 4–7, 13, 15–17, 21, 22, 24, 30, 69, 175, 177–179, 183, 185, 213, 214, 259, 260, 263, 264
Flexible, 179, 181, 182, 186, 213, 214, 232, 259, 260, 263, 264
FOM_{area}, 189, 205, 206, 209–211, 260
FOM_{DR}, 189, 191–197, 211, 255, 259, 260
$FOM_{eq,th}$, 189, 194–198, 201–205, 211, 255, 259, 260
FOM_{HD3D}, 189, 200–204, 211, 258

G
Gain, 33, 34, 48–50, 52, 64–67, 72, 75–77, 79–83, 87, 88, 101–105,

116, 117, 130, 169–171, 183–186, 217, 218, 233–235, 258
Gain accuracy, 79, 91, 101, 173
GSM receiver, 30, 33, 51, 61–63, 65, 70, 166–168

H
Harmonic distortion, xix, xx, 54, 55, 60, 108, 109, 113, 115, 193, 211, 215, 219–221, 229, 266
HD2, xix, 54, 109, 148, 200, 266
HD2D, xix, 57
HD3, xix, 54, 60, 108, 109, 148, 200, 266, 269
HD3D, xx, 108–110, 189, 197, 200–202, 204
Highly digitized receiver, 46, 69, 94, 116, 176, 177, 213, 214, 238, 249, 260

I
I, 215, 216, 219, 221, 222, 228, 229, 231, 232, 237, 266, 273, 274
IIP2, xx, 54, 57, 58, 68
IIP3, xx, 54, 57, 58
IIR, 81, 241, 242
IM2, 33, 55–59, 68, 220, 223, 229, 232, 245, 249
IM3, 55–59, 220, 223, 229, 232, 245, 249
Image rejection (IR), xiv, 33–38, 60, 62, 63, 221, 229–231, 236
Image rejection ratio (IRR), xiv, xx, 34–38, 62, 63, 148, 221, 274
Implementation, 1, 9, 15–17, 19, 20, 45, 50, 82, 83, 88, 89, 92, 94, 95, 189, 193, 208, 209, 213, 257, 258, 263, 264
Implementation loss (IL), xiv, xx, 33, 52, 199, 211
Input intercept point (IIP), xiv
Input stage, 36, 100, 105, 106, 108–111, 119, 122, 131, 132, 170, 171, 174, 176, 183, 195, 197, 202, 204, 217, 218
Intercept point (IP), v, xiv, xx, xxi, 7, 10, 11, 25, 92–97, 170, 175, 182, 187, 213, 240, 249, 250, 263
Intermediate frequency (IF), xiv, 27–29, 33–35, 37, 38, 40, 41, 44, 63, 166
Intermodulation (IM), xiv, 54–59, 108, 223, 229–231, 245, 246
Inter-symbol-interference (ISI), xiv, 112, 113, 131, 143, 218
IP2, 33, 54, 55, 57, 62, 64, 66, 67
IP3, 33, 54, 55, 57, 62, 64, 66, 67

L
Linearity, 28, 54, 66–68, 70, 79, 80, 86–88, 99, 100, 103, 107–109, 113, 114, 169, 170, 189, 202–204, 213, 214, 222, 223, 258, 259
Local oscillator (LO), xiv, 28, 35, 36, 63, 169, 177–179, 181, 182, 187
Loop delay, 91, 114, 127–130, 169, 242, 256
Loop filter, 9, 71, 73–76, 78, 82–84, 95, 97–99, 101, 103, 129, 130, 178, 179, 186, 215–217, 226, 233, 239

M
Micrograph, 228, 229, 236, 248, 254, 255
Moore's law, 1, 24
Multi-bit DAC, 79, 81, 92, 105, 114, 152, 164, 171, 172

Multi-mode, 2, 4, 175, 182, 186, 187, 214, 222, 228, 232, 238, 246, 249, 258–260, 263
Multi-standard, 2, 4–6, 10, 31, 69, 175, 176

N
Near zero IF (NZIF), xiv, 28, 35–39, 59, 61, 62, 68, 70
NMOS in N-well, 225
Noise, 8, 14, 16, 20, 23, 39, 48, 50–52, 105–108, 151–160, 166, 169, 170, 172, 173, 189, 193, 255–257
Noise benchmark, 189, 204
Non-linearity, 36, 103, 108–115, 176, 243, 266
Non-return-to-zero (NRTZ), xiv, xx, 134, 135

O
OIP2, 57
OIP3, 57
Output intercept point (OIP), xiv, xx
Over-sampling ratio (OSR), xiv, xx, 71–73, 77, 78, 81, 84, 88, 107, 117, 134, 164, 183, 185, 199, 216, 234, 235

P
Performance benchmark, 91, 190
Phase delay, 127
Phase locked loop (PLL), xv, 136, 139, 163, 168, 178, 179, 181, 182, 215, 216, 219, 221, 223, 228
Portability, xvii, 16, 24, 93, 94, 213, 249, 250, 256, 259, 260, 264
Portable, 13, 17, 21, 30, 92, 94, 199, 249, 256, 259, 260
Power consumption, 22, 30, 36, 38, 41, 45, 94, 98–100, 182, 189, 191, 193–195, 205, 206, 231, 232, 249, 250, 256, 257
Power efficiency, xix, 4, 8, 9, 24, 69, 189, 191, 193, 194, 197, 206, 211, 238, 255–257, 260

Q
Q, 53, 59, 83, 132, 153, 215, 216, 219, 221, 222, 228, 229, 231, 232, 237, 266, 273, 274
Quadrature, xiv, 29, 35, 53, 63, 159, 160, 215, 221, 229, 231
Quality indicator, v, 1, 6, 7, 10–16, 18, 20, 23, 24, 26, 27, 30, 67, 70, 213, 214, 257, 258, 261, 263, 264
Quantization noise, 71, 72, 74, 76, 77, 80, 81, 83, 84, 88, 103, 111, 132, 145, 148, 153–156, 180, 186, 215, 216, 241, 242
Quantizer, 72–74, 76, 78–84, 86, 98–102, 104, 111, 112, 116, 117, 127, 128, 132, 170, 171, 184–186, 216, 240, 243, 244, 253

R
RC integrator input stage, 105, 106, 111, 116, 122, 170, 171, 173, 174, 183, 200, 202–204, 211
Receiver architecture, v, 6, 9, 11–13, 27–31, 34, 39, 43, 46, 69, 71, 175, 177–180, 215, 223, 232
Re-configurability, xvii, 16, 264
Reliability, xviii, 16, 18
Reproducibility, xviii, 16
Return-to-zero (RTZ), xv, xxi, 112–114, 117–122, 131, 132, 134–138, 143, 144, 146, 149, 150, 171, 218, 219, 271

Re-usability, xviii, 223
RF front-end, 12, 31, 33, 34, 41, 42, 44, 47, 49, 50, 54, 62, 65–68, 70, 175–177
Robust, 6, 9, 13, 16, 18, 21, 25, 70, 92, 93, 95, 97, 98, 138, 174, 213, 259, 260, 263, 264
Robustness, v, xviii, 7, 13, 16, 21, 24, 69, 91, 93, 169, 175, 258, 259, 263

S
Scalability, xvii, xviii, 16
Scaling, 7, 9, 22, 75, 92, 93, 176, 177, 183, 185, 196, 199, 226, 264
SC DAC, 98, 106, 107, 112, 113, 116, 122, 131, 134, 136–138, 140–142, 146–148, 151–154, 157, 161–168, 171–173, 223, 259
SC feedback DAC, 10, 98, 99, 106, 122, 137, 138, 155, 161, 162, 167, 173–175, 183, 186, 263
Selectivity, 29, 33, 46, 58, 64, 215
Shannon, 1, 9, 50, 51
Shannon limit, 51
SI DAC, 106, 117, 119, 133–135, 137, 138, 143–145, 149–151, 154–156, 159–168, 170–173, 223, 273
SI feedback DAC, 105, 117, 154, 155
Sigma Delta converter, v, 1, 173
Signal-to-distortion ratio (SDR), xv, xxi, 58, 59
Signal-to-jitter-noise ratio (SJNR), xv, xxi, xxiii, 152–154, 172
Signal-to-noise ratio (SNR), xv, xxi, 9, 50, 52, 53, 61, 65, 66, 68, 107, 108, 134–139, 151, 152, 154–156, 159–161, 171, 172, 189–192, 244–247
Signal-to-quantization-noise ratio (SQNR), xv, xxi, xxiii, 71–74, 76–78, 80–82, 84, 86, 88, 99, 130, 183, 199, 224, 244, 251
Simplicity, xviii, 25, 133, 152, 165, 181
State-of-art, 12, 29–31, 189, 191, 193, 198, 202, 203, 205–211, 255
Switch, 117, 120–122, 124, 125, 127, 136, 171, 218
Switched capacitor (SC), xv, 9–11, 97, 98, 105, 113, 138, 141, 145, 190, 193, 203, 224, 232
Switched current (SI), xv, 10, 98, 105, 143
Switched resistor (SR), xv, 98, 105, 120, 123, 183
System, 14, 15, 61, 94, 214, 215, 223, 232, 263, 265, 266, 273
System quality, xvii, 13, 15, 23, 67

T
TAJE model, 133, 134, 136–139, 171, 172
Technology, 1–3, 10, 20, 21, 25, 28, 30, 31, 74, 76, 88, 91, 92, 170, 171, 181, 182, 190, 194, 198, 199, 249, 250
Technology scaling, 7, 8, 21, 22, 92–94, 170, 174, 182, 261
Telecommunication standards, 176
Testability, xviii, 17
Time delay, 127, 128, 171
Time jitter, xxiii, 10, 138, 139, 162
Time-to-amplitude-jitter-error (TAJE), xv, 133, 134, 136–139, 171, 172
Time-to-market, 5, 6, 24, 26, 92, 93, 96, 170, 187, 260

Time-to-phase-jitter-error (TPJE), xv
TPJE model, 133, 139, 146, 151, 156, 161, 162, 164, 166–168, 171–173, 259, 273
Transistor, xix–xxi, 3, 7–9, 20, 22, 30, 92, 93, 95, 109, 190, 193, 195–197, 200, 205, 217, 251–253

V
Variability, xviii

W
Wireless, 4, 10, 13, 18, 27, 95, 214, 232, 258, 260

Z
Zero IF (ZIF), 28, 35, 36, 38, 68, 70, 215

FSC
www.fsc.org
MIX
Papier | Fördert
gute Waldnutzung
FSC® C083411

Kolibri 360 GmbH
Ferdinand-Jühlke-Straße 7
99095 Erfurt, Deutschland
produktsicherheit@kolibri360.de